Lecture Notes in Earth Sciences 74

Springer-Verlag Berlin Heidelberg GmbH

Martin Beniston John L. Innes (Eds.)

The Impacts of Climate Variability on Forests

With 144 Figures and 37 Tables

Springer

Editors

Professor Martin Beniston
University of Fribourg, Institute of Geography
Pérolles, CH-1700 Fribourg, Switzerland

Dr. John L. Innes
Swiss Federal Institue for Forest, Snow and Landscape Research (WSL)
CH-8903 Birmensdorf, Switzerland

"For all Lecture Notes in Earth Sciences published till now please see final pages of the book"

Cataloging-in-Publication data applied for

Die Deutsche Bibliothek - CIP-Einheitsaufnahme

The impacts of climate variability of forests : with 37 tables /
Martin Beniston ; John L. Innes (ed.).

(Lecture notes in earth sciences ; 74)
DOI 10.1007/978-3-540-69107-5

ISSN 0930-0317

Originally published by Springer-Verlag Berlin Heidelberg New York in 1998
MyCopy version of the original edition 1998

Typesetting: Camera ready by author
SPIN: 10630132 32/3142-543210 - Printed on acid-free paper
www.springer.com/mycopy

Acknowledgements

We would like to convey our appreciation to all the contributing authors for their dedication and effort which has led to this book. We would also like to thank the reviewers for generously giving their time.

The financial support of the Swiss Agency for the Environment, Forests and Landscape and the Swiss National Science Foundation is gratefully acknowledged. This funding enabled several contributors to attend the Workshop on Past, Present and Future Climatic Variability and Extremes : The Impacts on Forests, which was held in September 1997 in Wengen (Bernese Alps, Switzerland). The papers which appear in this book are based on material presented at the Wengen-97 Workshop.

We would also like to give special thanks to Ms. Sylvie Bovel-Yerly who finalized the manuscripts in their camera-ready form, in an efficient and timely manner.

Foreword: research on the impacts of climate change and policy making

Richard Volz

The Framework Convention on Climate Change contains commitments for countries not only to reduce greenhouse gas emissions, but also to promote and cooperate in research on the climate system, on assessing the impacts of climate change, and on preparing to adapt to climate change. At the second Ministerial Conference on the Protection of Forests in Europe, which was held in June 1993 in Helsinki, Finland, European countries made four resolutions, as a joint response to many of the forest decisions taken in the United Nations Conference on Environment and Development in Rio de Janeiro in 1992. The Helsinki Resolution 4 (H4) "Strategies for a Process of Long-term Adaptation of Forests in Europe to Climate Change" (Ministerial Conference 1993) deals with the effects of climate change on forests and reinforces the importance of this issue.

In addition to international decisions, we also have national legal obligations preserve the forest and protect the population, buildings and installations against natural hazards. We need to know how to react, although it is uncertain how the risks will change and what measures will be necessary or appropriate in the future. Swiss forestry strategy emphasises conservation and biodiversity. We may wonder whether this is sufficient for adapting to climate change or whether we should advise the planting of trees suitable for a warmer climate. We are dependent on research to provide answers to such questions. However, the outcome of a meeting between scientists and representatives of the Agency for the Environment, Forests and Landscape was that we will only receive answers if we ask clear questions. As a first step, we have therefore prepared a paper detailing our questions on some selected topics such as "forests and forestry", "water fluxes and natural hazards" and "nature protection" (Volz et al. 1998).

An important role of the administration is to prepare decisions for politicians and afterwards to ensure that these decisions are implemented. Politicians feel that a prosperous economy is their responsibility and short-term decisions are therefore often taken over the economy. Long-term factors such as climate change are hardly taken into account when making economic decisions. In the tourist industry for instance, planning never extends more than ten years ahead. This short-term mentality influences policy. Consequently, government is often forced to act within a short time-scale and under pressure. Anything which seems to be urgent is taken more seriously than factors which should be long-term priorities for sustainability (Conseil du développement durable 1997). As a result, it is all the more important that the research community focuses attention on the long-term issue of climate change, which is affecting or will affect economy and society. It is also important that the results of this research are clearly communicated.

The complexity of the climate change process, and the difficulty in identifying specific cause–effect relationships, make climate change a difficult task in the political context, especially in relation to its potential impacts, as the effects of climate interact with many other environmental factors. Searching for and preparing a system for monitoring the forests, we thought that phenological observations should be a good indicator for the effects of global warming on trees. A study of phenological data collected by the Bern cantonal forestry service showed (as might be expected) that leaf break of deciduous trees now occurs earlier in the spring than it did during the nineteenth century (Vasella 1997). This cannot simply be explained by warmer spring temperatures, the most determining factor for this phenophase, as there has been no change in the needle flushing of *Picea abies*. Phenological observations are not very precise, which raises the question of whether the change could merely be the result of altered observation practices. Earlier leaf break indicates an earlier start of photosynthetic activity, which coincides with a general increase in increment of *Fagus sylvatica* (BRÄKER 1996). The growth increase of *Picea abies* seems however to be less distinct, which indicates differences between deciduous trees and conifers. We concluded that phenology is a useful indicator in a forest monitoring system, but that alterations which were expected to be an effect of climate may also be attributable to other causes. This makes it very difficult to establish convincing indicators for the impact of climate change and without impacts climate change itself would have no political significance. Therefore, we have to show clear indications of climate change impact to convince politicians and the population, or they will not accept any measures. This is a great challenge and efficient cooperation will be crucial.

References

Bräker OU (1996) Tree-Ring Growth Trends since the Swiss Forest Damage Inventory „Sanasilva 1984“: Results of a 1992 Pilot Study. Radiocarbon 363-370

Conseil du développement durable (1977) Sustainable Development, Action Plan for Switzerland. Swiss Agency for the Environment, Forests and Landscape, Bern; pp 31

Ministerial Conference on the Protection of Forests in Europe, 16-17 June 1993 in Helsinki: Documents. Ministry of Agriculture and Forestry, Helsinki

Vassella A (1997) Phänologische Beobachtungen des Bernischen Forstdienstes von 1869 bis 1882. In: Phänologie von Waldbäumen, Umwelt-Materialien 73:9-75

Volz R, Nauser, M, Schiess C, Küttel M (1998) Auswirkungen von Klimaänderungen, Fragen an die Forschung. Umwelt-Materialen; Bern, pp 33

Preface

Martin Beniston

The Wengen Workshops on Global Change Research take place on an annual basis in the mountain resort of Wengen in the Bernese Alps, Switzerland. These workshop series were intiated in 1995, and are under the responsibiltiy of the Department of Geography of the University of Fribourg, Switzerland. Each year, a theme of interest to the global change community is identified; the Workshops are then organized jointly with other specialized scientific institutions. The ultimate objective of the Wengen Workshop series is to provide a scientific forum which is at the leading edge of interdisciplinary science.

In the context of the 1997 Wengen Workshop, which took place in September 1997, the selected topic was *Past, Present and Future Climatic Variability and Extremes: The Impacts on Forests*. The meeting was jointly organized with the Swiss Federal Institute for Forest, Snow and Avalanche Research in Birmensdorf, Switzerland.

The rationale for the 1997 meeting focused on the fact that, while the debate on global climate change has considered primarily mean changes in variables such as temperature and precipitation, many environmental and socio-economic systems are at least as sensitive to climatic extremes and variability as to average climate. Natural ecosystems and forests are one typical example of systems which, while constrained within certain ranges of mean climate, can undergo rapid and often irreversible damage in the face of short-lived but intense extreme events. Based upon these considerations, the Wengen-97 Workshop focused on the following themes:

- Issues related to past and present variability and extremes from a climatological viewpoint, and the shifts which have been observed this century or reconstructed for a more distant past
- An assessment of recent progress in the simulation of both variability and extremes in a future, warmer climate, in particular on regional scales
- Observed impacts of the variability of climate and extreme events on forest growth, species composition, competition between species, forest health and forest distribution, based on current observations and reconstructions from dendrochronological, palynological, and other paeloecological proxies
- Assessment of the capability of forest models to simulate forest response to extreme events, using hypothetical climatological data or a range of plausible scenarios for climate extremes

The 1997 meeting was truly interdisciplinary, with expertise from scientists working in climate modelling, climate statistics, forest ecosystem research, forest modelling, dendrochronology, palynology, and other specialized fields. The selected papers in this book are a reflection of the scientific discussions which took place in Wengen.

List of contributors

Tomasso Anfodillo, Universita di Padova, Dipartimento Territorio e Sistemi Agro Forestali, Legnaro, Italia.

Martin Beniston, University of Fribourg, Institute of Geography, Fribourg, Switzerland.

Silvio Borella, University of Bern, Physics Institute, Bern, Switzerland.

Rudolf Brazdil, Masaryk University, Department of Geography, Brno, Czech Republic.

Gerd Bürger, Department of Global Change and Natural Systems, Potsdam Institute for Climate Impact Research, Potsdam, Germany.

Vinicio Carraro, Universita di Padova, Dipartimento Territorio e Sistemi Agro Forestali, Legnaro, Italia.

Marco Carrer, Universita di Padova, Dipartimento Territorio e Sistemi Agro Forestali, Legnaro, Italia.

Jean Combe, Antenne Romande, FNP Forêt - Neige – Paysage, Lausanne, Switzerland.

Michel Déqué, Météo Nationale, CNRM, Toulouse, France.

Francisco Javier Doblas-Reyes, Météo Nationale, CNRM, Toulouse, France.

Yves Durand, Météo Nationale, CNRM, Toulouse, France.

Joe Eastman, Colorado State University, Dept of Atmospheric Science, Fort Collins Colorado, USA.

Jean-Louis Edouard, Faculté des Sciences et Techniques de St. Jérôme, Institut Méditerranéen d'Ecologie et de Paléoécologie, Marseille, France.

Gary Funkhouser, University of Arizona, Laboratory of Tree-Ring Research, Tucson, USA.

Rüdiger Grote, BTU Cottbus, Potsdam Institute for Climate Impact Research, Cottbus, Germany.

Frédéric Guibal, Faculté des Sciences et Techniques de St. Jérôme, Institut Méditerranéen d'Ecologie et de Paléoécologie, Marseille, France.

Joel Guiot, Faculté des Sciences et Techniques de St. Jérôme, Institut Méditerranéen d'Ecologie et de Paléoécologie, Marseille, France.

Jarle I. Holten, Terrestrial Ecology Research, Buvika, Norway.

Karl Hönninger, Institute of Silviculture, University of Agricultural Sciences, Vienna, Austria.

Malcolm K. Hughes, University of Arizona, Laboratory of Tree-Ring Research, Tucson, USA.

John L. Innes, WSL (Swiss Federal Institut for Forest Snow and Landscape), Birmensdorf, Switzerland.

Trudy Kavanagh, University of Western Ontario, Dept. of Geography, London Ontario, Canada.

Thierry Keller, Faculté des Sciences et Techniques de St. Jérôme, Institut Méditerranéen d'Ecologie et de Paléoécologie, Marseille, France.

Christian Körner, Universität Basel, Botanisches Institut, Basel, Switzerland.

Petra Lasch, Department of Global Change and Natural Systems, Potsdam Institute for Climate Impact Research, Potsdam, Germany.

Manfred J. Lexer, Institute of Silviculture, University of Agricultural Sciences, Vienna, Austria.

Marcus Lindner, Department of Global Change and Natural Systems, Potsdam Institute for Climate Impact Research, Potsdam, Germany.

Lixin Lu, Colorado State University, Dept of Atmospheric Science, Fort Collins Colorado, USA.

Brian H. Luckman, University of Western Ontario, Dept. of Geography, London Ontario, Canada.

Eric Martin, Centre national de recherches météorologiques, Centre d'études de la neige, St Martin d'Heres, France.

David L. Peterson, University of Washington, Field Station for Protected Area Research, Seattle, USA.

Roger A. Pielke, Colorado State University, Dept of Atmospheric Science, Fort Collins Colorado, USA.

Martine Rebetez, Antenne Romande, FNP Forêt - Neige – Paysage, Lausanne, Switzerland.

Matthias Saurer, Paul Scherrer Institut, Villigen, Switzerland.

Fritz H. Schweingruber, WSL (Swiss Federal Institut for Forest Snow and Landscape), Birmensdorf, Switzerland.

Rolf Siegwolf, Paul Scherrer Institut, Villigen, Switzerland.

Felicitas Suckow, Department of Global Change and Natural Systems, Potsdam Institute for Climate Impact Research, Potsdam, Germany.

Lucien Tessier, Faculté des Sciences et Techniques de St. Jérôme, Institut Méditerranéen d'Ecologie et de Paléoécologie, Marseille, France.

Carlo Urbinati, Universita di Padova, Dipartimento Territorio e Sistemi Agro Forestali, Legnaro, Italia.

Thomas T. Veblen, University of Colorado, Department of Geography, Boulder, USA.

Pier Luigi Vidale, Colorado State University, Dept of Atmospheric Science, Fort Collins Colorado, USA.

Ricardo Villalba, Laboratorio de Dendrocronologia IANIGLA, Mendoza, Argentina.

Robert L. Walko, Colorado State University, Dept of Atmospheric Science, Fort Collins Colorado, USA.

Table of contents

The impact of climatic extremes on forests: an introduction

John L. Innes

Abstract. Important extreme events affecting forests include wind storms, frost, excessive snowfalls, glaze, drought and temperature anomalies (such as an unusually warm period in mid-winter). These tend to result in the mortality of individual trees, and may be highly selective of species or even individuals. Individuals also vary in their responses, although the relative importance of genotype and site characteristics is often difficult to separate. Extreme climatic events tend to be much more important in initiating change in forest ecosystems than average climatic conditions. The changes created by such disturbances provide the gaps which will enable species better suited to the current environmental conditions to become established. Consequently, a forest at any given time will consist of a mosaic of patches, each occupied by species that were favoured at the time of gap formation. Models of forest succession under a changing climate need to pay much greater attention to the role of extreme climatic events.

1. Introduction

Climate change presents a major challenge to forest ecologists studying the dynamics of natural and semi-natural forest ecosystems. Not only do they have to contend with the changes that all forests are constantly undergoing, but they have to try to separate this natural variation from changes induced by climate. Climate may bring about gradual, long-term changes in forest composition, or the changes may be sudden, triggered by climatic extremes. Alternatively, extreme events may initiate long-term, gradual trends which are only partially predictable (Oliver and Larson 1996). Extreme climatic conditions tend to result in the formation of gaps in the forest, with the size of the gap being determined by the nature of the climatic event, the nature of the forest and a variety of other factors. Such gaps form patches, and both primary and secondary forests are made up of a dynamic mosaic of different patches, with the patch scale being dependent on the ecological processes operating in the landscape. The composition, structure and functions of the ecosystem(s) in each patch is in a permanent state of flux (Froment and Tanghe 1967; Goovaerts et al. 1990; Koerner et al. 1997). Many forests have the added complication that management activities cause further alterations to the ecosystem. Consequently, determining the effects of climate change on this highly dynamic mosaic is a complex and difficult task.

While the distribution of biomes can be mapped and modelled with relative ease, the application of such models to what is actually happening on the ground at a particular site is much more difficult. The mosaic structure of forests means that such models are scale-dependent: As the scale becomes finer, so the importance of those local factors which determine the nature of the forest ecosystem generally increases (e.g. Luckman and Kavanagh, this volume; Peterson, this volume). While climatic factors clearly operate at local scales, their relative importance may be reduced in comparison to other determinants. For example, tropical species will be excluded from a particular temperate area primarily by the temperature and precipitation characteristics of the climate, but it is the local conditions which will determine which of a range of possible temperate species in that biogeographical region may be present. Even then, local variations in the environment may make it possible for species to survive in apparently hostile environments. Variation in local conditions can occur across spatial (both horizontal and vertical) and temporal gradients. For example, the susceptibility of a tree to an extreme climatic event can vary according to its location, height in relation to the rest of the canopy and age. Similarly, the marked vertical gradients that occur in forest climates means that the impact of a particular climatic event may differ between the forest floor and canopy.

Changes in mean climatic conditions may be accompanied by changes in climatic variability and the pattern of extreme events (Waggoner 1989; Rebetez 1996; Rebetez and Beniston, this volume). For example, a shift in mean temperature may conceal a marked change in the frequency distribution of extreme temperatures. There has been increasing interest in extreme events concurrent with the realisation that the disturbance dynamics of forests play an important role in determining their composition and the processes operating within them (c.f. Oliver and Larson 1996).

2. Climate and forest ecosystems

Climate affects forests in a variety of ways, as detailed in several recent reviews (e.g. Kirschbaum et al. 1996a, 1996b; Beniston et al. 1996), and in several chapters in this book. The focus in this introductory paper is on the importance of extreme events in comparison to changes in average conditions. In looking at such events, the whole forest ecosystem should be considered and not just the trees. This implies observing the forest over a number of different spatial and temporal scales. However, as illustrated by the contributions to this book, the majority of experimental work and models have been restricted to the impacts on trees, as these form the most important structural characteristics of forest ecosystems and, perhaps more importantly, trees have generally been viewed as the most important economic product of forest ecosystems.

Given the changes that have been recorded in climate and atmospheric chemistry over the past 100 years, it seems likely that some changes in forests should have already occurred. What evidence is there is for this? Many studies have indicated changes in tree growth rates at high latitudes and high altitudes. Some of these changes may be related to the improved nutritional status of intensively managed forests in central Europe (Spiecker et al. 1996), but other explanations have also been proposed. For example, in the southern hemisphere, an increase in the growth rates of *Lagarostrobus franklinii* in Tasmania since 1965 has been attributed to warmer conditions (Cook et al. 1991).

The trends in growth rates have been matched by increasing rates of seedling establishment at and just below the treeline (Rochefort et al. 1994; Luckman and

Kavanagh, this volume; Peterson, this volume). Many studies have documented increases in the altitude of the forest or tree limits in mountainous regions (Innes 1991), although the evidence for latitudinal extensions of the forest limit is less convincing. Such records indicate that both the composition and the extent of forests actively and rapidly respond to fairly short-term climatic fluctuations. These changes reflect both selective immigration and selective mortality of particular tree species at the altitudinal and latitudinal treelines.

3. Assessing climate change impacts on forest ecosystems

As the potential impacts are so varied, observations at different scales are needed to determine the effects of climate variability and extremes on forests (c.f. Carrer et al. this volume). Such systems range from ecophysiological studies of individual trees to modelling studies of the entire forest community. The themes are related through a variety of cross-scalar mechanisms. Much better integration of such studies is urgently required in view of the accumulating evidence that predictions, experimental evidence and observations made in the field are providing conflicting results (Körner 1995). For example, after identifying that only about 2% of the studies looking at the impacts of elevated CO_2 deal with natural vegetation, Körner (1995) listed several areas where improvements could be made when up-scaling from experiments to predictions. These included giving greater attention to:

- Non-linear system responsiveness
- Mineral nutrition
- Biologically effective CO_2 enrichment
- Interactions between individuals (both mono- and interspecific)
- Acclimation.

A major factor that has been very poorly addressed is the way in which forest ecosystems may resist change. The most basic form of resistance is the presence of within-plant plasticity (Kuiper and Kuiper 1988; Bell and Lechowicz 1994). Trees are long-lived organisms and therefore must be able to cope with fluctuating environmental conditions. Many species reach ages of >1000 years and continue to survive despite the substantial changes in environmental conditions that have occurred over the past 1000 years. This indicates that an individual tree can have considerable plasticity with respect to the environment, and it should not be assumed that a change in the environment will necessarily result in the mortality of trees. Instead, such mortality is often induced by extreme events, such as drought, wind or fire.

One reason for a tree's plasticity is that as it grows, it experiences very different environments. As a result, a number of characteristics may change. For example, the shade needles of *Sequoia sempervirens* are very different from the sun-exposed foliage. Over longer time scales, mesophyllous trees subjected to water stress may become increasingly sclerophyllous, as has been shown for *Fagus sylvatica* (Bussotti et al. 1995). Such changes are rarely, if ever, taken into account when modelling the response of a species to environmental stress.

While an individual may show considerable plasticity, as indicated by studies of clones (e.g. Kramer 1995), there is even greater plasticity within a species. This includes both phenotypic and genotypic variability among individuals. The tolerance of a species to a particular environmental stress as determined from a study of a small number of individuals is therefore likely to reveal highly misleading results. An example is provided by *Picea abies*, one of the most important timber trees in Europe. This species is widely distributed, occurring from Spain eastwards and from the Mediterranean to north of the Arctic Circle. While its optimal growth occurs under a restricted suite of environmental conditions, it is clearly capable of growing under a wider range. Field evidence suggests that it may not be capable of regenerating over such an extended range of conditions, but the environmental factors controlling the regeneration are only now being fully determined. They reflect a complex set of interactions which even in the heart of its geographical distribution may prevent regeneration on unsuitable sites.

At the same time, most individual plants appear to be adapted to their specific micro-site conditions, with plants of the same species introduced to a new site often being less successful than the individuals already there. Consequently, it appears that there is an intricate balance between specialisation and plasticity (Bell and Lechowicz 1994).

Changes within a forest may also occur more slowly than expected because of the resistance of the forest to change. Successful regeneration under a closed canopy is relatively rare, with most tree species requiring a canopy gap before they can assume a dominant position. The rate of gap formation is dependent on the incidence of natural disturbances, but may be low. For example, Peterken and Mountford (1996) document one stand in Lady Park Wood, England, in which no gaps were formed in the canopy between 1945 and 1992, despite the occurrence of severe droughts during the period. Clearly, in such situations, it would be very difficult for a new tree species to colonise the woodland.

Such factors need to be taken into account when examining the potential impact of climatic variability on forests. Various climatic phenomena can have a direct or indirect impact on forests, and some of these are examined below.

3.1 Temperature

The distribution of most major biome types, many plant communities and many individual species is correlated with mean annual temperature. However, this may be because mean annual temperature tends, under current climatic conditions, to be correlated with a number of variables that play a more decisive role in controlling the distributions of plants. For example, some species cannot tolerate frost, and a number of tropical and sub-tropical species show injury at temperatures between zero and 15°C (McWilliam 1983). The occurrence of a single frost event, while unlikely to have any effect on mean annual temperatures, may drastically change the species composition of a forest, particularly in relation to the composition of seedlings in the regeneration layer, which tend to be much more sensitive than mature individuals of the same species (e.g. Larcher and Mair 1969).

Freezing injury is particularly important for forest species. While many boreal and temperate species can withstand freezing provided that they are sufficiently acclimatized, major damage can be caused by unseasonal freezing. Some of the greatest damage occurs as a result of late spring frosts, when the new shoots or leaves have already started to extend and are therefore soft and susceptible to thermal shocks (Butin 1995). The extent of such damage depends on the species, the phenological stage of the tree and the nature of the frost, as well as the preceding weather conditions (e.g. Braathe 1995). Early in the spring, the damage may be restricted to buds, but later frosts may affect both the young foliage and the extending shoots (e.g. Cannell 1984). As the onset of growth is primarily determined by temperature, climatic extremes (either in the form of unusually warm conditions early in spring followed by a colder period with frost or an unusually late frost) can be particularly significant.

Early autumn frosts may also cause damage if they occur before the trees have hardened sufficiently. Such autumn damage tends to be more significant in evergreen trees than in deciduous ones, as the foliage of the deciduous trees has already completed much of its life cycle by this time. In conifers, the damage is characterised by the death of needles (Redfern and Cannell 1982). The onset of autumnal hardening is determined by a combination of temperature and photoperiod, and may therefore be less influenced by temperature variability than spring damage.

In winter, several different types of temperature-related injury may occur. Trees from frost-free environments that are introduced to areas experiencing such conditions appear to be particularly susceptible. For example, although *Nothofagus obliqua* shows considerable potential as a plantation species in Great Britain, frost cracking has restricted its use to areas free of frost. Extreme events are particularly important. For example, in January 1982, very low temperatures in northern Britain resulted in the death of many trees of southern origin, particularly *Pinus nigra* (Redfern and Rose 1984). Such factors may be particularly important when considering the introduction of new species or provenance as a response to changes in mean temperature (e.g. Hannerz 1994). The occurrence of winter sunscald may also be important. It is characterised by the formation of cracks on sun-exposed stems and is usually caused by the stem being exposed to strong sunlight during the day and then sub-zero temperatures at night.

Plants have a lowermost temperature that they can withstand. In some cases, these temperatures are much lower than would be found in nature. For example, many coniferous species from the western USA can withstand temperatures down to between -60°C and -80°C (Sakai and Weiser 1973). As with frost susceptibility, the tolerance of winter temperatures of most species varies according to provenance, with provenances from more extreme conditions being the most tolerant (e.g. Williams and MacMillan 1971). Generally, exceptionally cold conditions during mid-winter do not cause excessive damage to trees, primarily because they are already hardened. Instead, sudden changes, particularly when they cross the 0°C threshold, are more important.

Another form of injury occurs when the roots of some species are exposed to freezing. The extent of injury depends on the species and how low the temperatures get (Havis 1976). In many areas, a layer of snow protects the ground from freezing. However, if snowcover is low, then the freezing front may penetrate the rooting zone. Injury following reduced snowcover occurred in the mixed hardwood forests of southern Quebec in the early 1980s (Lachance 1985; Bernier et al. 1989) and in northern Finland in the winter of 1986–87 (Jalkanen 1993; Jalkanen et al. 1995). Such injury is often (but not always: c.f. Sundblad and Andersson 1995) related to the provenance, with individuals derived from areas with more extreme climates often being hardier than those derived from milder areas (e.g. Rehfeldt 1988; Aho 1994). This was illustrated in Les Landes, France, in 1962–1963, when 100,000 ha of *Pinus pinaster* derived from Portuguese sources was killed or severely damaged by frost, and again in 1985, when between 30,000 and 50,000 ha were severely impacted, whereas the local variety was much less severely affected (Le Tacon et al. 1994). Events of this magnitude are likely to have an impact on the choice of provenance in new plantings being made to anticipate future changes in climate.

An important form of injury occurring during winter is called winter desiccation (or winter drying), although there is some confusion surrounding the term. Winter desiccation is caused by heavy frosts inducing water loss from foliage (Jalkanen 1993), as opposed to direct freezing injury. The injury is believed to be a critical factor influencing the altitudinal and latitudinal extents of many species (Wardle 1971; Tranquillini 1979; Kozlowski 1982; Larcher 1995), although Körner (this volume) has offered an alternative explanation for the location of treelines.

The occurrence of warm winds during winter may also induce damage. This is evident in the "red belt" damage to conifers in mountainous areas of the western USA and Canada (e.g. Henson 1952; Bella and Navratil 1987). In this case, the damage is associated with the warm Chinook wind, although whether it is caused by needle desiccation or by sharply varying temperatures is unclear (Henson 1952, Jalkanen 1993). Similar damage has occurred in Europe and it is an important cause of injury in Great Britain, particularly amongst introduced North American species (Redfern 1993). Its importance in Scandinavia declines eastwards; it is frequent in Norway (Venn 1993) but tends to be relatively rare in Finland (Jalkanen 1993).

A related process has been termed freeze-drying (e.g. Jalkanen 1993), and occurs when air temperatures in spring rise sufficiently for dehardened needles to transpire while the ground remains frozen (Kozlowski 1982; Berg and Chapin 1994; Jackson and Grace 1996). A single sunny day in spring followed by a severe night frost can be sufficient to induce major injury (Jalkanen 1993). The process is accentuated by low atmospheric humidity, and may also be affected by increased transpiration following cuticle erosion by wind-driven snow and ice (Hadley and Smith 1986). In the most severe cases, substantial areas of forest can be killed (Robins and Susut 1974).

Cold conditions during the growing season, if persistent, can result in indirect damage to forests. Foliage may not be able assimilate sufficient carbon to produce the necessary defence compounds to protect the trees against attacks by pathogens. For example, the scleroderris canker (*Ascocalyx* (*Gremmeniella*) *abietina*) is particularly abundant on *Pinus sylvestris* in Finland following unfavourable summers (Jalkanen 1993).

Extremely hot temperatures may also cause injury to woody plants. One of the most common forms occurs when a stand is opened up, allowing sunlight to penetrate through the upper canopy. Many thin-barked species, such as *Fagus sylvatica*, are particularly susceptible to injury under such conditions (e.g. Woodcock et al. 1995). Living, hydrated tissues are generally killed at temperatures of 50 to 60°C (Kozlowski et al. 1991), although in most trees injury may occur at temperatures of 40 to 50°C (Larcher 1995), and gross photosynthesis generally reaches a plateau at 8–18°C (e.g. Tranquillini 1955). Seedlings may be particularly affected because of the high temperatures that can develop close to the ground surface. As with cold tolerance, many species show differential seasonal sensitivity, with the greatest tolerance occurring in mid-summer and the lowest in spring. There is also evidence that heat tolerance increases in particularly hot summers (Kozlowski et al. 1991), reducing the potential impact of such extremes.

3.2 Precipitation

Droughts (defined here as periods with insufficient soil moisture for plants) may play a critical role in determining species distributions. As with frost, a single drought may have a major effect on the composition of a forest (e.g. Hursh and Haasis 1931), although the effects may primarily be the result of secondary agents following the initial weakening of the trees by the drought. A drought such as that which occurred in parts of Great Britain in 1976 can initiate considerable mortality amongst larger overstory trees, reversing the normal trend for mortality to be largely restricted to the smaller trees within a stand (Elliott and Swank 1994; Peterken and Jones 1987). The trend for mortality to be greatest amongst the smaller size classes may be related to the initial density of the stand, the greater susceptibility of saplings and poles below the general canopy layer (Peterken 1996), or it may be a function of the dynamics of the drought. Mortality may occur at the time of the drought or, more commonly, in the following years (Peterken and Mountford 1996). Such mortality often affects some species more than others (e.g. Clinton et al. 1993), resulting in changes to the composition of the forest.

The 1976 drought in England resulted in long-term effects in some forests of *Fagus sylvatica*. Mortality continued for 15 years after the drought, and the growth rates of some larger trees appear to have permanently reduced. The effects varied between individuals, with adjacent trees responding differently (Peterken and Mountford 1996), although older trees were much more affected than younger ones. Substantial lag effects associated with drought impacts on *Abies alba* in the Vosges Mountains of France have also been noted (Becker 1989).

As a drought develops, the available soil water decreases as a result of uptake by plants, evaporation and drainage into the groundwater. There is evidence that as the drought develops, trees maintain the minimum water potential (Ψ_{min}) above the threshold potential for cavitation (e.g. Cochard et al. 1996; Lu et al. 1995, 1996), thereby reducing the risk of twig and branch mortality. The direct impact of such conditions is the wilting of foliage, caused by a loss of turgor in the cells. However, a number of effects may occur without any evidence of wilting. These include reductions in growth and changes in the carbon allocation patterns within trees. The latter may be evident in the form of changes in leaf size and shoot extension patterns. Often these effects may be most apparent in the year following a drought, and the occurrence of two sequential years with droughts may cause substantially different effects (Innes 1992a). The timing of the drought is also important, with droughts in the early part of the growing season having very different effects to those in the middle of the summer (Innes 1992a).

Drought appears to be an important factor determining the amount of foliage held by trees, and several studies have related the occurrence of drought to reductions in the crown density of trees in Great Britain (Innes 1993), France (Landmann 1995), and Norway (Strand 1997). In addition, the spatial variation in crown density is, for some species, related to the water-holding capacity of the soil (Webster et al. 1996). Some of the defoliation that in the late 1970s and early 1980s was attributed to air pollution may actually have been the direct and indirect response of trees to the severe drought in 1976 (Innes 1994).

A particularly important type of damage associated with precipitation occurs during glaze or ice storms. These involve rainfall or fog that freezes either on entering a layer of super-cooled air near the ground or on contact with trees and other surfaces. The damage occurs because the weight of ice that accumulates on the branches leads to severe mechanical damage, particularly if strong winds occur when the trees are laden with ice. Minor events result in small amounts of damage that the trees usually recover from easily. However, severe events can cause major damage, with a particularly damaging event occurring in January 1998 in the northeast USA and southeast Canada. Damage during glaze storms may create canopy gaps that encourages of sensitive species by others (Carvell et al. 1957; Whitney and Johnson 1984; de Steven et al. 1991).

3.3 Wind

Another important factor that needs to be considered when looking at climate influences on forests is wind. Wind as a chronic cause of stress is most important in coastal and mountain areas. In the latter, it may have an impact on the altitudinal treeline. Foliage damage is particularly important in coastal areas, where it is often (but not always) associated with salt injury (Wilson 1980). Damage may also occur as a result of persistent strong winds, which leads to higher transpiration and eventually to plasmolysis of cells, marginal necrosis and premature foliage loss (e.g. Raitio 1991).

While wind pruning, involving the removal of leaves and foliage-bearing branchlets, may be significant in certain circumstances (Grier 1988), the most important impact of wind occurs in the form of extreme events. These can result in damage to large areas of forest. For example, hurricane "Hugo" destroyed 70 million m^3 of timber in South Carolina, USA, in September 1989 (Weeden 1990) and in Germany, 60 million m^3 of timber was blown by severe gales in early 1990 (Geipel 1992). The impact of wind is dependent on the age structure of forests. Young, vigorous forests are much more resistant than older, mature or over-mature forests (Grayson 1991). Trees on shallow soils seem to be more susceptible than those on deeper soils (Savill 1983), although exceptions to this generalisation occur. Trees in recently thinned forests are more susceptible than those sheltered in dense stands. The impact of a particular storm is also dependent on the preceding conditions; the wind force alone is not a particularly good indication of either the extent or the nature of the damage that may be caused (Kozlowski et al. 1991, Combe, this volume). In particular, the duration of the storm may be important as swaying of trees results in the progressive breakage of the supporting roots, making the trees more susceptible to windthrow. Such damage may also result in increased susceptibility to root damage and may decrease the ability of trees to take up water. Trees with roots infected by pathogenic fungi are often more susceptible to windthrow (e.g. Landis and Evans 1974; Shaw and Taes 1977). In addition, a severe storm may create an initial opening in a dense stand which is then enlarged by less severe storms (Robertson 1991). During severe storms, there seems to be a tendency for the damage to occur in a mosaic of different-size patches, rather than large areas of forest being completely flattened (e.g. Pierce 1921; Lorimer 1989; Nelson et al. 1994).

3.4 Fire

Other forms of disturbance associated with climate include floods, landslides and other mass movements and fire. Of these, by far the most important is fire. Fire and climate are intricately linked, with the magnitude and frequency of fires being related to the available fuel wood, its flammability, and the weather conditions. Fire is one of the most important causes of forest disturbance when considered at a global scale (Oliver and Larson 1996). There are optimum climatic conditions for the occurrence of fires: Too dry and there is insufficient production of the woody material needed to keep the fire burning; too wet and the moisture content will remain too high for the fuel to burn. Several other factors influence the occurrence of forest fires, mostly related to the provision of suitable fuel wood. For example, insect and disease outbreaks can predispose a forest to burning (Schowalter et al.

1981), as can the occurrence of large numbers of downed stems following a windstorm (Lorimer and Frelich 1994).

The natural frequency of fire was probably closely related to the occurrence of lightning. However, it is difficult to determine natural fire frequencies, because man has had a major influence on fire frequency for so long. The influence of man is frequently underestimated, and many forests today are the result of thousands of years of disturbance induced by fires set by man. Even in areas where fire is generally considered to be of minor importance, such as in many moist temperate forests, anthropogenic fires may have had an important influence on the structure and composition of the forest. The frequency of fires is also affected by short-term climatic variations, and Villalba and Veblen (1997) have documented a relationship between the establishment of *Austrocedrus chilensis* woodland and the absence of fire associated with cooler and wetter conditions.

In the pine and mixed forests of the southwestern USA, surface fires occur under natural conditions at a frequency of at least once every 10–15 years (Weaver 1951; Wagener 1961; Kilgore 1973), although this figure is dependent on a variety of factors, such as local topography. Many forests are adapted to less frequent, more severe fires, and the diversity of the original forest in the Great Lakes – St. Lawrence area has been attributed to such fires (Heinselman 1981). Some temperate coniferous forests, such as the *Pseudotsuga menziesii* forests of the Pacific Northwest (Romme 1982), and much of the boreal forest are also adapted to less frequent, more severe fires (Bonan and Shugart 1989). Much (>60%) of the *P. menziesii* forests of the central Cascade Mountains, for example, appear to have developed following a catastrophic fire in the 15th century (Agee and Huff 1980, Perry 1994). Calculating the frequency of fire at a particular site is often difficult as individual fires burn very unevenly across a landscape, leaving unburned patches and patches with varying levels of burning intensity.

Much research on the occurrence of forest fires has been concentrated in the USA and in the boreal zone (see Johnson 1992). The impact of fire in other forest ecosystems has been less thoroughly studied. However, fire can be a significant factor affecting such forests, as for example in the temperate rain forests of southern Chile (Innes 1992b; Veblen 1985; Veblen et al. 1983). Permanent changes from one vegetation type to another following fire has been documented in a number of areas, including boreal forest (changing to tundra) in northern Quebec, Canada (Sirois and Payette 1991) and dry tropical forest (changing to shrubland) in Zambia (Trapnell 1959).

Tropical rain forests are also susceptible to fire, as demonstrated by the deliberate use of fire to clear them for agriculture. In some parts of the world, dry conditions associated with El Niño events may predispose rain forests to burning (Lopez-Portilla et al. 1990), as was clear in the case of the fires in Indonesia during the second half of 1997. The presence of charcoal in the Amazonian soils indicates past fires, possibly coinciding with periods of extended drought (Saldarriaga and West 1986). The extent of fire in the tropical rain forest region remains largely undocumented, but it appears to be dependent on fuel moisture levels (Uhl et al.

1988). These may increase significantly if the canopy is opened up by some form of disturbance, with Kauffman and Uhl (1990) showing that a 50% reduction in canopy cover has the potential to increase average temperatures in the forest by 10 °C and to decrease relative humidity by 35%.

Boreal forests are considered to be highly susceptible to fire, and the available evidence suggests that large-scale fires, involving over 100,000 ha, are fairly frequent, occurring about once every 100 years (Johnson 1992). One such fire burned in 1950 in northern British Columbia and Alberta, Canada, destroying 1.4 million ha of forest. Another burned from 6 May to 2 June 1987, destroying 870 000 ha of forest in the Daxinganling Mountains of China, and killing about 200 people (Goldammer and Di 1990). The same year, Goldammer and Di (1990) estimate that 1.5 million ha of forest in neighbouring parts of Russia was burnt. In 1915, an estimated 14 million ha of forest in Siberia was burnt (Shostakovitch 1925). However, the extent to which these are determined by climatic extremes is unclear.

3.5 Interactions with carbon dioxide

When looking at the potential impacts of climate change on forests, the predicted changes in carbon dioxide concentrations should also be considered. For example, some experimental work suggests that higher atmospheric carbon dioxide concentrations may increase the water use efficiency of plants, making them less susceptible to drought. However, this hypothesis is based on extrapolations of experimental studies to field situations. Such up-scaling using linear models is extremely questionable (Körner 1995). In particular, global mean carbon dioxide concentrations bear little relationship to those experienced by plants in a forest. Carbon dioxide concentrations show marked variations in both space and time within a forest and may not be clearly coupled with atmospheric concentrations. Marked diurnal and seasonal variations occur, and concentrations also vary horizontally and vertically within the forest matrix, although these variations tend to be much less marked during daylight hours within the canopy itself. For example, in mid-summer, CO_2 concentrations at 0.05 m and 12 m above the ground in a forest may differ by more than 100 $\mu L\ L^{-1}$ (Bazzaz and Williams 1991) and similar vertical gradients have been documented by Buchmann et al. (1996).

4. Future developments in impact studies of extreme climatic events

It is clearly important to use multiple-scale observation systems when considering the impacts of climate on forest ecosystems. For example, drought tolerance needs to be examined at the scale of individual trees, using several observation points within the tree's rooting zone. Stand-level effects require detailed and time-consuming observations of all the individuals within the stand, often with the extrapolation of some parameters (e.g. meteorological conditions) across the stand area. Extrapolating such observations to a regional scale requires the use of appropriate models, and the verification of the models may be able to utilise satellite observations (e.g. Myeni et al. 1997).

Recent developments in forest ecosystem monitoring provide a particularly good opportunity to determine the impacts of climate change and in particular extreme climatic events on forest ecosystems. For the first time outside a very few specialised areas such as Hubbard Brook, USA, a range of different types of data are being collected simultaneously at the same site (De Vries et al. 1997). The nature and quality of the data are obviously dependent on the programme, but some programmes are now installing a range of equipment that will enable a number of different aspects of the ecosystem to be linked together. For example, towers with meteorological equipment at several different heights enable vertical differences in climate to be assessed within a stand. These reveal major differences to the climatic data recorded by standard meteorological stations located according to WMO guidelines. Simultaneously, relevant climatic data, such as soil temperature and soil water potential may be collected and linked to for example, the data derived from dendrometers, which continuously measure the diameters of trees. In addition, recent developments in sap flow measurements mean that it is possible to link all of these to water movement within trees. When combined with isotope studies within the trees, the potential to understand the links between climate and water stress in trees becomes very large (c.f. Valentini et al. 1994; Livingston and Spittlehouse 1996; Panek and Waring 1997).

5. Conclusions

Climatic extremes affect forests in many different ways. Impacts may be direct, as in the case of a wind storm, or indirect, as with fire. The impacts operate over a range of different scales, making assessment of ecosystem impacts difficult. In particular, what may be considered as a climatic extreme by a climatologist may not be an extreme event for a forest ecologist, particularly one with a historical perspective (c.f. Hughes 1998, this volume). Most studies concentrate on the impacts on trees, but these represent only a fraction of the species that can be affected by climatic extremes. To understand better the impacts, it will be necessary to change the way that we look at forests, which until now has frequently been from the point of view of the forester looking at the impact on the production of timber. Future studies need to assess the impacts from the viewpoint of the organisms being affected. For example, how a flightless insect perceives a disturbance will be very different to the perception of a raptor. Such differences are primarily related to scale (Peterken 1996).

Current observation systems are mostly based at one particular scale, and there is an urgent need to find appropriate methods for transcending these scalar limitations. Modelling provides one such method, but the various down-scaling and up-scaling methods that are available require further improvement. Combined observation and modelling offers considerable potential, not only in terms of calibrating models with site-specific data, but also in relation to checking the accuracy of the model predictions. Such a process involves long-term observations, which are costly and may not be of immediate value in the short-term. The time involved may be much greater than hitherto realised: the composition and dynamics of forests today have very often been determined to a greater or lesser extent by anthropogenic disturbances over the past 200 years.

Natural disturbances have also played a role, and all forests should be seen as being in a state of continual change. The long-term studies done at sites such as Hubbard Brook in the USA and Solling in Germany are beginning to indicate the value of such investigations for understanding the basic ways in which forest ecosystems respond to stresses such as extreme climatic events. Given the time-scales involved, it is important that observational studies are complemented by historical studies, and Quaternary paleoecologists have a major role to play in this respect. If forest ecosystems are to managed in a sustainable way, then a combination of palaeoecological studies, long-term observations and modelling will provide much of the information that is essential for such a management approach.

6. References

Agee JK, Huff MH (1980) First year ecological effects of the Hoh fire, Olympic Mountains, Washington. In: Proceedings of the Sixth Meteorology Conference, April 22–24, 1980, Society of American Foresters, Washington D.C., pp 175–181

Aho M-L (1994) Autumn frost hardening of one-year-old *Pinus sylvestris* (L.) seedlings: Effect of origin and parent trees. Scan J For Res 9:17–24

Bazzaz FA, Williams WE (1991) Atmospheric CO_2 concentrations within a mixed forest: Implications for seedling growth. Ecology 72:12–16

Becker M (1989) The role of climate on present and past vitality of silver fir forests in the Vosges mountains of northeastern France. Can J For Res 19:1110–1117

Bell G, Lechowicz MJ (1994) Spatial heterogeneity at small scales and how plants respond to it. In: Caldwell MM, Pearcy RW (eds) Exploitation of environmental heterogeneity by plants. Ecophysiological processes above- and belowground. Academic Press, San Diego, pp 391–414

Bella IE, Navratil S (1987) Growth losses from winter drying (red belt damage) in lodgepole pine stands on the east slopes of the Rockies in Alberta. Can J For Res 17:1289–1292

Beniston M, Fox DG, Adhikary S, Andressen R, Guisan A, Holten JI, Innes JL, Maitima J, Price MF, Tessier L, Barry R, Bonnard C, David F, Graumlich L, Halpin P, Henttonen H, Holtmeier F-K, Jaervinen A, Jonasson S, Kittel T, Kloetzli F, Körner Ch, Kräuchi N, Molau U, Musselman R, Ottesen P, Peterson DL, Saelthun N, Shao X, Skre O, Solomina O, Spichiger R, Sulzman E, Thinon M, Williams R (1996) Impacts of climate change on mountain regions. In: Watson RT, Zinyowera MC, Moss RH (eds) Climate change 1995. Impacts, adaptations and mitigation of climate change: scientific-technical analyses. Cambridge University Press, Cambridge, pp 191–213.

Berg EE, Chapin FS III (1994) Needle loss as a mechanism of winter drought avoidance in boreal conifers. Can J For Res 24:1144–1148

Bernier B, Paré D, Brazeau M (1989) Natural stresses, nutrient imbalances and forest decline in southeastern Quebec. Water Air Soil Poll 48:239–250

Bonan GB, Shugart HH (1989) Environmental factors and ecological processes in boreal forests. Ann Rev Ecol Syst 20:1–28

Braathe P (1995) Birch dieback – caused by prolonged spring thaws and subsequent frost. Norw J Agric Sci, Suppl. 20:1–59

Buchmann N, Kao W-Y, Ehleringer JR (1996) Carbon dioxide concentrations within forest canopies – variation with time, stand structure, and vegetation type. Global Change Biol 2:421–432

Bussotti F, Bottacci A, Bartolesi A, Grossoni P, Tani C (1995) Morpho-anatomical alterations in leaves collected from beech trees growing in conditions of natural water stress. Environ Exp Bot 35:201–213

Butin H (1995) Tree diseases and disorders. Causes, biology, and control in forest and amenity trees. Oxford University Press, Oxford

Cannell MGR (1984) Spring frost damage on young *Picea sitchensis*. I. Occurrence of damaging frosts in Scotland compared with western North America. Forestry 57:159–175

Carvell KL, Tryon EH, True RP (1957) Effects of glaze on the development of Appalachian hardwoods. J Forestry 55:130–132

Clinton BD, Boring LR, Swank WT (1993) Canopy gap characteristics and drought influences in oak forests of the Coweeta Basin. Ecology 74:1551–1558

Cochard H, Bréda N, Granier A (1996) Whole tree hydraulic conductance and water loss regulation in *Quercus* during drought: evidence for stomatal control of embolism? Ann Sci For 53:197–206

Cook E, Bird T, Peterson M, Barbetti M, Buckley B, D'Arrigo R, Francey R, Tans P (1991) Climatic change in Tasmania inferred from a 1089-year tree-ring chronology of Huon pine. Science 253:1266–1268

De Vries W, Vel EM, Reinds GJ, Deelstra HD (1997) Intensive monitoring of forest ecosystems in Europe. Technical Report 1997. European Commission, Brussels and UN Economic Commission for Europe, Geneva

Elliott KJ, Swank WT (1994) Impacts of drought on tree mortality and growth in a mixed hardwood forest. J Veg Sci 5:229–236

Froment A, Tanghe M (1967) Répercussion des formes anciennes d'agriculture sur les sols et la composition floristique. Bull Soc Roy Bot Belg 100:335–352

Geipel R (1992) Naturrisken, Katastrophenbewältigung im sozialen Umfeld. Darmstadt, Wissenschaftliche Buchgesellschaft

Goldammer JG, Di X (1990) Fire and forest development in the Daxinganling Montane-Boreal coniferous forest, Heilongjiang, Northeast China – a preliminary model. In: Goldammer JG, Jenkins MJ (eds) Fire in Ecosystem Dynamics. Mediterranean and northern perspectives. SPB Academic Publishing, The Hague, pp 175–184

Goovaerts P, Frankart R, Gérard G (1990) Effet de la succession de différentes affectations sur les propriétés chimiques de pédons en Fagne de Chimay (Belgique). Pédologie 40:179–194

Grayson AJ (1991) The 1987 storm: impacts and responses. Forestry Commission Bulletin 87, HMSO, London

Grier CC (1988) Foliage loss due to snow, wind, and winter drying damage: its effects on leaf biomass of some western conifers. Can J For Res 18:1097–1102

Hadley JL, Smith WK (1986) Wind effects on needles of timberline conifers: Seasonal influences on mortality. Ecology 67:12–19

Hannerz M (1994) Predicting the risk of frost occurrence after budburst of Norway spruce in Sweden. Silva Fenn 28:243–249

Havis JR (1976) Root hardiness of woody ornamentals. HortScience 11:385–386

Heinselman ML (1981) Fire and the disturbance and structure of northern ecosystems. Gen. Tech. Rep. WO – U.S. Forest Service GTR-WO-26, pp 7–57

Henson WR (1952) Chinook winds and red belt injury to lodgepole pine in the Rocky Mountain parks area of Canada. For Chron 28:62–64

Hursh CR, Haasis FW (1931) Effects of 1925 summer drought on southern Appalachian hardwoods. Ecology 12:380–386

Innes JL (1991) High-altitude and high-latitude tree growth in relation to past, present and future global climate change. The Holocene 1:168–173

Innes JL (1992a) Observations on the condition of beech (*Fagus sylvatica* L.) in Britain in 1990. Forestry 65:35–60

Innes JL (1992b) Structure of evergreen temperate rain forest on the Taitao Peninsula, southern Chile. J Biogeog 19:555–562

Innes JL (1993) Forest health: Its assessment and status. CAB International, Wallingford

Innes JL (1994) Climatic sensitivity of temperate forests. Environ Pollut 83:237–243.

Jackson GE, Grace J (1996) Field measurements of xylem cavitation: are acoustic emissions useful? J Exp Bot 47:1643–1650

Jalkanen R (1993) Abiotic and biotic diseases of the northern boreal forests in Finland. In: Jalkanen R, Aalto T, Lahti M-L (eds) Forest pathological research in northern forests with a special reference to abiotic stress factors. Paper 451, Finnish Forest Research Institute, Rovaniemi, pp 7–21

Jalkanen R, Aalto T, Derome K, Niska K, Ritari A (1995) Lapin neulaskato. Männyn neulaskatoon 1987 johtaneet tekijät Pohjois-Suomessa. Loppuraportti. Metsäntutkimuslaitoksen tiedonantoja 544, 1–75.

Johnson EA (1992) Fire and vegetation dynamics. Studies from the North American boreal forest. Cambridge University Press, Cambridge

Kauffman JB, Uhl C (1990) Interactions and consequences of deforestation and fire in the rainforests of the Amazon Basin. In: Goldammer JG (ed) Fire in the tropical and subtropical biota. Springer-Verlag, Berlin, pp 117–134

Kilgore BM (1973) The ecological role of fire in Sierra conifer forests: Its application to national park management. Quatern Res 3:496–513

Kirschbaum MUF, Bullock P, Evans JR, Goulding K, Jarvis PG, Noble IR, Rounsevell, M, Sharkey, TD, Austin MP, Brookes P, Brown S, Bugmann HKM, Cramer WP, Diaz S, Gitay H, Hamburg SP, Harris J, Holten JI, Kriedemann PE, Le Houerou HN, Linder S, Luxmoore RJ, McMurtrie RE, Pitelka LF, Powlson D, Raison RJ, Rastetter EB, Roetter R, Rogasik J, Sauerbeck DR, Sombroek W, van de Geijn SC (1996a) Ecophysiological, ecological, and soil processes in terrestrial ecosystems: a primer on general concepts and relationships. In: Watson RT, Zinyowera MC, Moss RH (eds) Climate change 1995. Impacts, adaptations and mitigation of climate change: scientific-technical analyses. Cambridge University Press, Cambridge, pp 57–74.

Kirschbaum MUF, Fischlin A, Cannell MGR, Cruz RVO, Galinski W, Cramer WP, Alvarez A, Austin MP, Bugmann HKM, Booth TH, Chipompha NWS, Ciesla WM, Eamus D, Goldammer JG, Henderson-Sellers A, Huntley B, Innes JL, Kaufmann MR, Kräuchi N, Kile GA, Kokorin AO, Körner Ch, Landsberg J, Linder S, Leemans R, Luxmoore RJ, Markham A, McMurtrie RE, Neilson RP, Norby RJ, Odera JA, Prentice IC, Pitelka LF, Rastetter EB, Solomon AM, Stewart R, van Minnen J, Weber M, Xu D (1996b) Climate change impacts on forests. . In: Watson RT, Zinyowera MC, Moss RH (eds) Climate change 1995. Impacts, adaptations and mitigation of climate change: scientific-technical analyses. Cambridge University Press, Cambridge, pp 95–129.

Körner Ch (1995) Towards a better experimental basis for upscaling plant responses to elevated CO_2 and climate warming. Plant Cell Environ 18:1101–1110

Koerner W, Dupouey JL, Dambrine E, Benoît M (1987) Influence of past land use on the vegetation and soils of present day forest in the Vosges mountains, France. J Ecol 85:351–358

Kozlowski TT (1982) Water supply and tree growth. Part I. Water deficits. Forestry Abstracts 43:57–95

Kozlowski TT, Kramer PJ, Pallardy SG (1991) The physiological ecology of woody plants. Academic Press, San Diego

Kramer K (1995) Phenotypic plasticity of the phenology of seven European tree species in relation to climatic warming. Plant Cell Environ 18:93–104

Kuiper D, Kuiper PJC (1988) Phenotypic plasticity in a physiological perspective. Acta Oecol Plant 9:43–59

Lachance D (1985) Répartition géographique et intensité de dépérissement de l'érable à sucre dans les érablières au Québec. Phytoprotection 66:83–90

Landis TD, Evans AK (1974) A relationship between *Fomes applanatus* and aspen windthrow. Plant Disease Reports 58:110–113

Landmann G (1995) Forest decline and air pollution effects in the French mountains: a synthesis. In: Landmann G, Bonneau M (eds) Forest decline and atmospheric deposition effects in the French Mountains. Springer-Verlag, Berlin, pp 407–452

Larcher W (1995) Physiological plant ecology. Ecophysiology and stress physiology of functional groups. Springer-Verlag, Berlin

Larcher W, Mair B (1969) Die Temperaturresistenz als ökophysiologisches Konstitutionsmerkmal. 1. *Quercus ilex* und andere Eichenarten des Mittelmeergebietes. Oecologia Plantarum 4:347–376

Le Tacon F, Bonneau M, Gelpe J, Boisseau T, Paradat P (1994) Le dépérissement du pin maritime dans les Landes de Gascogne à la suite des introductions de graines d'origine ibérique et des grands froids des années 1962–63 et 1985. Rev For Fr 46:474–484

Livingston NJ, Spittlehouse DL (1996) Carbon isotope fractionation in tree ring early and late wood in relation to intra-growing season water balance. Plant Cell Environ 19:768–774

Lopez-Portilla J, Keyes MR, Gonzales A, Cabrera C, E, Sanchez O (1990) Los incendios de Quintana Roo: catastrofe ecologica o evento periodico? Ciencia y Desarrollo 16:43–57

Lorimer CG (1989) Relative effects of small and large disturbances on temperate hardwood forest structure. Ecology 70:656–567

Lorimer CG, Frelich LE (1994) Natural disturbance regimes in old-growth northern hardwoods: implications for restoration efforts. J Forest 92:33–38

Lu P, Biron P, Bréda N, Granier A (1995) Water relations of adult Norway spruce (Picea abies (L) Karst) under soil drought in the Vosges Mountains: water potential, stomatal conductance and transpiration. Ann Sci For 52:117-129

Lu P, Biron P, Granier A, Cochard H (1996) Water relations of adult Norway spruce (Picea *abies* (L) Karst) under soil drought in the Vosges mountains: whole-tree hydraulic conductance, xylem embolism and water loss regulation. Ann Sci For 53:113–121

McWilliam JR (1983) Physiological basis for chilling stress and the consequences for crop production. In: Raper CD, Kramer PJ (eds) Crop reactions to water and temperature stresses in humid, temperate climates. Westview Press, Boulder, Colorado, pp 113–132

Myeni RB, Keeling CD, Tucker CJ, Asrar G, Nemani RR (1997) Increased plant growth in the northern high latitudes from 1981 to 1991. Nature (London) 386:698–702

Nelson BW, Kapos V, Adams JB, Oliveira WJ, Braun OPG, do Amaral IL (1994) Forest disturbance by large blowdowns in the Brazilian Amazon. Ecology 75:853–858

Oliver CD, Larson BC (1996) Forest stand dynamics. Update edition. John Wiley & Sons, New York

Panek JA, Waring RH (1997) Stable carbon isotopes as indicators of limitations to forest growth imposed by climate stress. Ecol Appl 7:854–863

Perry DA (1994) Forest ecosystems. Baltimore, The John Hopkins University Press

Peterken GF (1996) Natural woodland. Ecology and conservation in northern temperate regions. Cambridge University Press, Cambridge

Peterken GF, Jones EW (1987) Forty years of change in Lady Park Wood, the old-growth stands. J Ecol 75:477–512

Peterken GF, Mountford EP (1996) Effects of drought on beech in Lady Park Wood, an unmanaged mixed deciduous woodland. Forestry 69:125–136

Pierce FR (1921) Tornado destroys great forest. Popular Mechanics, May 1921, 659–662

Raitio H (1991) Problem hos lövträd. Skogsaktuellt 1991(3):24–25

Rebetez M (1996) Seasonal relationship between temperature, precipitation and snow cover in a mountainous region. Theor. Appl Climatol 54:99–106

Redfern DB (1993) Climatic injury as a problem of introduced species in northern Britain. In: Jalkanen R, Aalto T, Lahti M-L (eds) Forest pathological research in northern forests with a special reference to abiotic stress factors. Paper 451, Finnish Forest Research Institute, Rovaniemi, pp 44–49

Redfern DB, Cannell MGR (1982) Needle damage in Sitka spruce caused by early autumn frosts. Forestry 54:39–45

Redfern DB, Rose DR (1984) Winter cold damage to pines. Report on Forest Research 1984. HMSO, London, pp 33–34

Rehfeldt GE (1988) Ecological genetics of *Pinus contorta* from the Rocky Mountains (USA): A synthesis. Silv Gen 37:131–135

Robertson A (1991) Some effects of wind on northern forests in a changing climate. Commonw For Rev 70:47–55

Robins JK, Susut JP (1974) Red belt in Alberta. Canadian Forest Service Information Report NOR-X-99

Rochefort RM, Little RL, Woodward A, Peterson DL (1994) Changes in sub-alpine tree distribution in western North America: a review of climatic and other causal factors. The Holocene 4:89–100

Romme WH (1982) Fire and landscape diversity in subalpine forests of Yellowstone National Park. Ecol Monog 52:199–221

Sakai A, Weiser CJ (1973) Freezing resistance of trees in North America with reference to tree regions. Ecology 54:118–126

Saldarriaga JG, West DC (1986) Holocene fires in the northern Amazon Basin. Quatern Res 26:358–366

Savill PS (1983) Silviculture in windy climates. For Abstr 44:473–488

Schowalter TD, Coulson RN, Crossley DA (1981) Role of southern pine beetle and fire in maintenance of structure and function of the southeastern coniferous forest. Environ Entomol 10:821–825

Shaw CG, Taes EHA (1977) Impact of *Dothistroma* needle blight and *Armillaria* root rot on diameter growth of *Pinus radiata*. Phytopathology 66:1319–1323

Shostakovitch VB (1925) Forest conflagrations in Siberia. With special reference to the fire of 1915. J For 23:365–371

Sirois L, Payette S (1991) Reduced postfire tree regeneration along a boreal forest–forest tundra transect in northern Quebec. Ecology 72:619–627

Spiecker H, Mielikäinen K, Köhl M, Skovsgaard J (eds) (1996) Growth trends in European forests. Studies from 12 countries. Springer Verlag, Berlin.

de Steven D, Kilne J, Matthiae PE (1991) Long-term changes in a Wisconsin *Fagus—Acer* forest in relation to glaze storm disturbance. J Veg Sci 2:201–208

Strand G-H (1997) Effects of early summer drought on the crown density of Norway spruce. Forestry 70:157–160

Sundblad LG, Andersson B (1995) No difference in frost hardiness between high and low altitude *Pinus sylvestris* (L.) offspring. Scand J For Res 10:22–26

Tranquillini W (1955) Die Bedeutung des Lichtes und der Temperatur für die Kohlensäureassimilation von *Pinus cembra* Jungwuchs an einem hochalpinen Standort. Planta 46:154–178

Tranquillini W (1979) Physiological ecology of the alpine timberline. Springer-Verlag, Heidelberg

Trapnell CG (1959) Ecological results of woodland burning experiments in northern Rhodesia. J Ecol 47:129–168

Uhl C, Kauffman JB, Cummings DL (1988) Fire in the Venezuelan Amazon 2: Environmental conditions necessary for forest fires in the evergreen forest of Venezuela. Oikos 53:176–184

Valentini R, Anfodillo T, Ehleringer JR (1994) Water sources and carbon isotope composition ($\delta^{13}C$) of selected tree species of the Italian Alps. Can J For Res 24:1575–1578

Veblen TT (1985) Stand dynamics in Chilean *Nothofagus* forests. In: Pickett STA, White PS (eds) The ecology of natural disturbance and patch dynamics. Academic Press, Orlando, pp 35–51

Veblen TT, Schlegel FM, Oltremari JV (1983) Temperate broad-leaved evergreen forests of South America. In: Ovington JD (ed) Ecosystems of the World 10. Temperate broad-leaved evergreen forests. Elsevier, Amsterdam, pp 5–31

Venn K (1993) Red belts in boreal forests. In: Jalkanen R, Aalto T, Lahti M-L (eds) Forest pathological research in northern forests with a special reference to abiotic stress factors. Paper 451, Finnish Forest Research Institute, Rovaniemi, pp 50–54

Villalba R, Veblen TT (1997) Regional patterns of tree population age structures in northern Patagonia: climatic and disturbance influences. J Ecol 85:113–124

Wagener WW (1961) Past fire incidence in Sierra Nevada forests. J For 59:739–748

Waggoner PE (1989) Anticipating the frequency distribution of precipitation if climate change alters its mean. Agric For Meteorol 47:321–337

Wardle P (1971) An explanation of the alpine treeline. N Z J Bot 9:371–402

Weaver H (1951) Fire as an ecological factor in the southwestern ponderosa pine forests. J For 49:93–98

Webster R, Rigling A, Walthert L (1995) An analysis of crown condition of *Picea*, *Fagus* and *Abies* in relation to environment in Switzerland. Forestry 69:347–355
Weeden P (1990) The aftermath of Hurricane Hugo. Wildfire News Notes 4:1–5
Whitney HE, Johnson WC (1984) Ice storms and forest succession in southwestern Virginia. Bull Torrey Bot Club 111:429–437
Williams GJ, MacMillan C (1971) Frost tolerance of *Liquidambar styraciflua* native to the United States. Can J Bot 49:1551–1558
Wilson J (1980) Macroscopic features of wind damage to leaves of *Acer pseudoplatanus* L. and its relationship with season, leaf age, and windspeed. Ann Bot (Lond) New Series 46:303–311
Woodcock H, Vollenweider P, Dubs R, Hofer R-M (1995) Crown alterations induced by decline: a study of relationships between growth rate and crown morphology in beech (*Fagus sylvatica* L.). Trees: Struct Func 9:279–288

Meteorological extremes and their impacts on forests in the Czech Republic

Rudolf Brázdil

Abstract. Meteorological extremes in the Czech Republic (CR) cause considerable damage to forest stands. The effects of such extremes has increased conspicuously in the latter half of the present century, with salvage felling due to meteorological factors accounting in some years for more than half of the total timber cut in the CR. The most important reason for this salvage felling is damage due to wind (61 %), followed by damage due to snow (16 %), drought, air pollution and ice deposits. Using data from four professionally-maintained weather stations and one special station, time series for maximum wind gusts are analysed as well as the frequencies of days with ice deposits, maximum mass of ice, heights of new snow ≥ 10 cm and, for areal precipitation series from Bohemia and Moravia, precipitation sums for the year, the summer half-year and the frequencies of occurrence of dry months. The problems of measuring these characteristics and their homogeneities are discussed. Their annual distribution and their long-term changes (fluctuations, trends) are studied. The main forest disasters of the 20th century attributable to the identified meteorological extremes are described. The analysis does not, however, permit reliable conclusions about the future behaviour of those extremes and their impact on forests under conditions of global warming.

1. Introduction

The present distribution of forest stands in the Czech Republic (CR) is the result of the interaction of geographical, climatic and anthropogenic factors. Geographical factors include those associated with the locality (soil, slope, exposition, etc.). Man's activities in the forests of the CR are reflected (a) in a change in the species diversity of stands that began in the latter half of the 19th century, with a change towards the economically most required and profitable wood species (Kupčák 1994) and (b) in the degradation of forest stands by industrial air pollution, particularly SO_2. This is especially evident in *Picea abies* forests in the borderland mountain regions of the northern part of the Czech Republic, from the Krušné hory Mountains to the Moravskoslezské Beskydy Mountains (Palát et al. 1994).

The most important climatic factors affecting the growth of trees in the CR are air temperature and precipitation. The response functions calculated for tree-ring chronologies from the CR indicate a significant growth response by trees to air temperatures in those mountain regions where precipitation does not limit tree growth (see e.g. Sander et al. 1995 for the Šumava Mountains or Brázdil et al. 1997 for the Krkonoše Mountains). At lower altitudes, where air temperatures during the summer half-year are sufficient for tree growth, precipitation plays the dominant role (see for example the dendrochronology of *Abies* sp. from south Moravia, covering the period 1498–1996 – Kyncl and Kyncl 1997).

In the case of air temperature, damage to forests arises principally as a consequence of sudden temperature changes (early autumn and late spring frosts), but long-lasting severe winter temperatures or high summer temperatures can also be important (Forst et al. 1985). For example, damage due to frost was extremely extensive during the exceptionally severe winter of 1928–29 (particularly

February). The damage or death of forest stands throughout the CR was due to a sudden temperature reversal between 31 December 1978 and 1 January 1979 (Report on Forestry of the Czech Republic 1995) when, after the passage of a cold front, air temperature dropped from about 10°C to between -15 and -20°C (the greatest drop occurred at Třeboò in south Bohemia: 29.8°C in 24 h – for details see Rein and Štekl 1981). The greatest damage occurred in stands of *Fagus sylvatica* in the Moravskoslezské Beskydy Mountains, where there was the combined impact of the severe temperature drop and high pollution levels (Musil 1982; Förchtgott 1986). With precipitation, damage is not only due to drought, but also to intense rainfall, floods and hailstorms (Forst et al. 1985).

In addition, damage or complete destruction of individual trees, groups of trees in a stand and whole stands can occur due to strong winds or to a large amount of wet snow or thick ice deposits (the term ice deposits is used here to include hard and soft rime, clear ice and glaze). The least dangerous form of tree damage is the snapping of branches. This mainly occurs in deciduous or old coniferous stands. In coniferous stands, the top part of the tree often snaps (termed top-break – Fig. 1). More vigorous trees can overcome this damage without great difficulty if the broken part does not exceed one-third of the green crown. If a greater part is broken off (termed crown-break), the tree may die. The most severe form of stem damage is the snapping of any part of the tree below the green crown (termed stem-break), which may sometimes be partly attributable to rot weakening the stem. In young and middle-aged stands the tree can be bent. Most leaning coniferous trees remain so and they soon die or are snapped in the following years by wind or snow. With wind breakage, trees are often blown over, and the tree stem may then be broken. The falling broken stems may cause mechanical damage to the remaining standing trees. Wind and snow breakage can also have secondary consequences, such as subsequent breakage in blown-down stands, and soil water-logging and erosion (Vicena et al. 1979).

Global warming due to anthropogenic intensification of the greenhouse effect may have important effects on the state of forests, as shown by the applications of alternative regional scenarios of climatic change for the CR (see Vinš et al. 1997). Potential changes in the frequencies of occurrence and intensity of meteorological extremes have not been clarified. Small changes in the mean climate or its variability may be accompanied by great changes in the frequencies of extreme phenomena, so it is necessary to study selected meteorological extremes and their impacts on forests at the local and regional scales. This may have important applications for the implementation of adaptation measures in the forest economic sector. Such adaptations must be planned well in advance because of the long lead-time involved in forest productivity.

This contribution deals with selected aspects of the analysis of wind, ice deposits, snow and drought, and their impacts on forests in the CR.

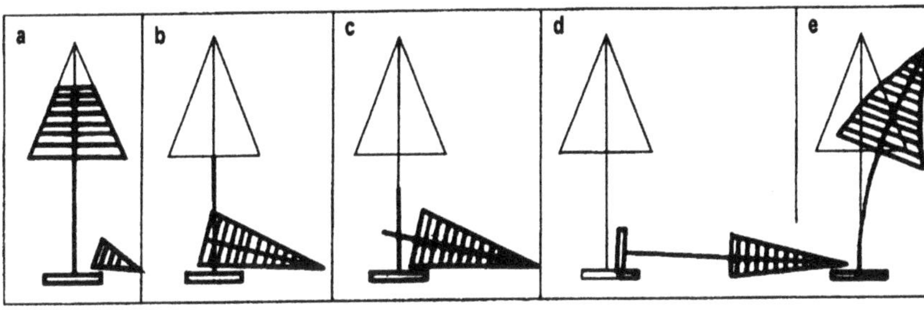

Fig. 1. Possible kinds of tree damage during wind, snow and ice deposit disasters: a) top-break, b) crown-break, c) stem-break, d) wind-fall, e) wind-lean (adapted after Vicena 1992)

2. Impacts of meteorological extremes on the forest economy in the Czech Republic

Meteorological extremes are often responsible for severe forest damage; this is recorded by the amount of salvage felling. Disasters affect the growth, renewal and planning of the economy and often require economic intervention, with adverse impacts on the overall economy.

Although biotic effects (primarily insects) were mostly responsible for forest disasters from the early 18th century until the mid-20th century, in the latter half of the 20th century abiotic factors have became prevalent, principally meteorological extremes and air pollution (Report on Forestry of the Czech Republic 1996). As a result, damage due to wind, ice deposits and snow accounted for 43.8% of salvage felling of timber in 1900–1950, whereas in 1951–1980 it was as much as 73.0%. Damage due to drought and air pollution grew from 1.8% to 18.1% (Forst et al. 1985). The historical causes for the rise in this damage are the extensive conversion of the original natural stands into monocultures which, when they were first established, appeared to be economically attractive. Later it became apparent that the monocultures had little resistance to wind and snow. The impaired and weakened stands then became victims of the impacts of further stress factors. In a situation of considerable forest damage, economic problems, particularly the neglect of economic measures, were important for the stability of stands, especially the preparation of young stands (Report on Forestry of the Czech Republic 1995).

Salvage felling[1] as a percentage of the total timber harvest in the CR increased during the period 1963–1995 (Fig. 2). The highest proportion was reached in 1985 (82.4 %), followed by 1993 and 1994 (77.8 and 77.7 %, respectively). Up to 1992,

[1] Information about the species diversity of the salvage felling was obtained from reports from individual forest districts concerning the occurrence of harmful factors in some forests. Some salvage felling can sometimes be postponed to the following year because of the occurrence of a disaster (such as gale).

meteorological factors were primarily responsible for the salvage felling, but in 1993–1995 biotic factors (insects, fungal diseases, damage due to game) were more important. The increasing proportion of salvage felling is considered by Kupka (1995) to be a clear symptom of the growing instability of forest ecosystems in the CR.

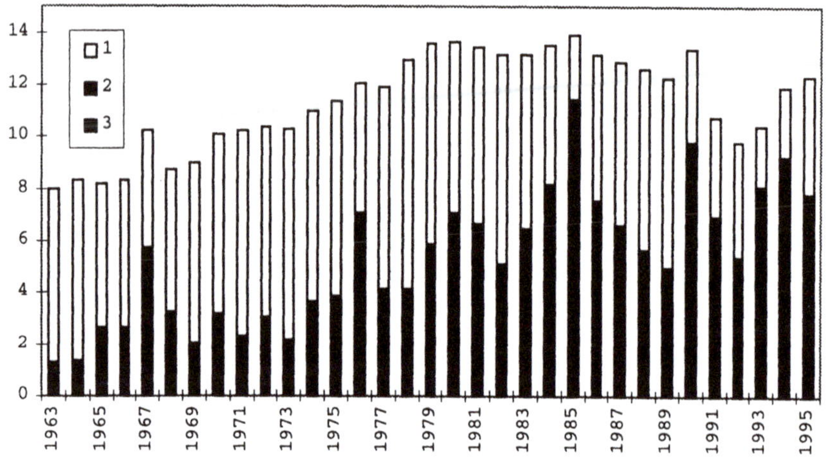

Fig. 2. Timber felling in the CR (millions m^3) in the period of 1963–1995: 1) total felling, 2) total salvage felling, 3) salvage felling due to meteorological extremes. Data sources: FGMRI; Report on Forestry of the Czech Republic 1995.

The amount of salvage felling due to meteorological factors in the CR between 1963 and 1995 increased until 1990 (Fig. 2), when the maximum was reached (9.2 million m^3 of timber, i.e. 69 % of the whole timber cut in the CR in that year). The values were lower in the 1990s. Salvage felling due to meteorological factors accounted for more than one-half of the total timber cut in 1967 (52.9 %), 1984 (52.2 %) and 1985 (59.8 %). Damage due to wind was prevalent (Fig. 3), reaching a maximum in 1990 (8.4 million m^3 of timber). Only in 1969, 1970, 1979 and 1980 (3.4 million m^3 of timber) were they exceeded by damage due to snow and in 1996 by damage due to ice deposits (2 million m^3 of timber). Salvage felling due to drought was greatest in the years 1993–1995 (1.5 million m^3 of timber each in 1994 and 1995) and by pollution in 1980 and 1981 (more than 1 million m^3 of timber in each year). However, these two factors may not always be clearly distinguished in statistics about salvage felling (Jancarík et al. 1993). Salvage felling between 1963 and 1996 accounted for 134.7 million m^3 of timber, with the following percentages related to individual causes: wind 61.1 %, snow 15.9 %, drought 9.7 %, industrial pollution 9.5 % and ice deposits 3.8 %.

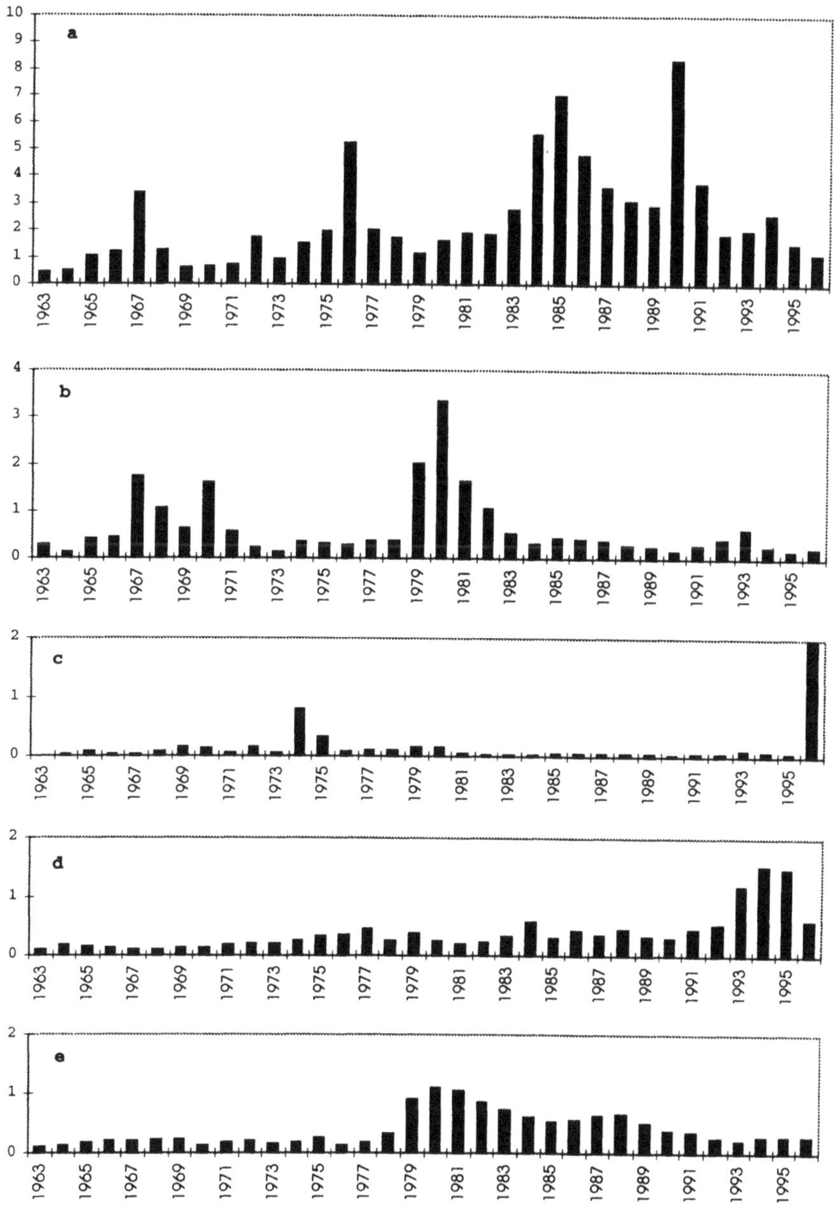

Fig. 3. Salvage felling in the CR (millions m^3) due to meteorological factors in the period of 1963–1996: a) wind, b) snow, c) ice deposits, d) drought, e) air pollution. Data sources: FGMRI

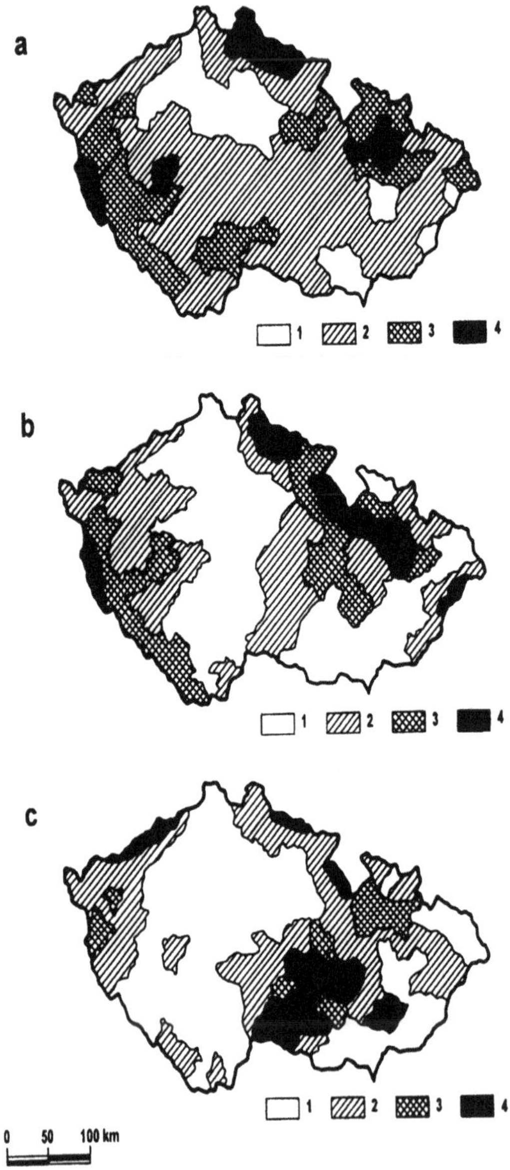

Fig. 4. Location of damage due to (a) wind, (b) snow and (c) ice deposits in the CR in the period of 1963–1975 (Vicena et al. 1979). Degree of damage according to the mean amount of salvage felling in $m^3\ ha^{-1}$: a) 1 – negligible (≤ 0.10), 2 – weak (0.11–0.50), 3 – medium (0.51–1.00), 4 – strong (≥ 1.01); b) 1 – negligible (≤ 0.10), 2 – weak (0.11–0.25), 3 – medium (0.26–0.60), 4 – strong (≥ 0.61); c) 1 – negligible (≤ 0.010), 2 – weak (0.011–0.100), 3 – medium (0.101–0.200), 4 – strong (≥ 0.201)

When examining the mean annual salvage felling per hectare in the period of 1963–1975, the location of wind and snow breakage can be specified in the CR by the degree of damage in the stands (Vicena et al. 1979; Parez 1980a, 1980b). In the case of wind (Fig. 4a) 9% of the stand area was affected by severe damage (in the Jizerské hory Mountains, the Krkonoše Mountains, part of the Ceský les Mountains, the Brdy Mountains and the Bohemian-Moravian Highlands, the Hrubý Jeseník Mountains, and the western part of the Beskydy Mountains) and 24% by medium damage. Ice deposits caused severe damage to 8% of the stand area of the CR (particularly the Bohemian-Moravian Highlands, the highest positions of Mt. Kralický Snìsník, the Orlické hory Mountains and the Krkonoše Mountains), and approximately the same extent was classified as having a "medium degree of damage" (Fig. 4c). Snow breakage also accounted for severe damage in 8% of the stand area (in the Jizerské hory Mountains, the Krkonoše Mountains, the Orlické hory Mountains, the Ceský les Mountains, the Hrubý Jeseník Mountains, and the Moravskoslezské Beskydy Mountains), and there was "medium degree of damage" to 23%, similar to that from wind breakage (Fig. 4b). Meteorological extremes thus affect largely the same regions.

The occurrence of wind and snow breakage is also a function of altitude. The destructive effects of wind are mainly concentrated in a zone at 400–800 m (Vicena et al. 1979); wind damage at higher altitudes is not exceptional, but the areas affected are smaller. Snow breakage mainly occurs at elevations from 450 to 900 m (at lower altitudes the snow is often replaced by rain whereas at higher altitudes the falling snow is often powdery, resulting in less damage). Damage due to ice deposits is limited to specific altitudinal zones, determined partly by exposure, ranging from 600–800 m in the Bohemian-Moravian Highlands to 1000–1250 m in the region of Mt. Kralický Snìsník.

3. Instrumental records of meteorological extremes

For the analysis of meteorological extremes we have selected monthly series of maximum wind gusts, days with ice deposits, maximum mass of ice deposits, depths of new snow ≥ 10 cm and sums of precipitation. Because of the likelihood that such phenomena are temporally and locally limited, testing of the relative homogeneity of their series can be a problem. For this study, it was therefore necessary to use observations from stations for which there are no conspicuous impairments in data homogeneity. Conclusions about the possible homogeneity of such series will have a rather qualitative character and we therefore only used professionally-maintained meteorological stations (Fig. 5).

Mt. Milešovka (H = 835 m) is a typical mountain station situated on an isolated phonolite cone in northwest Bohemia with measurements since 1905. The history of the station, including metadata, is documented in detail in the papers by Štekl and Zacharov (1993) and Štekl and Podzimek (1993). The station is ranked as a reference station of the WMO.

Fig. 5. Location of meteorological stations on the territory of the CR used in this study

Prague–Karlov (H = 232 m) is a typically urban station situated on the roof platform of the former Institute of Meteorology and Climatology of Charles University, 27 m above the ground, with measurements since 1921.

Brno–Tuřany (H = 241 m) is the airport station on the Tuřany terrace, southeast of the city of Brno. Until 15 April 1958, the station was situated about 3.5 km north-west of the present position. Measurements started in 1946.

Kostelní Myslová (H = 569 m) is a rural station representing the lower part of the Bohemian-Moravian Highlands, with measurements since 1946.

Series of areal sums of precipitation for Bohemia (Jílek 1957) and Moravia (Brázdil et al. 1985) were used to characterise drought. The Bohemian series (1875–1996) was first calculated by the isohyet method, then by the squares method and later as an arithmetic mean from all rain-gauge stations. The Moravian series (1881–1988) was calculated as a weighted arithmetic mean of 143 stations, using their altitudes above sea level as weights. Although the change in the calculation method of the areal means may be a source of inhomogeneity, no breaks were indicated when the monthly series were tested using the Alexandersson method (Alexandersson 1986).

4. Analysis of selected meteorological extremes in the Czech Republic

4.1 Wind

Mean hourly wind speed and the maximum gusts are measured by anemographs at professionally-maintained meteorological stations in the CR. The quality of measurements is adversely affected by several factors, including interruptions due to ice deposit formation, lightning strikes to the sensor or the adjustment or replacement of the instrument, such as with the introduction of more sensitive anemographs (Metra) in the early 1960s (Slabý 1993; Štekl 1997). At stations without anemographs the wind force was assessed according to the 13-point Beaufort scale, and a statistic was derived comprising days with strong (6 to 7 on the Beaufort scale) or stormy (8 or more) winds. This scale divides storms into gales (9–11) and hurricanes (12). A sudden short-term increase of wind speed with gusts of >30–40 m s^{-1} is often accompanied by squalls, which are signs of rapidly ascending and descending movements of air, occurring in cumulonimbi along fronts and instability lines (Kameník 1986; Štekl 1996). Observations at weather stations may of course fail to record the occurrence of tornadoes, although such events can cause severe damage along transects in forests (see e.g. Fikar 1950; Šálek 1994; Setvák et al. 1996).

The destructive effects of wind on forest stands (snapping, windfall) mainly materialises after reaching some limiting threshold, which is the product of the momentary wind speed and the force necessary to overcome the mechanical resistance of the stand. As this force is determined by the stand properties, it is practically constant at a given moment and the damage therefore depends on the maximum wind speed or gusting (Nekovár and Valter 1996). Damage to trees arises at a wind speeds of more than 60 km h^{-1}, and severe damage at wind speeds of more than 90 km h^{-1} (Vicena et al. 1979). Consequently, the following evaluation deals with the characteristics of maximum wind gusts. Analyses for the period 1961–1987 for the stations Prague–Ruzynì, Milešovka and 40 stations in the CR have been made by Slabý (1990a, 1990b, 1993). He considered the main reason for the gusts to be the intensified pressure gradient in the cold half-year.

Wind gusts recorded at the stations Prague–Karlov and Milešovka have been analysed here. At the Prague–Karlov observatory, gusts have been measured since 1921 using a Dines pressure tube anemograph (Munro, London) installed on a spire structure 8.5 m above the platform, 33.8 m above the ground (height above sea level 269 m). The only break in the records occurred from 24 January to 28 February 1947, when repairs to the damaged anemograph tubing were carried out. In 1960 the instrument was replaced by a universal anemograph (Metra) and on 30 October 1964 by its more sensitive version (Slabý 1993). A further replacement occurred in 1996. Unfortunately, these interventions affected the homogeneity of observations, so the series of maximum wind gusts has been divided into two parts (1921–1959 : 0.39% missing days, 1960–1996 : 0.21% missing days). In the latter case, only the 1965–1995 period can be considered as homogeneous. Anemograph measurements have been undertaken since the establishment of the Milešovka station in 1905, but data for maximum wind gusts at the station are only available since 1956 (up to that time the values of gusts in m s^{-1} were converted to the Beaufort scale). The universal anemograph Metra is installed 4.5 m above the spire of the observatory, 22.5 m above the ground. Unlike Prague–Karlov, the more exposed Milešovka site has more frequent measurement breaks (1.44 % missing days). Series of monthly maximum wind gusts and monthly numbers of days with wind gusts of ≥ 17 m s^{-1} (V17), ≥ 20 m s^{-1} (V20), ≥ 25 m s^{-1} (V25), ≥ 30 m s^{-1} (V30), ≥ 35 m s^{-1} (V35) and ≥ 40 m s^{-1} (V40) were compiled for both stations.

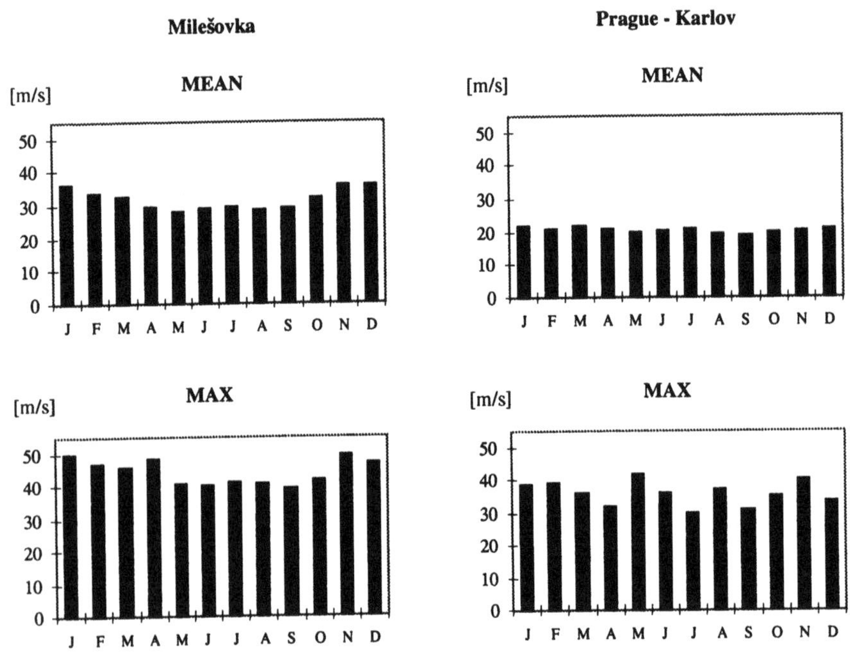

Fig. 6. Annual variation of means of maximum monthly wind gusts (MEAN) and absolute monthly wind gusts (MAX) at Milešovka (1961–1990) and Prague–Karlov (1965–1994)

The basic characteristics of the two stations are presented for the thirty-year homogeneous periods 1961–1990 (Milešovka) and 1965–1994 (Prague–Karlov). The annual variation is better seen at Milešovka (Fig. 6), where the means of maximum wind gusts in the winter half-year months exceed the means of the summer half-year months by as much as 8 m s^{-1} (maximum in January, minimum in May). The maximum wind gust was measured at Milešovka at 0445 on 14 January 1967, coming from the NW direction and having a value of 50 m s^{-1} (i.e. the upper measurement limit of the anemograph, so that the actual gust speed may have been even higher). At Prague–Karlov, in a balanced annual variation, the monthly means of maximum wind gusts are in the range of 20 m s^{-1}, also with a January maximum, but with the minimum in September. The strongest wind gust (41.7 m s^{-1}) was measured on 11 May 1992 at 1214 h, from the WSW direction. An examination of the annual variation of days with gusts exceeding the threshold limits (Fig. 7) indicated that the two stations record maximum frequencies in November–January and in March, but that they differed in the minima (Milešovka: May and August; Prague–Karlov: September). At Milešovka, the maximum monthly gust was less than 17 m s^{-1} in only 2 months (0.6 %), whereas at Prague–Karlov it was below 17 m s^{-1} in 28 months (7.7 %). For other wind speed categories, the corresponding percentages at Milešovka were substantially higher (for V20: 98.1 and 60.8 % of all months, respectively, for V25: 85.3 and 28.2 %, respectively, for V30: 53.1 and 9.1 %, respectively, for V35: 31.7 and 1.9 %, respectively and for V40: 12.2 and 0.6 %, respectively). The prevailing directions for maximum gusts are WNW, NW and W at Milešovka; for speeds of V30 and higher, more frequent gusts come from the NW than from WNW (Fig. 8). At Prague–Karlov the most frequent V17 gusts are from the WSW, WNW and W, but the directions of the V20 and V25 gusts are the same as at Milešovka (W, WNW and NW). This corresponds with the findings of Slabý (1993) about the prevalence of gusts from the western quadrant in Bohemia, while in Moravia and Silesia gusts generally come from the S and SE.

The fluctuation of annual maximum wind gusts and the frequencies of days with gusts exceeding specific thresholds were analysed for the period of 1956–1996 at Milešovka and for the periods 1921–1959 and 1965–1995 at Prague–Karlov (Figs. 9, 10). At Milešovka, the two characteristics show a significant increase (t-test, p = 0.05) only for the V17 frequencies (9 days in 10 years). At Prague–Karlov, there was a reduction in the frequencies (with the exception of the V30 frequencies), but this was only significant for the V17 frequencies in the years 1921–1959 (-2.6 days in 10 years).

The most severe damage due to wind in the CR in the period 1900–1980 occurred in January (22.1%), followed by July (15.3%) and November (13.2%) (Table 1). The most significant wind disasters of the 20th century are listed in Table 2. The event of 23–24 November 1984 is denoted as the disaster of the century which, in total, involved 12 million m^3 of timber (for a meteorological analysis of this event, see Setvák and Strachota 1986). Sprátek (1990) considered this event to be comparable to that of 1 March 1990, when an estimated 11 million m^3 of timber was blown.

	J	F	M	A	M	J	J	A	S	O	N	D
Wind	22.1	6.6	2.4	6.9	5.7	4.2	15.3	10.0	0.6	4.1	13.7	8.4
Snow	0.5	5.4	3.9	23.9	.	.	.	.	.	23.7	31.9	10.7
Ice deposits	60.2	7.6	21.3	0.5	.	.	.	.	.	2.6	0.2	7.6

Table 1. Monthly variation in the percentage of disaster damage due to meteorological factors in the CR in the period of 1900–1980 (Forst et al. 1985)

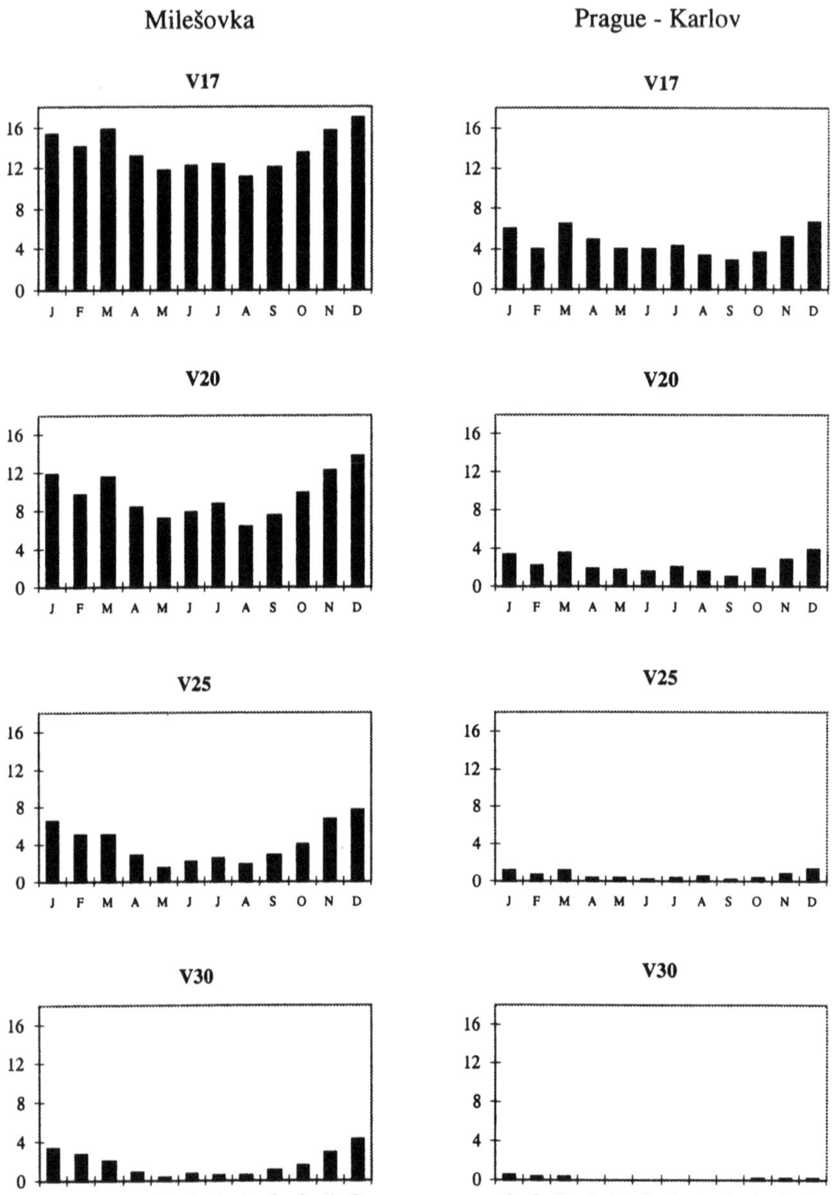

Fig. 7. Annual variation of numbers of days with gusts exceeding specific thresholds at Milešovka (1961–1990) and Prague–Karlov (1965–1994)

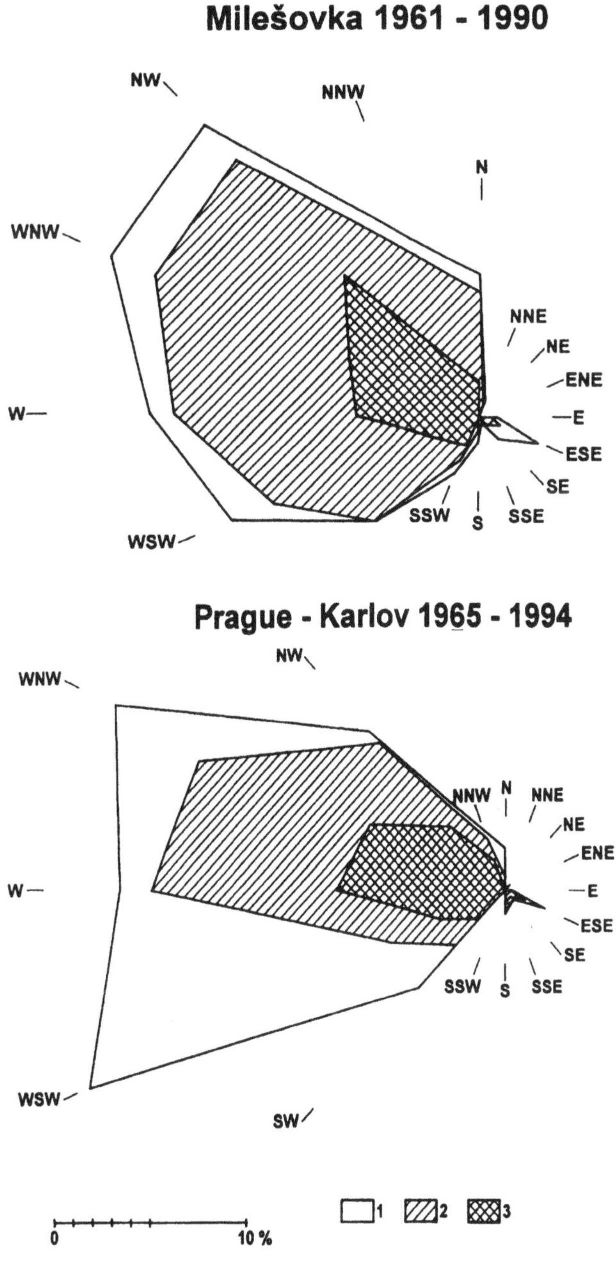

Fig. 8. Wind roses of maximum monthly wind gusts at Milešovka (1– V17, 2 – V25, 3 – V35) and Prague–Karlov (1 – V17, 2 – V20, 3 – V25)

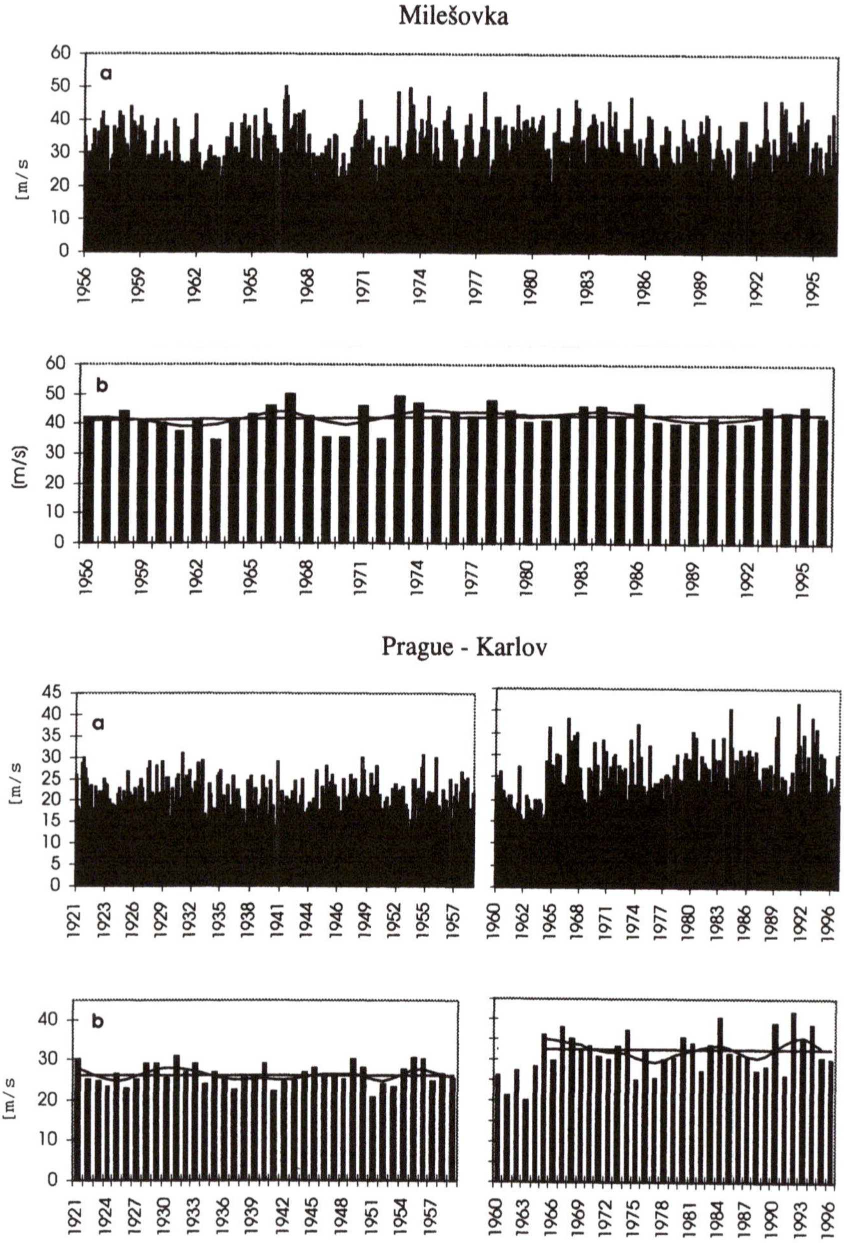

Fig. 9. Fluctuation of maximum monthly (a) and annual (b) wind gusts at the stations Milešovka (1956–1996) and Prague–Karlov (1921–1996). Values (b) smoothed by the Gauss filter (10 years) and completed by the linear trend

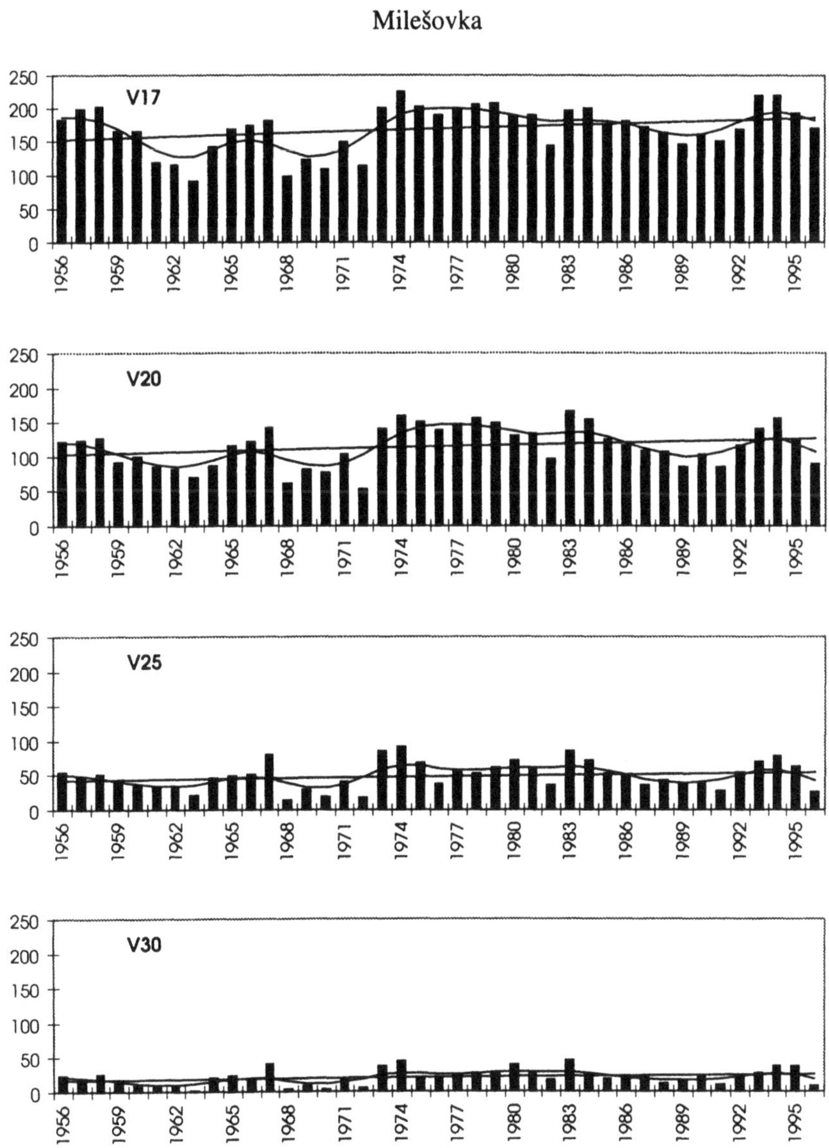

Fig. 10. Annual fluctuations in the numbers of days with maximum wind gusts exceeding specific thresholds at Milešovka (1956–1996) and Prague–Karlov (1921–1996). Smoothed by the Gauss filter (10 years) and completed by the linear trend

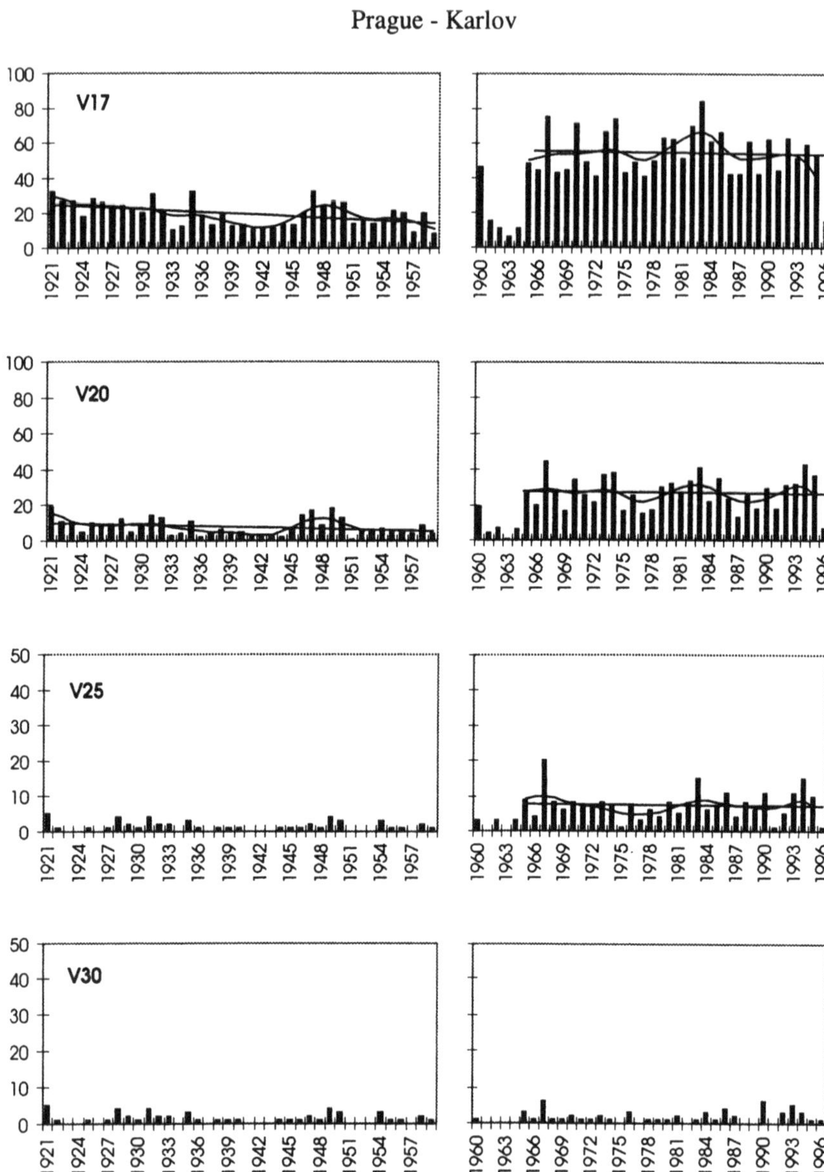

Fig. 10. – continued

Wind	4 July 1929; 16–17 January 1955[a]; 21, 23, 28 February 1967; 2–4 January 1976; 23–24 November 1984; 1 March 1990
Ice deposits	1904–1905; 1908; 1911; 1916; 1933; 1969; 1970; 1972; 1973/74; 1979; 1980; 1995/96; 1996/97
Snow	1916; 25–30 October 1930[b]; 1933[c]; 6–7 December 1939; 15–16 November 1941; 28–29 October 1956; 21 February – 1 March 1967; 25 November 1969 – 10 March 1970; 31 December 1978 – 1 April 1979[d]; 24 April 1980[d]

[a] meteorological analysis see Gregor (1955)
[b] 7 million m^3 of snow breakage timber
[c] together with glaze and ice
[d] for detail see Parez (1982)

Table 2. The most significant wind, ice deposits and snow disasters during the 20th century in the CR (Parez 1980b; Report on Forestry of the Czech Republic 1995; Vicena et al. 1979; Vicena and Vokroj 1991)

4.2 Ice deposits

Wet snow and ice deposits on symmetric tree crowns intensify the stress on the stem, causing buckling which, in combination with the lateral pressure of wind (or only the crown's eccentric centre of gravity), often causes the top to break close to the crown base (Sereda 1994).

The different types of ice deposits recorded in observations at weather stations in the CR include soft rime (hoar frost), hard rime and clear ice. Ice deposits are formed at various temperatures by the freezing of fog or cloud droplets onto trees. Freezing of super-cooled rain drops or drizzle on frozen surfaces leads to ice riming. When recording ice deposit phenomena, the times of their start and end are indicated, as are their duration and their intensity and the number of days with ice deposits. These observations are closely linked to the accuracy with which the observer works, which limits useful data to those from professionally-maintained meteorological stations. The proper measurement of the mass of ice deposits is carried out at special stations during the ice deposit cycle (i.e. the period of ice deposit formation), by means of a 1 m measuring pole with a diameter of 60 or 30 mm. However, long series of such observations are the exception (for details, see Popolanský 1996b).

The problems of ice deposits have been studied in the CR particularly with respect to their impact on energetics (see especially the papers by B. Hrudi ka in the 1930s, in Popolanský 1996b and more recent work, e.g. Vrána 1980, 1986). Thanks to this work, what may be the longest continuous series in the world for the maximum mass of ice deposits has been obtained at Studnice (H = 800 m) in the Bohemian-Moravian Highlands, where measurements at a height of 5 m above the ground began in the winter of 1940–41 (Popolanský 1996a, 1996b). A homogeneous series of maximum masses of ice deposits is presented in Fig. 11. The values fluctuate greatly from year to year. The minima occurred in the 1940s and in 1975–1992, whereas the maxima were in the 1950s–1960s and the 1990s. Absolute maxima were recorded in the winters of 1956–57 (16.5 kg m^{-1}), 1962–63 (16.8 kg m^{-1}), 1967–68 (15 kg m^{-1}) and 1995–96. The last event, which occurred in December 1995 and involved 21.6 kg m^{-1} has been estimated by Popolanský (1996b) to correspond to a 100-year return period, although such statistics are notoriously unreliable because they are based on a short series. Indeed, in the winter of 1996–97, an even greater value (24.9 kg m^{-1}) was recorded (Popolanský 1997). In relation to the effects of ice deposits on trees it is important to note that the mass of the ice deposits increases with the height above the ground. Thus, at 10 m height it is a multiple of 1.32 and at 20 m the multiple is 1.76, based on values measured at 5 m above the ground (Popolanský 1996b).

The data on maximum values of ice deposits at Studnice correlate well (statistically significant correlation of 0.60) with the values of salvage felling caused by ice deposits in the CR in the period of 1963–1996. This can be related (among other things) to the fact that the Bohemian-Moravian Highlands are among those regions most heavily affected by ice deposits.

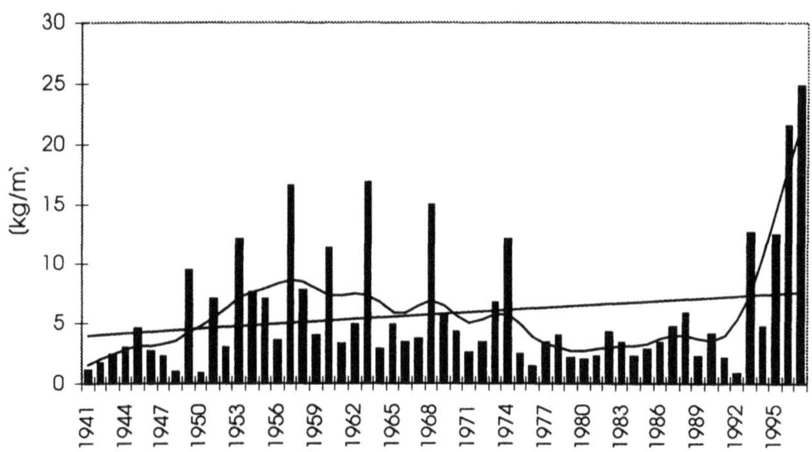

Fig. 11. Variation of maximum annual masses of ice deposits at Studnice in the period of 1940/41–1996/97 (data after Popolanský 1996b, 1997). Smoothed by a Gaussian filter (10 years) and completed by a linear trend. Years in the graph are marked according to the year in which the month of January falls

Problems associated with ice deposit phenomena in the CR in the period of 1954–1973 have been dealt with by Procházka (1976), who quoted a number of characteristics for 5 selected stations between 725 m and 835 m altitude. Rime occurs from October to May, and heavy ice deposits (layer thickness more than 3 cm) from November to March (or until April at Milešovka) with a December or January maximum. The ice deposits last on average 3–4 days, but in 1963 a 50-day ice deposit period was recorded in the Bohemian-Moravian Highlands (2 January – 20 February). Of the total number of days with ice deposits at Svratouch

(H = 737 m) in the Bohemian-Moravian Highlands, 34.3% had heavy ice deposits. Glaze occurs from October to April, with the maximum in some of the winter months.

At Milešovka, icing phenomena have been studied in detail by Procházka (1996). However, the time series for the number of days with ice deposits at Milešovka is dependent on the activity of the observer. In 1928 the observatory was included in the reporting network of the Czechoslovak aeronautical weather service, which made it possible to employ a worker with a university education, whereas before then only one observer had been present. Another decisive moment occurred in 1958, when the number of observers increased and a round-the-clock service was introduced. Therefore, in Fig. 12, the variations in the ice deposits have also been divided into the three partial sectors in which the homogeneity is believed to be reasonable. The high fluctuation of values during the war is explained by Procházka (1996) as being due to the frequent changing of observers, often after one month (see Štekl and Zacharov 1993). Thus there is an unrealistic increase of 8.7 days/10 years in the number of days with rime when the three partial periods are combined (Fig. 12). An analogous procedure was also used for glaze, the observation of which seems to have been less affected by the above problems. The number of days with glaze appears to be increasing significantly at a rate of 1.2 days/10 years (further significant trends were 3.6 days/10 years in 1905/06–1927/28 and -3.1 days/10 years in 1928/29–1957/58).

The most significant disasters due to ice deposits in the 20th century in the CR are listed in Table 2 (for a meteorological analysis of selected strong icing situations in the CR see for example Vrána 1980, 1986; Štekl, 1997). As with wet snow, strong winds can also accentuate the damage caused by ice deposits. In the period 1900–1980, disastrous damage due to ice deposits in the CR were concentrated mostly in January (60.2%) and March (21.3%) (Table 1).

4.3 Snow

The harmful effects of snow in forest stands depend mainly on the snow density. Powder snow with a specific density 100–200 kg m^{-3} does not cause direct damage. Moist to wet snow with a specific density 300–500 kg m^{-3} or greater is more dangerous. This causes damage in cases when a great amount of snow falls in a short time, when it accumulates in the tree crowns over a long period of time, or when it is combined with strong winds. Crowns snap (according to the moisture of the snow) with snow accumulation depths of 25–40 cm and greater (Vicena et al. 1979).

The depth of new snow is measured at weather stations by a snowmeter plate and a snowmeter stake at 0700 h, mean local time. The methodology of these measurements has hardly changed since they began. Moist or wet snow can be judged only indirectly from cases when there was snowfall at temperatures above 0°C, even though, according to Procházka (1976) it can also fall at air temperatures below 0°C.

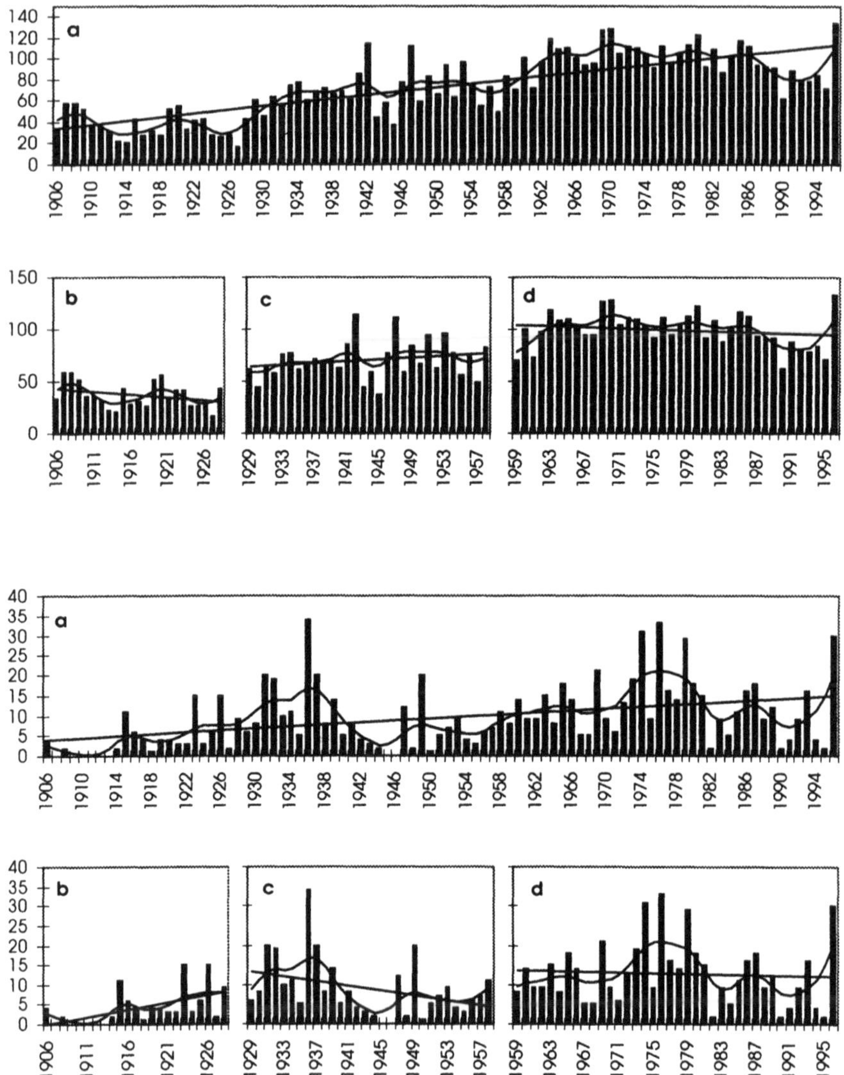

Fig. 12. Fluctuations in the number of days with rime (above) and glaze (below) at Milešovka in the periods (a) 1905/06–1995/96, (b) 1905/06–1927/28, (c) 1928/29–1957/58 and (d) 1958/59–1995/96. Smoothed by the Gauss filter (10 years) and completed by the linear trend. Years in the graph are marked according to the year in which the month of January occurs

For this analysis, the occurrence of snow days with new snow depths of ≥ 10 cm was selected. Over the year, at lower altitudes the largest number of these days falls in January, although they occur from November to March (see Brno-airport; Fig. 13). At medium elevations, they occur from October to April, and their frequency in the months of the winter half-year increases (Kostelní Myslová in January to March with the maximum in January, Milešovka from November to March with the maximum in February).

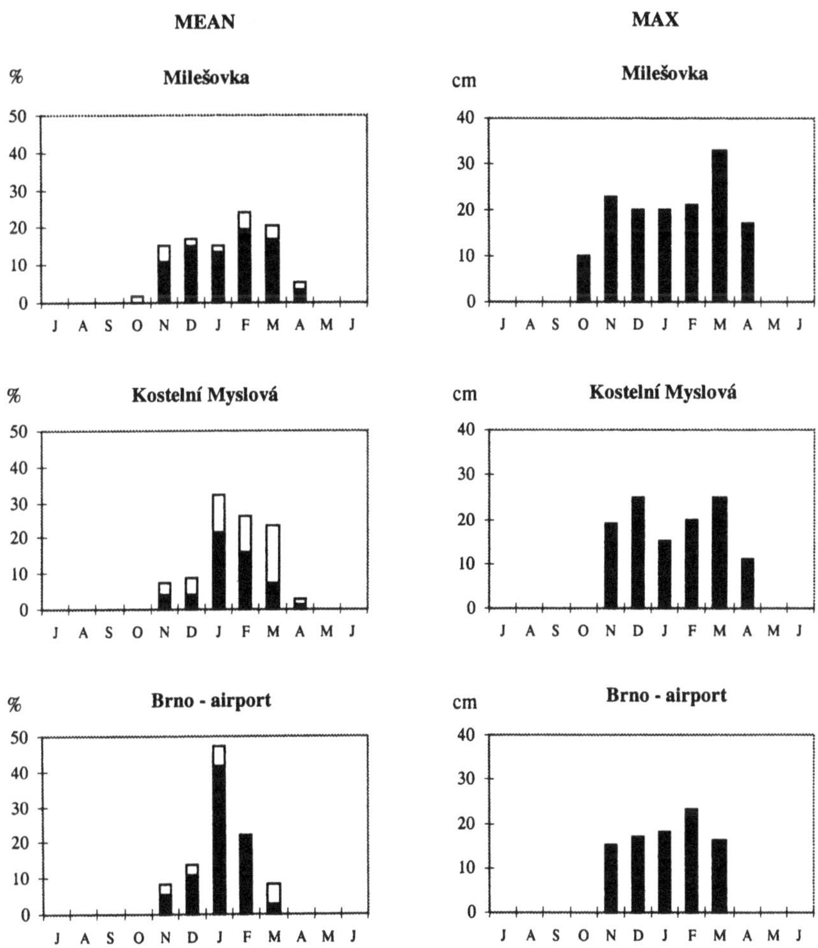

Fig. 13. Monthly variation in the frequency of days with new snow ≥ 10 cm (MEAN) with a marked proportion of wet snow (white columns) and absolute daily maxima of new snow ≥ 10 cm (MAX) at Milešovka (1945/46–1995/96), Kostelní Myslová and Brno-airport (both 1946/47–1996/97)

Cases were taken into account when temperatures were >0° for at least one of the records between 0700–0700 h (mean local time) in order to assess the possibility of wet snow falling. Such cases can occur in practically all months at stations recording a sum of new snow ≥ 10 cm (see Fig. 13). The total percentage of such days in the "days with new snow ≥ 10 cm" is relatively low (Brno-airport 16.7 %, Milešovka 19.8 %); an exception is Kostelní Myslová (44.9 %).

The fluctuation in the annual number of "days with new snow ≥ 10 cm" and "maximum heights of new snow ≥ 10 cm" each year is indicated for Brno-airport and Kostelní Myslová for the period of 1946/47–1996/97 and for Milešovka for the period of 1945/46–1995/96 (Fig. 14). There were long periods when new snow ≥ 10 cm was not measured, particularly at lower elevation stations (e.g. the latter half of the 1940s and the 1990s for Brno-airport), but even at medium elevations (Kostelní Myslová) new snow ≥ 10 cm is not an annual occurrence.

The most significant snow disasters during the 20th century in the CR are listed in Table 2. In the period of 1900–1980 the main damage due to wet snow was concentrated in November (31.9 %), April (23.9 %) and October (23.7 %) (see Table 1).

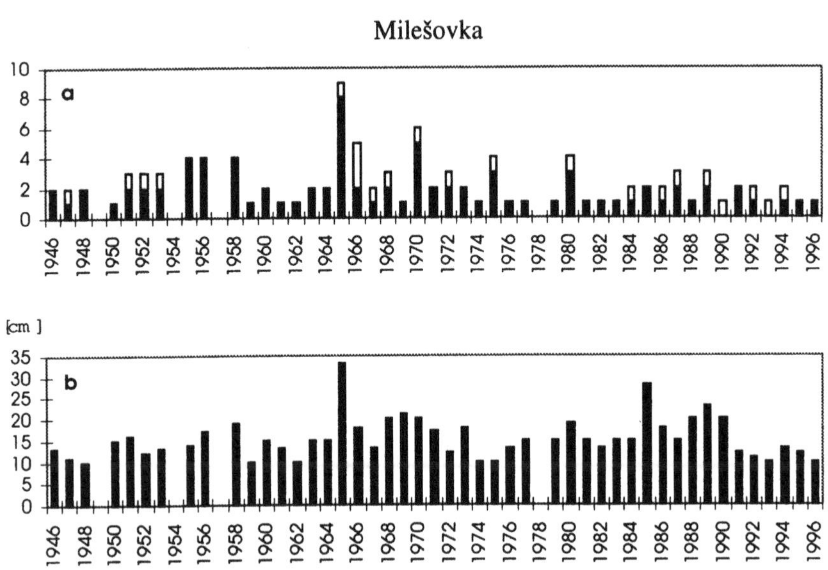

Fig. 14. Fluctuations in the numbers of days with new snow ≥ 10 cm (a) and the maximum height of new snow ≥ 10 cm (b) at Milešovka (1945/46–1995/96), Kostelní Myslová and Brno-airport (both 1946/47–1996/97). Days with new snow at positive temperatures (wet snow) are marked by empty columns. Years in the graph are marked according to the year in which the month of January occurs

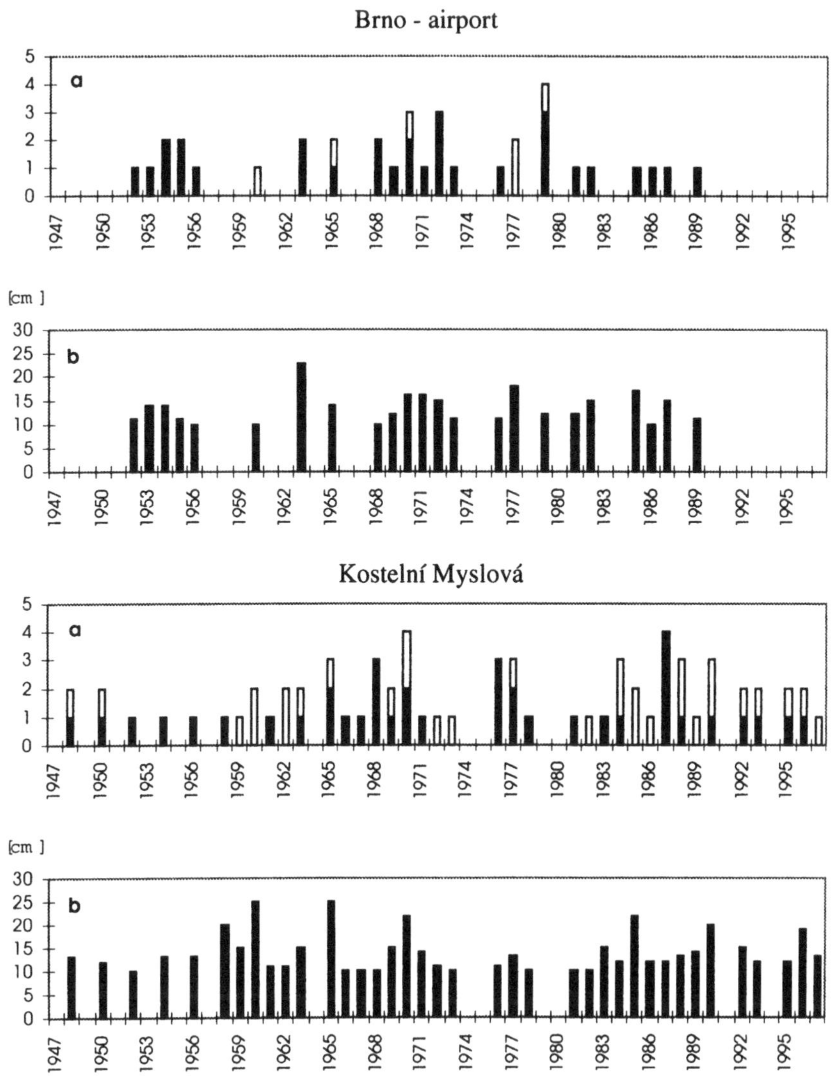

Fig. 14. – continued

4.4 Drought

Drought can cause severe damage to seedlings, transplants and mature trees. Both individual trees and whole groups of trees can die (dead standing trees). Indirectly, the consequences of drought appear in the form of increased forest fire hazard and in the rapid increase of insect pest populations (e.g. bark beetles after the 1992 drought – Mrkva 1993). Damage due to drought mainly affects lower-altitude forests (up to 500 m above sea level), but the effects can have large areal extents (Forst et al. 1985).

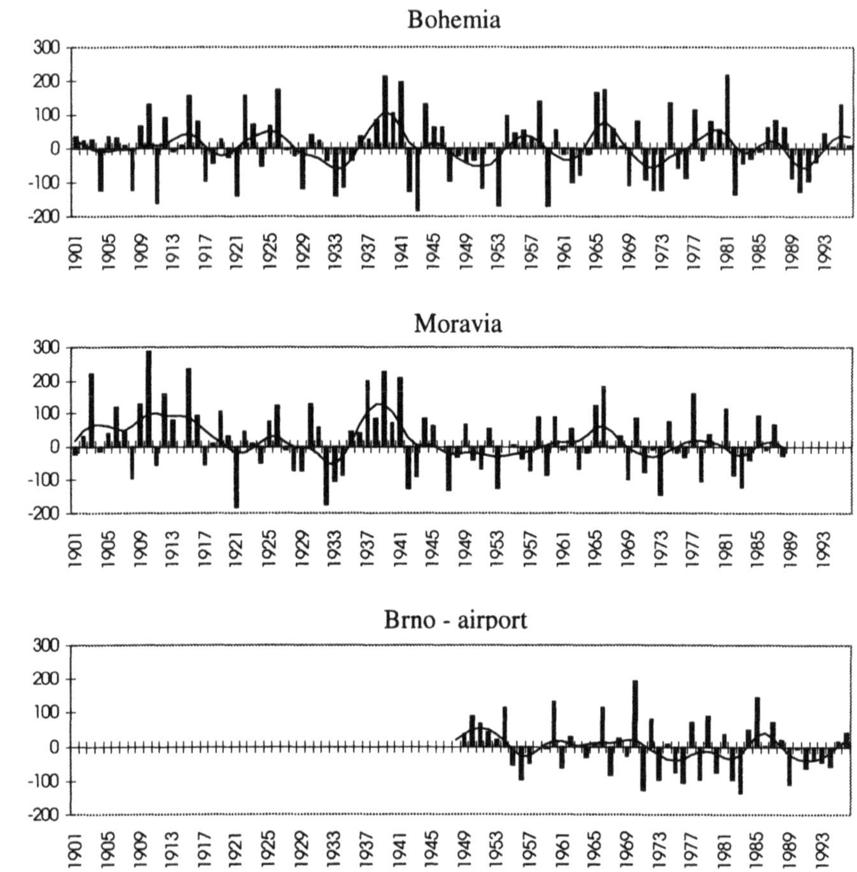

Fig. 15. Fluctuations in the annual precipitation sum anomalies (mm) in Bohemia (1901–1996), Moravia (1901–1988) and at Brno-airport (1948–1996). Smoothed by a Gaussian filter (10 years), reference period 1951–1980

The meteorological specification of a drought is difficult. It can be determined both from the point of view of precipitation sums and by including further data concerning air temperature and evaporation (e.g. using different indices). At the weather stations of the CR the amount of precipitation is measured by rain-gauges (Metra) with an orifice area of 500 cm^2 at a height of 1 m above the ground. The measured values are not corrected for systematic errors (Sevruk 1985, 1989) and such corrections have not been undertaken in the CR (unlike in a number of other

minima is more than twice this value (4.2°C/ century). The corresponding decrease in diurnal temperature range is therefore quite marked, i.e., over 2°C.

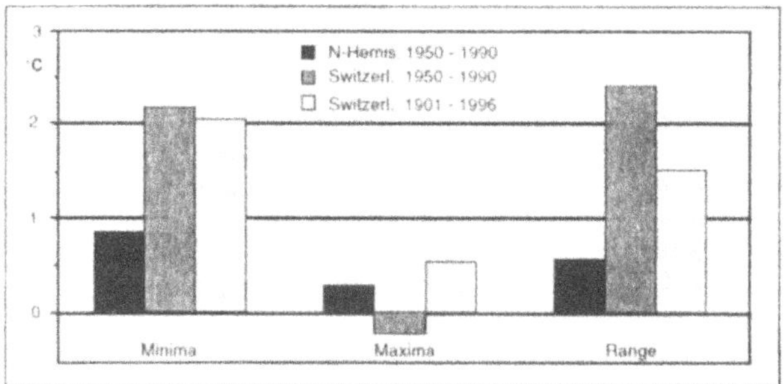

Fig. 2. Comparison of trends (per century) in diurnal temperature range between the Northern Hemisphere for the period 1950–1990 (Karl et al. 1993) and for Switzerland. Swiss trends are based on the periods 1950–1990 and 1901–1996.

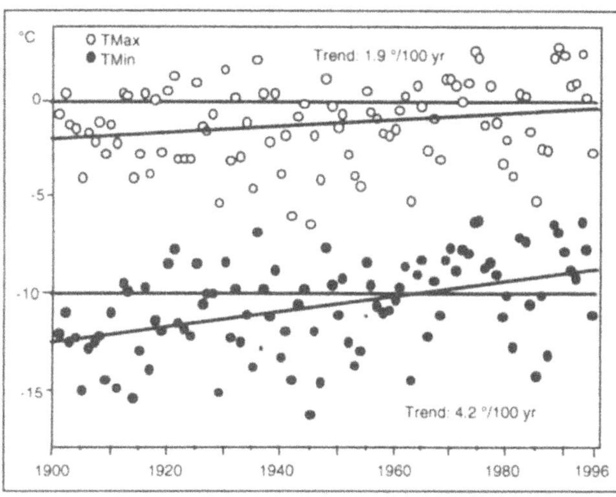

Fig. 3. Minimum and maximum January temperatures from 1901 to 1996 in Davos.

3.2 Monthly temperature trends

Not all months exhibit trends in minimum and maximum temperatures as substantial as those shown in the previous example. As highlighted in figures 4 and 5, for both sets, the trends are greatest in the autumn and winter, and are weakest in the late spring and early summer. Trends in maxima are negative in May and June, (-0.82°, resp. -0.68°C/ century), and maximum in October, (+1.44°C/ century). Minimum temperature trends range from +1.1°C/ century in May and June to +2.5°C/ century in January and February.

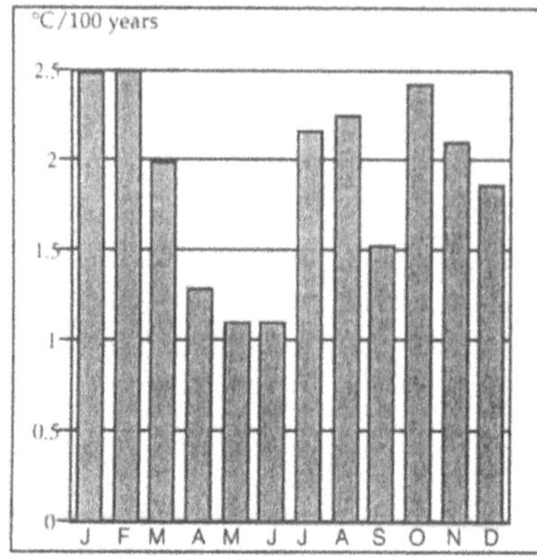

Fig. 4. Monthly mean minimum temperature trends (5 stations)

These strong trends can be explained to a good degree by changes in the behaviour of the North Atlantic Oscillation Index (a measure of the strength of the synoptic flow over the North Atlantic). Periods of high NAO index are closely linked to anomalously high wintertime temperatures and below average precipitation in the Alpine region (Beniston et al. 1994; Beniston 1997). Since the early 1970s, the NAO Index has been increasing and has been persistently above its long-term average value, until its sudden reversal in early 1996. This has influenced the large-scale pressure field and associated meteorological characteristics over much of the western and central European region. This period of anomalously high NAO values, spanning a quarter of a century, has certainly contributed to the accelerated warming trends observed over this time interval. The influence of the NAO index on the latter part of the temperature record is dominant; in the absence of such long-term forcing, the temperature trends would have been much lower.

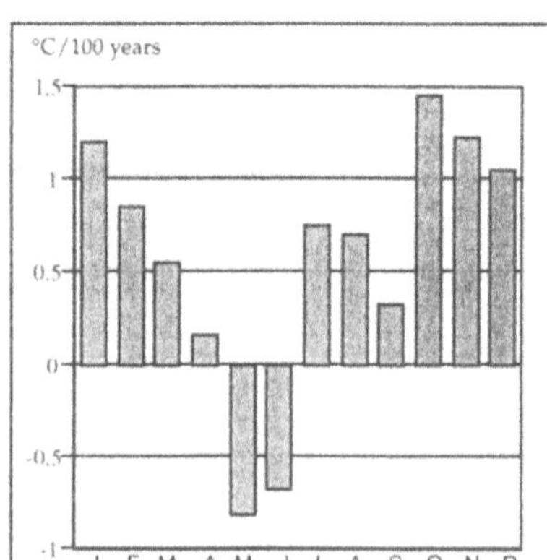

Fig. 5. As Figure 4, except for monthly mean maximum temperature trends

3.3 Variability

Figure 6 illustrates the nature of the relationship between the monthly mean values and the variance of the corresponding daily values for minimum temperatures, based on the period 1901–1996. There is a strong negative correlation from November to March, which implies that warmer nocturnal temperatures in winter may be associated with a reduction in their variability. During thc warmer part of the year, the strength of the relationship between means and variances abates. Positive correlation coefficients never reach the 90% significance level.

The situation for maximum temperatures is less clear. Figure 7 illustrates the nature of the relationship between the mean values and the variance of the corresponding daily values for maximum temperatures. As in the case of minimum temperatures, there is also a negative correlation in winter, but a less homogenous one than for the minima. In July, there is also a significant negative correlation for a couple of stations. In September and, to a lesser extent in August, there is a positive correlation coefficient, but only at lower elevations. The two mountain sites show no significant correlation for these two months.

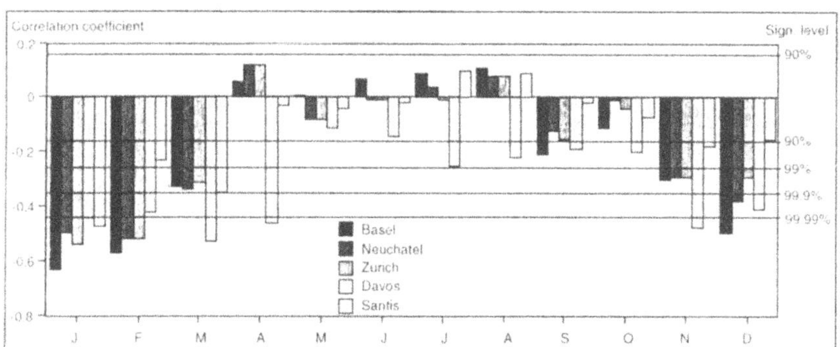

Fig. 6. Correlations between the monthly mean values and the variance of the corresponding daily values for minimum temperatures, based on the period 1901 - 1996.

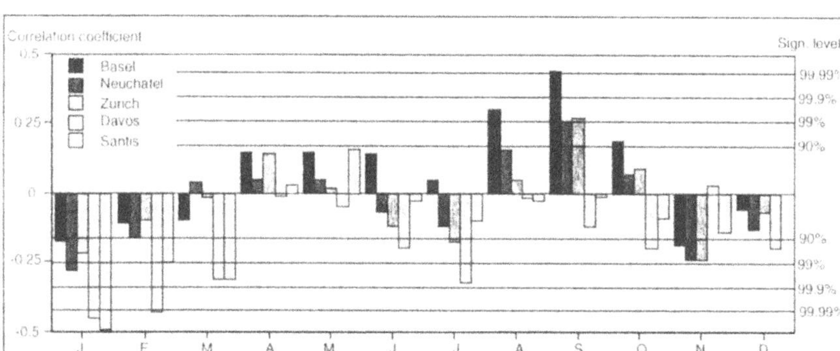

Fig. 7. As Figure 6, except for maximum temperatures

Reduced variance can also to some extent be interpreted as increased persistence. The high correlations between means and variance observed during the autumn, winter and early spring reflect the persistence of anticyclonic patterns over much of western and central Europe in these periods. Associated with these persistent pressures and occasional blocking highs, particularly during high NAO Index modes, are large positive temperature departures from their mean values, thereby establishing the link between persistence – or reduced variance – and increasing temperatures. On the other hand, the months of May and June are well known for their advective character, when variability is inherently high. They are also months when the warming tendency has been smallest this century. Under such circumstances, the mean-to-variance relationship cannot be expected to be particularly high.

Such considerations extend to the behaviour of the individual stations (Figures 6 and 8); during periods of persistent high pressure and mild temperatures, both the high-elevation sites and those at low altitudes are under the influence of the same, uniform weather pattern, and therefore the stations generally exhibit the same broad mean-to-variance relationships. During the more perturbed months of May and June, however, regional and local characteristics can lead to very different behaviour when a frontal system is present. Such characteristics include, for example, orographically-enhanced cloudiness and precipitation (which may be absent in the lowlands), the presence of snow at high elevations but not in the lowlands, and the direction of the winds associated with the frontal system. These may lead to different responses at sites located north or south of the Alps, at high or low elevations, or for those with different exposure to the perturbed atmospheric flows. The mean-to-variance correlations differ greatly from station to station, with changes in sign occurring for certain sites in a given month (e.g. June, as illustrated in Figure 6).

4. Conclusions

During the period 1950–1990, a decrease in the daily temperature range of 2.4°C/century has been measured in Switzerland, which is a trend more than four times greater than that for the Northern Hemisphere. For the period 1901–96, the reduction in the daily temperature range reaches 1.5°C. The discrepancy between the Swiss observations and the figures for the Northern Hemisphere confirms earlier findings regarding the extreme sensitivity of the Alpine region to large-scale climatic forcing.

Trends in minimum and maximum temperatures are greatest in the autumn and winter and weakest in the late spring and early summer. Trends in maxima are negative in May and June.

Warmer nocturnal and diurnal temperatures in winter are correlated with a reduction in their variability, highlighted by the negative correlation in the relationship between the monthly mean and the variance of temperatures from November to March.

During the warmer part of the year, there is no significant change in the daily variability, except for August and September afternoon temperatures, at lower elevations, the variability of which tends to increase in a warmer climate.

The general trend highlighted in this paper shows that during the 20th century, a warming climate in Switzerland has been linked to a strong winter night warming and has been accompanied during that season by a general decrease in variability, both when considering nocturnal and diurnal temperatures together, i.e. the daily temperature range, and when considering them separately. There has been a reduction in daily temperature range, due to a stronger increase of nocturnal temperatures, and there has also been a decrease in the variability of winter night-time temperatures as well as in the variability of winter diurnal temperatures when considered separately.

This means that the strong warming tendency observed in Switzerland this century is linked to the reduction in the number of cold winter days and particularly nights, i.e. the loss of the lowermost extremes in the temperature probability distribution. It would now be useful to have more detailed analyses of winter night temperatures in order to improve the assessment of the extent of the reduction in the lower extreme of minimum temperatures.

In terms of the summer period, warming trends are much smaller compared to the winter, and trends in variability are mostly inexistent, except for the maxima in August and September. This exception is worth further analyses as it could mean that there is also an increase in the upper range of the maximum temperature probability density function, although summer average temperatures have by no means increased in the same proportions as winter temperatures.

For vegetation, and forests in particular, the results discussed here principally imply a lengthening of the vegetation period, a reduction of the frost period and an increase of the possibility for vegetation to live at higher altitudes, to greater extents than the increase in average temperatures could let us expect. This is very important for combined vegetation – climate models which should be able to take these trends in temperature extremes into consideration. The impact of temperatures described here is of course considered independently from other climatological parameters. Trends in precipitation or sunshine duration, and particularly in the extremes of these variables, could modulate the expected impacts of trends in extreme temperatures described in this paper. Further work is therefore necessary to analyse in greater detail the behavior of other climatic parameters this century.

Acknowledgements. This work has been supported in part by the Forest Investigation Programme, a joint project between the Swiss Federal Office of the Environment, Forests and Landscape (BUWAL, Bern) and the Swiss Federal Institute for Forest, Snow and Landscape Research (WSL, Birmensdorf). We are grateful to MeteoSwiss for providing the meteorological data and to Michèle Kaennel for assistance with the literature survey for this article.

5. References

Bantle H (1989) Programmdokumentation Klima-Datenbank am RZ-ETH Zürich, Swiss Meteorological Institute, Zürich Bengtsson, L., M. Botzet, and M. Esch, 1995: Hurricane-type vortices in a general circulation model: Part 1. Tellus 47A:175–196

Bengtsson L, Botzet, M, Esch M (1995) Hurricane type vortices in a general circulation model. Tellus 47a:175-196

Beniston M (ed) (1994) Mountain Environments in Changing Climates. Routledge Publishing Co, London and New York

Beniston M (1997) Variations of snow depth and duration in the Swiss Alps over the last 50 years: links to changes in large-scale climatic forcings. Climatic Change 36:281–300

Beniston M, Rebetez M (1996) Regional behavior of minimum temperatures in Switzerland for the period 1979 – 1993. Theor Appl Clim 53:231–243

Beniston M, Rebetez M, Giorgi F, Marinucci MR (1994) An analysis of regional climate change in Switzerland. Theor Appl Clim 49:135–159

Beniston M, Diaz HF, Bradley RS (1997) Climatic change at high elevation sites: A review. Climate Change 36:233–251

Giorgi F, Mearns LO (1991) Approaches to the simulation of regional climate change. Rev Geoph 29:191–216

IPCC (1996) Climate Change. The IPCC Second Assessment Report. Cambridge University Press, Cambridge

Karl T, Jones P, Knight R, Kukla G, Plummer N, Razuvayev V, Gallo K, Lindseay J, Charlon R, Peterson T (1993) A New Perspective on Recent Global Warming: Asymetric Trends of Daily Maximum and Minimum Temperature. Bull Am Meteorol Soc 74, no 6

Katz RW, Brown BG (1992) Extreme events in a changing climate: variability is more important than averages. Climate Change 21:289–302

Kunz P, Rey J-M (1995) Intempéries du 23 septembre 1994 dans le Haut-Valais (Massif du Simplon). Bull. de la Murithienne 1995:1–13

Mearns LO, Katz RW, Schneider SH (1984) Extreme high-temperature events: Changes in their probabilities with changes in mean temperature. J Clim Appl Meteorol 23:1601–1613

Pfister C (1995) Scientific report to the Swiss National Science Foundation, National Program on Climate and Natural Hazards, Bern, Switzerland

Rebetez M (1996) Seasonal relationships between temperature, precipitation and snow cover in Switzerland. Theor Appl Clim 54:99–106

Rebetez M, Lugon R, Baeriswyl PA (1997) Climatic warming and debris flows in high mountain regions: The case study of the Ritigraben Torrent (Swiss Alps). Climate Change 36:371–389

Wagner D (1996) Scenarios of extreme temperature events. Climate Change 33:385–407

Evaluation of the $2xCO_2$ impact on European climate variability with a variable resolution GCM

Michel Déqué and Francisco Javier Doblas-Reyes

Abstract. This paper analyses climate variability in a numerical simulation of the impact of a doubling of atmospheric carbon dioxide concentrations. The sea surface temperatures (SST) come from a 240-year simulation with the Hadley Centre coupled model. Two 10-year time slices are simulated with a variable resolution version of the ARPEGE-IFS atmosphere general circulation model (AGCM). This model has a maximum horizontal resolution over Europe (60 to 100 km mesh size). The impact over Europe is analysed through the daily minimum and maximum temperature, and precipitation. Minimum and maximum temperatures increase whatever the season. Precipitation increases during the colder half of the year. The impact on temporal variability is analysed separately for three parts of the spectrum. Seasonal variability increases for the three variables. The intraseasonal variability of temperature decreases, whereas the intraseasonal variability of precipitation increases in winter, with more heavy rainy days and less dry days. The impact on the interannual variability is not significant. The extreme temperatures are warmer, but no significant impact is found on extreme precipitation.

1. Introduction

One of the major components of climate equilibrium is greenhouse gas absorption. Anthropogenic emissions of carbon dioxide (CO_2) and other gases since the beginning of the last century have led to an increase in the greenhouse effect (IPCC 1995), which was hidden by the natural variability of the atmosphere and by heat absorption in the oceans. Over the last 20 years, progress in numerical climate simulations has enabled some insights into possible climate changes in the next century. The reference greenhouse effect change is the doubling of the CO_2 concentration with respect to the 1980 decade, for which the observed climate is well documented. According to extrapolations based on recent decades (1% per year increase), this doubling might occur in the 2060 decade.

It is relatively well established that the atmosphere will warm by about 2 K on a global average (if all other parameters remain constant, in particular aerosols). Uncertainties remain over the regional distribution of the warming (which could in some places be a cooling), and on the impact on the hydrological cycle. More recently, the question of the impact on climate variability has been raised (Barrow and Hulme 1996), as a change in the frequency of heavy rains, long dry periods, cold waves, etc., could be more damaging to human activities than a 10% increase or decrease in precipitation or a 2 K mean warming.

The coupled ocean–atmosphere GCMs which are involved in such studies have a horizontal resolution of more than 250 km due to the computation cost of centennial integrations (see for example Murphy and Mitchell 1995). These GCMs are not sufficiently accurate to account for regional orographic features (e.g. the Alpine bow or the Apennines chain) responsible for temperature or precipitation patterns over horizontal scales smaller than 1000 km.

The solution consists of performing time slice integrations with a more accurate and more expensive model. The ocean is assumed to be fairly insensitive

to the use of another atmosphere GCM. In fact, the forcing on a time-scale of a few years is exerted much more from the ocean toward the atmosphere than from the atmosphere toward the ocean. This technique is similar to the Atmosphere Model Intercomparison Project (AMIP), but the SSTs come from a model instead of observations. It has been used by Stephenson and Held (1993) with the same atmosphere model, and generalised by Mahfouf et al. (1994). More recently, Timbal et al. (1997) used and compared two different SST forcings.
Time-slice (also called snapshot) experiments can be performed with two kinds of models:

i) regional models. The simulation also uses atmospheric variables from the coupled ocean–atmosphere simulation, and is sometimes called a downscaling simulation. The most traditional method consists of using a limited area model (LAM) forced at its boundaries (e.g. Marinucci and Giorgi 1992; Jones et al. 1997). Another method, proposed by Goyette and Laprise (1996), consists of recomputing the physical parameterisations on a finer grid, while the dynamic terms are inferred from the global GCM integration.
ii) global models. They can be high resolution models (Wild et al. 1995), or variable resolution models, as in the present paper.

Our approach is based on a variable resolution GCM. The advantage is that with the same amount of computation as with a global high resolution model, one can get a much higher resolution (but in practice no more than a factor of four for climate simulation) in a domain of interest, at the expense of the remaining part of the globe, where the resolution is reduced. Of the different methods available to vary the resolution over the globe, we use the one described in Courtier and Geleyn (1988), which maintains the same scaling factor in the x- and y- directions (isotropy) at any point on the globe.

A version of the ARPEGE-IFS atmosphere model has been developed for regional studies over Europe. This model is described and validated in Déqué and Piedelievre (1995). Two 10-year time-slice simulations have been performed with a standard and a doubled CO_2 rate. The SSTs come from a 240-year scenario simulation with the UKMO coupled ocean–atmosphere model (Mitchell et al. 1995). The models and experiments are described below. A more comprehensive description and the presentation of the mean impacts, including a comparison of the other impacts found in the literature, may be found in Déqué et al. (1998). The present paper is devoted to the impact of a CO_2 doubling on the climate variability on different time scales. Below, we describe the most prominent variability mode, the annual cycle. Some results of Déqué et al. (1998) are reported. We also address the intraseasonal variability, present the impact on the interannual variability and the examine the expectation for extreme events.

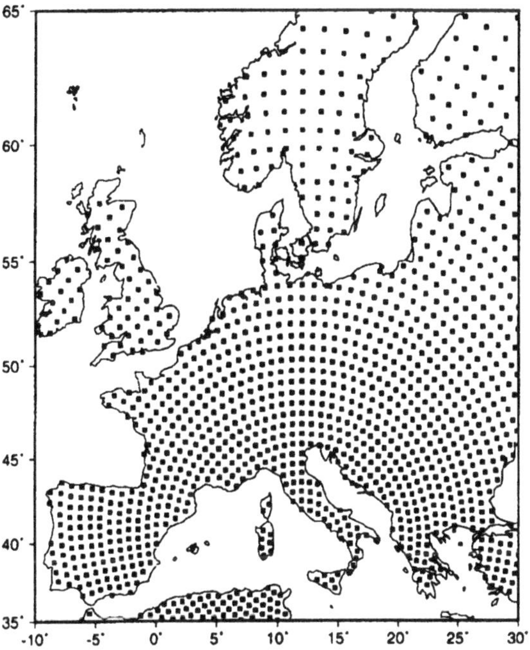

Fig.1. Distribution of the land grid-points of the numerical model over Europe

2. The models and experiments

The ARPEGE-IFS model is an atmospheric model developed by the French Meteorological Service Météo-France and the European Centre for Medium-range Weather Forecasts (ECMWF) for operational short- and medium-range forecasting. A climate version of this model has been developed at Météo-France. The base version is described by Déqué et al. (1994).

A variable resolution version, including additional developments, is described in Déqué and Piedelievre (1995). This version uses a spectral T63 truncation (the base version uses a T42 one). The associated grid has 96 pseudo-latitudes and 192 pseudo-longitudes. The pole of the new system of coordinates is located in the Mediterranean sea. As mentioned above, the equations are discretized on an isotropic grid (□x=□y). This cannot be true in the vicinity of the poles, since the Gaussian grid, which reduces the aliasing in the spectral truncations, must have a minimum number of longitude points as a function of latitude (Courtier and Naughton, 1994). Thus the grid cannot be isotropic for the physical parameterisations in the vicinity of the poles. A way to bypass this problem is to set the pole in an homogeneous area (a sea), where the non-isotropy is not detrimental. The antipode is located in the southern Pacific ocean. A stretching factor is applied as a function of the pseudo latitude. The maximum value is 3.5 at the Mediterranean pole, the minimum is 1/3.5 at the opposite pole. Thus the resolution over Europe is between 60 and 100 km, which is sufficient to capture coarsely the orographic features of the European continent. Figure 1 shows the grid point distribution over Europe. In Déqué and Piedelievre (1995), this version is integrated with observed SST in a 10-year simulation. The simulated climate over Europe is improved with respect to versions with homogeneous resolutions (T42 and T106).

Some improvements, described by Déqué et al. (1998), have been brought to this variable resolution version, which reduce some systematic errors over Europe. Two 10-year simulations have been carried out with the improved version. The first one uses a CO_2 concentration of 354 ppmv (the concentration observed in 1990), the second one a doubled value. In order to have a SST consistent with the radiative forcing, we have used values from a long coupled integration with a lower resolution.

The HADCM2 model (Johns et al., 1997) is a climate version of the UKMO unified model which includes an ocean GCM. The integration used here is a 240-year simulation starting with the CO_2 value of 1860. The first 130 years of the simulation are driven by the observed radiative forcing of the 1860–1990 period, so that the delay of the ocean due to its large thermal inertia is correctly taken into account. The fluxes exchanged between the atmosphere and the ocean undergo a constant correction, so that no drift appears in the SST, and the simulated 1990 values are very close (error less than 1 K) to the observed ones. Beyond 1990, the equivalent CO_2 increases by 1% per year so that it doubles by 2060 with respect to 1990.

Two 10-year time slices of monthly SST are extracted from this coupled simulation. Deep soil temperatures have also been extracted, since the bottom condition of the soil model used in ARPEGE-IFS is a fixed temperature. The corresponding forcing is weak, however (about 1 W/m^2, compared to a radiative forcing ten times greater). There is a small difference between the forcings used in HADCM2 and ARPEGE-IFS. In HADCM2, an equivalent CO_2 is used to represent the different greenhouse gases (except water vapour). In ARPEGE-IFS, only CO_2 is doubled, the other gases being kept constant. As a result, the radiative forcing is weaker in ARPEGE-IFS. However, this difference is negligible since we find a similar impact when we suppress the radiative forcing, keeping only the SST forcing (Déqué et al., 1998).

3. The annual cycle

In mid-latitudes, the largest variability mode is the annual cycle. In meteorology, the word variability often refers to fluctuations about this cycle. A section in this paper on the annual cycle is an occasion to recall the basic impact of a CO_2 doubling on the winter and summer climates. A change in this variability will imply a change of the amplitude of the annual cycle, which may have important consequences, e.g. in agriculture. A more detailed study of the winter and summer simulated impacts can be found in Déqué et al. (1998). In this section, additional results concerning the model validation are presented.

We restrict ourselves to three fields: the daily minimum and maximum temperatures, and the daily accumulated precipitation. Observational values for such fields are generally available in climatological databases. Here the daily simulated values are computed from 15 min by 15 min model outputs. These fields are often used in the literature on the numerical simulation of climate change, although the daily mean temperature (which is calculated as, or corresponds closely to, the halved sum of the minimum and the maximum) is preferred to the first two fields. Indeed, the mean impact on minimum and maximum temperature is similar, as we shall see below. However, it is worthwhile studying the minimum and maximum separately if the extremes and the variability are of concern: the greenhouse effect acts on the long-wave radiation which is the major component in the determination of minimum temperatures.

The European area is defined as the land grid-points situated between 35°N and 65°N, 10°W and 30°E. As in Déqué and Piedelievre (1995), four boxes of approximately the same area have been used (NW, NE, SW, SE), with 14°E and 48°N as divides.

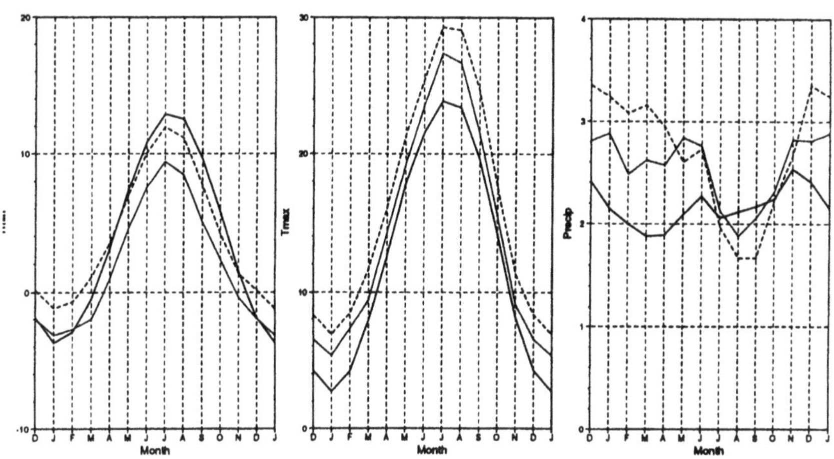

Fig. 2. Monthly means of minimum temperature (left: °C), maximum temperature (centre: °C) and precipitation (right: mm/day) over Europe; observation (solid), reference simulation (dot), $2xCO_2$ simulation (dash)

Figure 2 presents, for the three fields, the monthly means over Europe. The reference is the Climate Research Unit (CRU) observational database (Hulme et al., 1995). The results are 10-year averages. For the maximum and minimum temperatures, the warming due to the CO_2 doubling is not dependent on the season. For precipitation, two separate periods can be identified: an extended winter period (December through April) during which the precipitation increases, and an extended summer period (May through November) when it decreases. In order to simplify our analysis of the impacts and to get better statistical estimates by increasing the sample size, we will consider two phases of the annual cycle: the cold phase (NDJFMA) and the warm phase (MJJASO). They will be referred to as extended winter and extended summer.

Table 1 provides a quantitative approach of the model systematic error and of the impact through the spatial averages in the European boxes. The annual average temperature over Europe is 8.9°C in the simulation and 9.0°C in the observation. This apparently high accuracy masks the fact that the model overestimates the amplitude of the diurnal cycle and underestimates the amplitude of the annual cycle. The annual mean impact of a CO_2 doubling is 2.2 K (2.0 K in winter and 2.4 K in summer). The impact is slightly larger for the minimum (2.3 K) than for the maximum (2.1 K) temperature. For both fields the impact is larger in summer than in winter, and larger in the South than in the North. The model overestimates winter precipitation. Precipitation is too high in the northern boxes and too low in the southern ones. The annual mean impact of a CO_2 doubling on precipitation is negligible: the winter increase (0.3 mm/day) compensates for the summer decrease (0.2 mm/day).

		extended winter			extended summer			s.d. annual cycle		
		CRU	1xCO_2	2xCO_2	CRU	1xCO_2	2xCO_2	CRU	1xCO_2	2xCO_2
	NW	-0.8	-1.2	0.8	8.3	5.6	7.5	5.4	4.1	4.1
	NE	-4.6	-4.2	-1.9	8.2	5.4	7.2	7.6	5.8	5.6
Tmin	SW	2.5	0.4	2.7	11.7	6.9	10.0	5.3	4.0	4.5
	SE	0.1	-1.1	1.4	11.4	7.1	10.0	6.7	5.1	5.5
	EU	-0.7	-1.5	0.7	9.9	6.2	8.7	6.3	4.8	4.9
	NW	5.4	6.6	7.9	17.0	16.8	18.7	6.9	6.2	6.4
	NE	2.2	4.0	5.7	17.9	18.2	19.6	9.3	8.5	8.2
Tmax	SW	10.9	12.7	14.9	22.7	26.6	29.9	7.0	8.4	9.0
	SE	8.2	11.8	13.6	22.8	28.0	30.5	8.7	9.7	10.2
	EU	6.6	8.7	10.5	20.1	22.3	24.6	8.1	8.3	8.5
	NW	2.4	3.4	3.9	2.5	3.3	3.1	0.7	0.7	1.1
	NE	1.3	2.2	2.6	2.1	2.7	2.7	0.6	0.6	0.8
Prec	SW	2.6	3.1	3.2	2.0	1.9	1.5	0.8	1.4	1.6
	SE	2.4	2.1	2.5	2.0	1.4	1.2	0.9	0.9	1.3
	EU	2.1	2.7	3.1	2.2	2.3	2.1	0.8	1.0	1.2

Table 1. Means over the two extended seasons (NDJFMA and MJJASO) for daily minimum temperature (°C), daily maximum temperature (°C), and daily precipitation (mm/day) for the observations (CRU) and the two model simulations in the 4 boxes and the whole European area (EU). The last 3 columns give the standard deviation of the annual cycle

The amplitude of the annual cycle can be evaluated through the difference between the summer and winter values. However, this measure may be inadequate, as the cycle is not necessary sinusoidal, and the phase may be lagged. We therefore measured it by the standard deviation of the 12 monthly means. The last 3 columns of Table 1 show the corresponding values. The spatial averages are calculated with the variances in order to maintain unbiased estimates. The impact on temperature is a small increase in the seasonal variability, but this increase is not systematic in the different boxes. The variability of maximum temperature increases over France, whereas the variability of minimum temperature increases over Spain. However, the model underestimates the variability of minimum temperatures compared with observations. The impact on precipitation is more dramatic as the standard deviation is doubled. When comparing the reference simulation with observations, the variability is satisfactory in the North but too large in the South. Figure 3 shows the geographical distribution of the impact on precipitation. There is a decrease in the variability over Scandinavia.

4. The intraseasonal variability

In addition to the seasonal cycle, the weather fluctuations can be separated into the day-to-day variability, which can produce, due to the persistence of the atmosphere, longer periods of constant anomaly (e.g. blocking), and the year-to-year variability. The former, known as the intraseasonal variability, is addressed in this section. A 10-year sample contains sufficient independent events to use statistical tests for this type of variability. The latter, known as the interannual variability, is due to both the residual of the intraseasonal variations (the unpredictable part), and the forcing from the ocean (the predictable part). It also includes inter-decadal variability and longer climate trends. With only 10 years, statistical tests are not practicable: with 10 independent data, an F-test requires a factor of 3 between the variances to ensure a 95% significance, which implies huge changes. However, we will address the interannual variability below as an outlook on this part of the spectrum is of interest, at least to document the behaviour of the model.

It is not surprising that a doubling of CO_2 has an impact on the variability of precipitation (Gregory and Mitchell 1995; Fowler and Hennessy 1995). Due to the Clausius–Clapeyron relation, a warmer atmosphere will contain more water, and thus will be able to produce more precipitation. Over Europe, all models agree over an increase of precipitation, at least in winter when the soil drying out is not a limiting factor. Since the precipitation rate is bounded by zero, an increase of the mean drives an increase of the variance.

A simple way to isolate the intraseasonal variability from the annual and interannual components is to apply a time filter. A convenient and powerful filter can be constructed using the Dolph–Chebyshev filter (Lynch 1997) as a convergence factor for a least-square approximation to an ideal low-frequency filter. The variance is then calculated as the mean squared filtered data, and variance ratios can be tested against noise with an F-test. With a 2-10 day window, we have about 20 independent events per extended season, so that the critical ratio at the 95% confidence level is 1.3 (for two 10-year samples). We can also investigate the lower part of the spectrum with a 8-90 day window, where the critical ratio becomes 1.7 (with the assumption of four independent events per extended season).

Before looking at the impact of a CO_2 doubling on the filtered variance, we must validate the simulated values in the reference run against observation data. It is more difficult to obtain daily observed data than monthly means, and we will restrict our validation to the French climate network. If the model variability is not unrealistic for France, then the assumption should hold for the rest of Europe. On a daily basis, a single station value does not represent the same variable as a model value, in particular for precipitation. We have therefore computed a mean value involving nine meteorological stations. Thus, the mesoscale effects are reduced. These nine stations are spread uniformly over France, so that the average corresponds to a spatial average. For the model, we have taken the closest nine grid points to these stations.

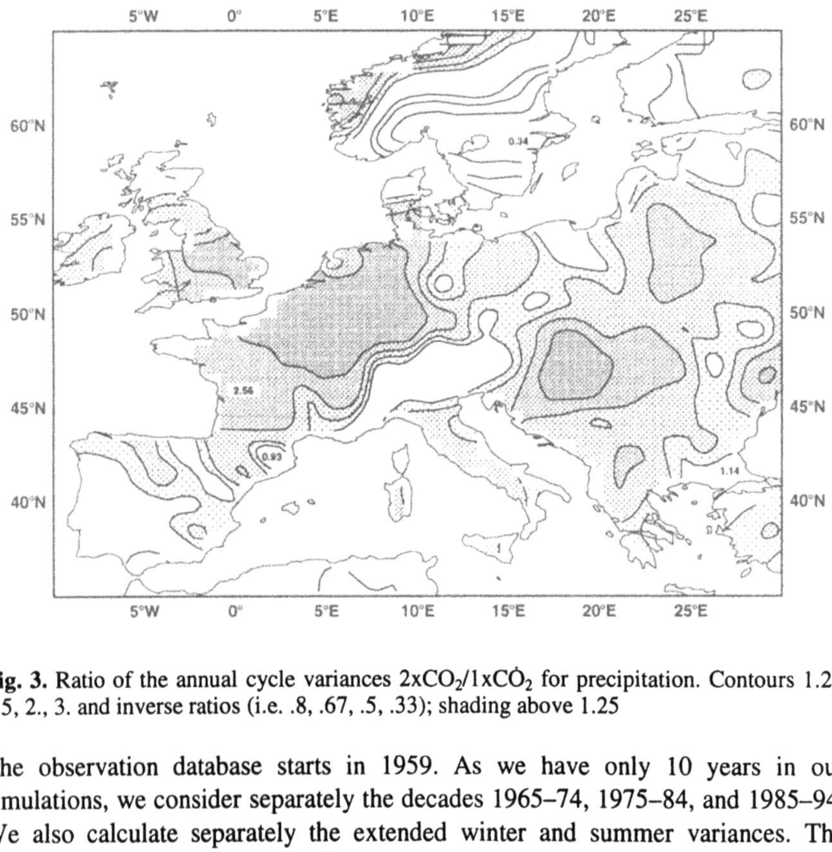

Fig. 3. Ratio of the annual cycle variances $2xCO_2/1xCO_2$ for precipitation. Contours 1.25, 1.5, 2., 3. and inverse ratios (i.e. .8, .67, .5, .33); shading above 1.25

The observation database starts in 1959. As we have only 10 years in our simulations, we consider separately the decades 1965–74, 1975–84, and 1985–94. We also calculate separately the extended winter and summer variances. The standard deviations are statistically stable with a 10-year sample. For winter 2-10 day minimum temperature variability, the standard deviations are 1.4, 1.4 and 1.4 K for the 3 decades. With the 8-90 day window, we get 1.9, 2.1 and 2.1 K. The model exaggerates the variability in the higher part of the spectrum (1.9 K), but is fairly good in the lower part (2.1 K). In summer, the model variability is larger than observed for both windows, but the order of magnitudes are respected. The maximum temperature variability shows a different behaviour: the higher part of the spectrum is satisfactorily simulated, and the lower part is underestimated (1.7 against 2.1 K in winter and 2.0 against 2.1 K in summer). On the scale of France, the simulated temperature intraseasonal variability is acceptable. With precipitation, we get for the three observed decades as good a statistical stability as we obtained for temperature. The model behaviour is also satisfactory. In winter the 2-10 and 8-90 windows provide simulated standard deviations of 2.1 and 1.7 mm/day (2.0 and 1.7 mm/day for the observations). In summer the simulated values are 1.9 and 1.5 mm/day respectively (1.7 and 1.4 mm/day for the observations). For temperature, the variance is higher in the lower part of the spectrum, whereas we get the opposite for precipitation. The "redness" of the spectrum (i.e. the persistence) is greater for temperature.

Having verified that the intraseasonal variability produced by the model is not completely unrealistic, we can investigate the impact of the CO_2 increase at the scale of the model grid point. Tables 2 and 3 summarise the results in the 2-10 and 8-90 day windows. The values are larger than in the last paragraph since the latter are averages of nine grid points. The variances are averaged in the four boxes and for the whole European domain. There is a slight reduction in the variability for temperature, the impact being larger for the minimum temperature. The impact is larger for precipitation, but restricted to winter. Figure 4 shows the geographical distribution for the 2-10 day window in each extended season. The patterns for the 8-90 day window (not shown) are very similar, but with slightly lower ratios. Comparing Table 2 with Table 3 shows that the redness of the spectra (estimated by the variance ratio of the higher to the lower parts) decreases when the CO_2 is doubled. This phenomenon is best illustrated by the winter precipitation: the increase in mean precipitation is accompanied, as expected, by an increase in variance. This increase is larger (in relative value) in the 2-10 day window than in the 8-90 day window. The precipitation events become less regular in time.

		extended winter		extended summer	
		$1xCO_2$	$2xCO_2$	$1xCO_2$	$2xCO_2$
Tmin	NW	2.9	2.6	2.7	2.7
	NE	3.3	3.0	2.9	2.8
	SW	2.9	2.7	2.7	2.5
	SE	3.2	3.0	2.7	2.6
	EU	3.0	2.8	2.8	2.6
Tmax	NW	1.9	1.7	2.0	2.1
	NE	2.3	2.1	2.3	2.4
	SW	2.3	2.3	2.4	2.4
	SE	2.3	2.4	2.5	2.6
	EU	2.2	2.1	2.3	2.4
Precip	NW	3.5	4.0	3.9	4.1
	NE	2.9	3.6	3.6	3.9
	SW	4.3	4.5	3.7	3.5
	SE	3.7	4.3	3.3	3.3
	EU	3.6	4.1	3.7	3.7

Table 2. Averages over the two extended seasons (NDJFMA and MJJASO) and in the five European boxes of the intraseasonal standard deviation of daily minimum temperature (K), daily maximum temperature (K), and daily precipitation (mm/day) for the two model simulations: window 2–10 days

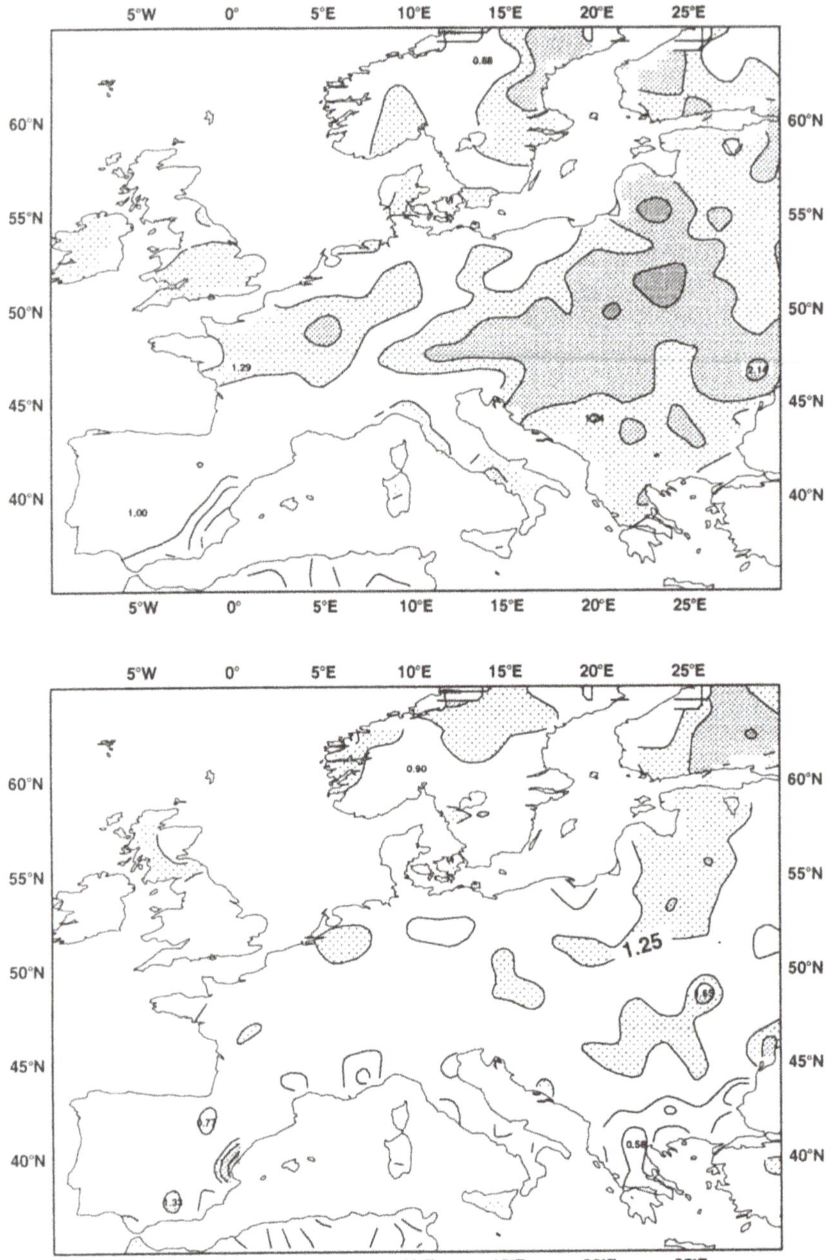

Fig. 4. Ratio of the intraseasonal (2–10 day window) variances $2xCO_2/1xCO_2$ for precipitation in extended winter (top) and summer (bottom). Contours 1.25, 1.5, 2., 3. and inverse ratios; shading above 1.25

		extended winter		extended summer	
		$1xCO_2$	$2xCO_2$	$1xCO_2$	$2xCO_2$
Tmin	NW	3.0	2.8	2.6	2.3
	NE	3.7	3.4	2.9	2.7
	SW	2.6	2.5	2.6	2.3
	SE	3.1	3.0	2.8	2.6
	EU	3.2	2.9	2.7	2.5
Tmax	NW	2.2	2.0	2.0	2.1
	NE	3.0	2.7	2.5	2.5
	SW	2.2	2.3	2.3	2.4
	SE	2.7	2.6	2.7	2.7
	EU	2.6	2.4	2.4	2.4
Precip	NW	2.3	2.4	2.4	2.3
	NE	1.8	2.1	2.1	2.2
	SW	2.9	3.0	2.4	2.3
	SE	2.4	2.7	2.1	2.2
	EU	2.4	2.6	2.3	2.3

Table 3. As Table 2 for the window 8–90 days

Another way to investigate the day-to-day variability of precipitation is to count the number of days in the two tails of the distribution, i.e. the dry daysand the intense rainy days (see for example Jones et al. 1997). In winter, 48% of the days receive less than 0.5 mm over Europe in the $1xCO_2$ simulation. This frequency decreases to 45% in the $2xCO_2$ simulation. In summer, the frequency of dry days increases from 54% to 59%. Figure 5a shows a wide area extending North of 45°N where the number of dry winter days decreases by 2 days/month, whereas an equivalent increase is found over Spain. In summer (Fig. 5b), such an increase is found over an area extending from France to Germany. Jones et al. (1997) found that the frequency of dry days increases by about 3% in winter (no impact in summer); in our simulations we get an increase about twice this value in summer, and a decrease in winter, so the impact is rather different in the two experiments.

At the other end of the distribution, the frequency of days with more than 10 mm increases from 6% to 8% in winter, and remains at 5% in summer. On closer inspection, it appears that over southern Europe the frequency of summer intense events is reduced. Jones et al. (1997) found that the frequency of intense events increases by about 30% in both seasons, which is similar to our result for winter. The different behaviour in summer is explained by the fact that their experiment exhibits an increase in precipitation in this season, whereas our experiment exhibits a decrease.

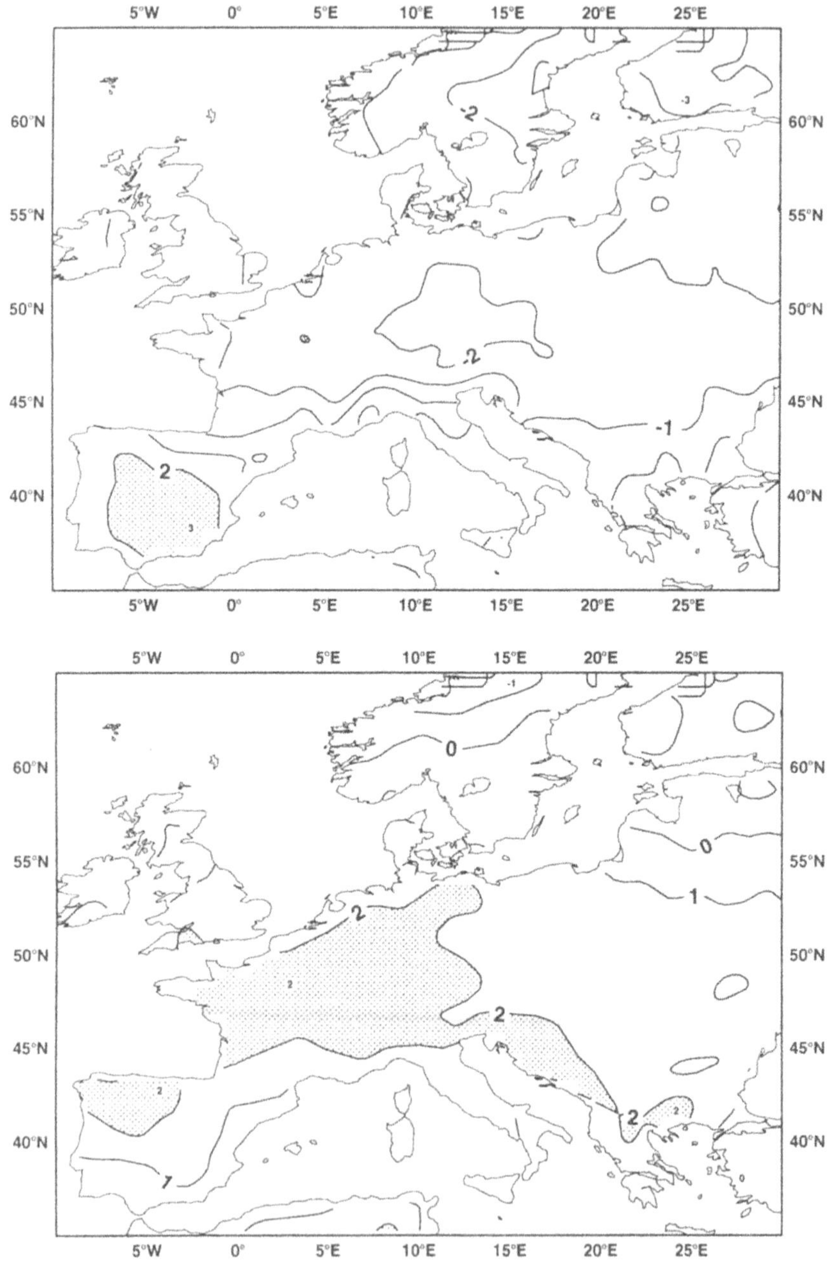

Fig. 5. $2xCO_2$-$1xCO_2$ differences in the number of day per month with less than 0.5 mm precipitation in extended winter (top) and summer (bottom). Contour interval 1 day/month; shading above 2 day/month

5. The interannual variability

As stated above, two 10-year samples are very short to study the interannual variability: the noise level in the ratio of the two variances is large, and we discard the decadal variations which occur in the SST. The last point is not critical, as we are studying a transient climate, but we need to keep in mind that any signals that are labelled significant may be due to either the radiative forcing or to the decadal SST forcing.

In order to increase the statistical stability, we calculate for each calendar month an unbiased estimate of the variance (based on 10 values), and we average the variances for each extended season. As in the last section, we start with a validation of the variance over France in the reference simulation. For the three decades, the observations yield less statistical stability than for the intraseasonal variability. For temperature, we have good stability in summer with 1.0, 0.9 and 1.0 K (Tmin) and 1.5, 1.6 and 1.6 K (Tmax). The model overestimates the standard deviation of minimum temperature (2.6 K) and underestimates that of the maximum temperature (1.2 K). In winter, the behaviour of the model is more satisfactory: the standard deviation of minimum temperature is 1.8 K (1.4, 1.5 and 1.8 K in the observations) and that of maximum temperature is 2.0 K (1.5, 1.5 and 2.0 K in the observations).

The model also has a satisfactory behaviour for precipitation in winter with a standard deviation of 1.0 mm/day (0.7, 0.9 and 1.0 mm/day in the three observed decades), as well as in summer with 0.9 mm/day (compared with 0.7, 1.0 and 0.9 mm/day). Thus, with the exception of temperature in summer, we can be quite confident in the interannual variability produced by the model. However we need to be cautious over the statistical significance of a small increase or decrease.

The results for Europe and its boxes are summarised in Table 4. The interannual variability generally increases, except for minimum temperature in winter. The impact is weak, however, if we have in mind the decade-to-decade variations in the observations over France. Figure 6 shows the impact in winter. The interannual variability over northeastern Europe is enhanced by a CO_2 doubling. There is a similarity between the intraseasonal and interannual precipitation variability (Fig. 4a and Fig. 6c).

6. The extremes

Estimating a climate extreme is a difficult statistical exercise, as adding or removing one daily value from among several thousands may change it completely. With 10-year samples, we will restrict the analysis to maximum or minimum daily values for the whole year, so that we can assume that the extreme is obtained by sorting a large number of independent values. For maximum temperature and precipitation, we select the maximum (the minimum precipitation is zero), whereas for minimum temperature, we select the minimum.

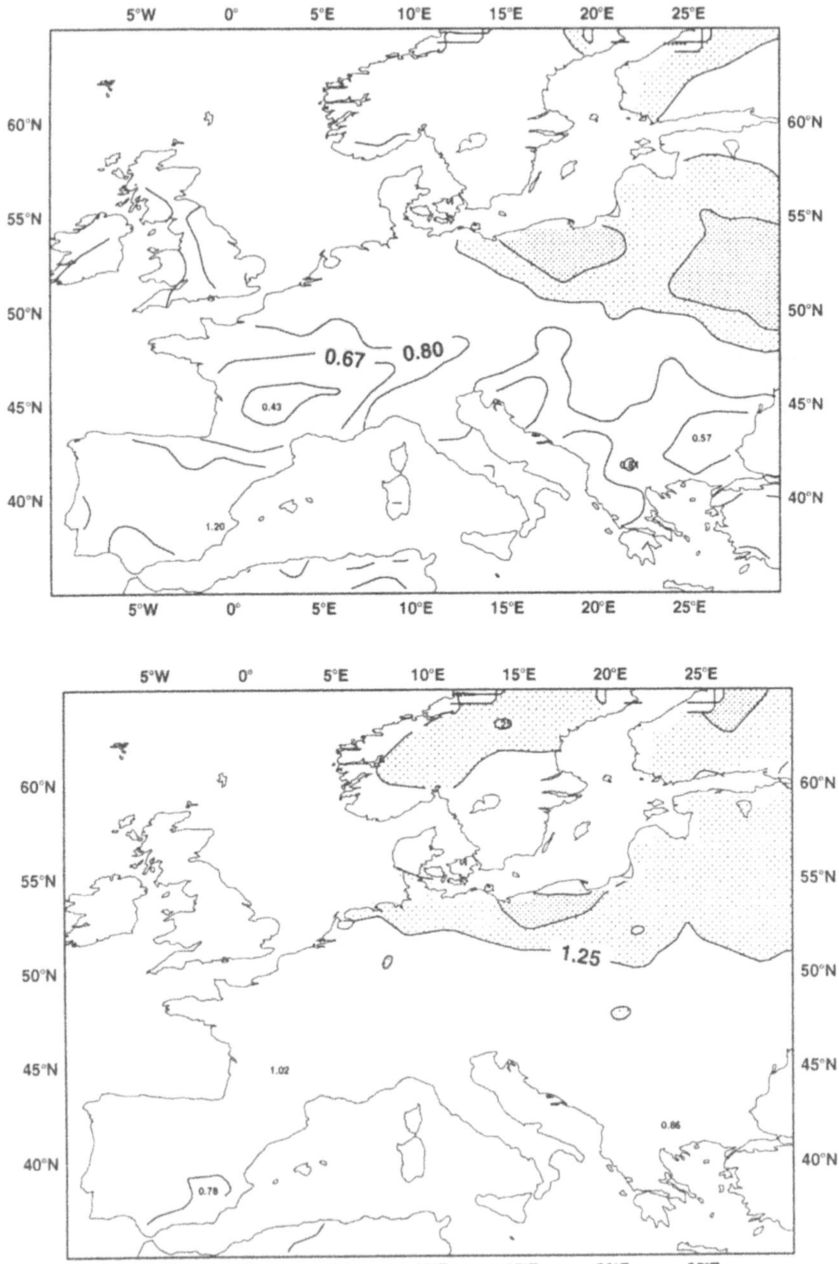

Fig. 6. Ratio of the interannual variances $2xCO_2/1xCO_2$ for the extended winter: minimum temperature (top), maximum temperature (bottom), and precipitation (next page). Contours 1.25, 1.5, 2., 3. and inverse ratios; shading above 1.25

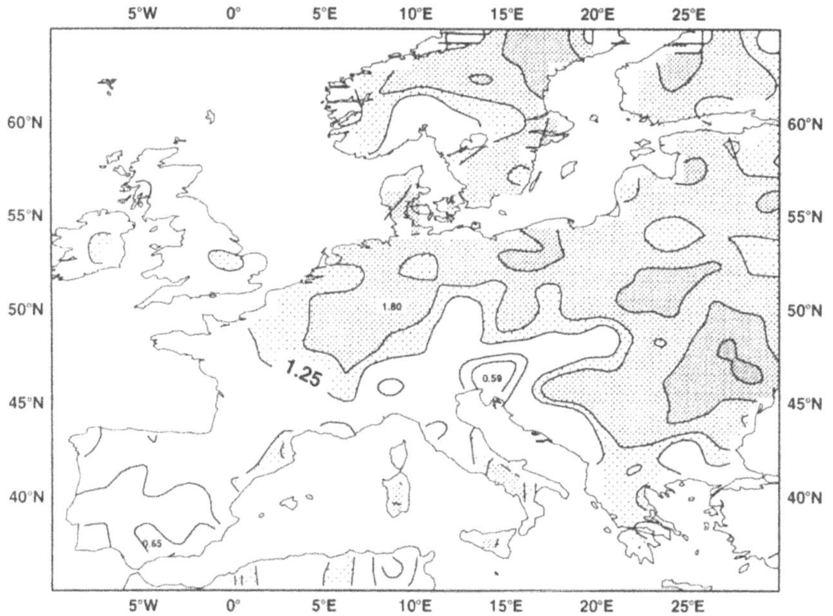

Fig. 6. (continue)

		extended winter		extended summer	
		$1xCO_2$	$2xCO_2$	$1xCO_2$	$2xCO_2$
Tmin	NW	2.2	2.1	1.5	1.5
	NE	2.4	2.8	1.9	1.9
	SW	2.1	1.9	1.6	1.6
	SE	2.2	1.9	1.8	2.0
	EU	2.3	2.2	1.7	1.8
Tmax	NW	2.0	2.1	2.3	2.7
	NE	2.6	3.0	2.8	3.1
	SW	2.3	2.2	2.9	2.6
	SE	2.5	2.5	3.1	3.4
	EU	2.4	2.5	2.8	3.0
Prec	NW	1.2	1.4	1.2	1.2
	NE	0.8	1.0	1.1	1.2
	SW	1.7	1.6	1.2	1.1
	SE	1.4	1.4	1.0	1.0
	EU	1.3	1.4	1.1	1.1

Table 4. As Table 2 for the interannual standard deviation

A climate extreme is a local value, therefore we cannot use an average over France for the validation against the observations. Instead, we will consider the observation series of Toulouse (located in the southern part of the country) in the following analysis.

When long series are available, it is possible to estimate the maximum for a longer series. The Gumbel distribution (Resnick, 1987) gives an asymptotic law to calculate probabilities, and return periods. This theorem also provides an estimate of the expectation of the maximum over a period of length *kn* and of its standard deviation as:

$$m_{kn} \approx m_n + \mu\sigma_n\left(\frac{1-k^{-\theta}}{\theta}\right) \tag{1}$$

$$\sigma_{kn} \approx \sigma_n\, k^{-\theta} \tag{2}$$

where m_n and σ_n are the mean and standard deviation of the maximum over a period of length n, and θ and μ are parameters depending on the shape of the distribution of the variable. As n tends to infinity, the right and left members become identical. The classical Gumbel distribution (double exponential law) corresponds to θ=0. Then Eqs. (1) and (2) reduce to:

$$m_{kn} \approx m_n + \mu\sigma_n \ln k \tag{3}$$

$$\sigma_{kn} \approx \sigma_n \tag{4}$$

We have four coefficients to estimate. We must take n large enough to be close to the asymptote, but we need several n-samples to estimate the means and standard deviations. With 10 years, we take n=365. We thus have 10 independent estimates of the maxima, and we calculate m_n and σ_n. If we want to extrapolate to several years, we need to estimate θ and μ. For this, we estimate m_{2n} and σ_{2n} by grouping pairs of years. Using a 5-sized sample gives a poor estimate of the mean and variance, but we can repeat the operation 100 times after scrambling the 10 years and average the 100 estimates to improve the accuracy. In order to test the method, we have used 10 years to build a statistical model of the observed daily series at Toulouse, and compared with direct estimates based on the whole 37-year database.

Figure 7 shows, for the 3 variables, the simplest estimate obtained by random sub-samplings of the 37 years (thick line), and estimates based on Eq. (1) using the decades 1965–74, 1975–84 and 1985–94 (thin solid lines). Extrapolating to 30 years is not safe with only 10 years of input data. In fact the standard deviation given by Eq. (2) indicates a large uncertainty about the extrapolated values. The extrapolation to 30 years from 10-year samples is associated with a standard deviation of 7 K for minimum temperature, 1 K for maximum temperature, and 15 mm/day for precipitation. The random sub-sampling using only a decadal series (dotted lines), without using Eq (1), is more accurate (closer to the thick solid curve) for the maximum temperature, but presents a larger divergence with 10 years for the minimum temperature and precipitation. There is therefore no clear preference between the two methods in terms of statistical stability; however the former has the advantage of providing a standard deviation and an extrapolation to any sample size.

The extreme values at the model grid point closest to Toulouse with the 10 years of the $1xCO_2$ simulation are rather different from the observed extremes, but remain acceptable. The coldest daily minimum temperature is -18°C, which is colder than observed. The warmest maximum temperature is 45°C, which is warmer than observed. The most intense daily precipitation is 53 mm/day. This is less than observed, but corresponds to the maximum observed with 5-year samples. The model thus fails to produce very rare precipitation events, and exaggerates the amplitude of the diurnal cycle of temperature.

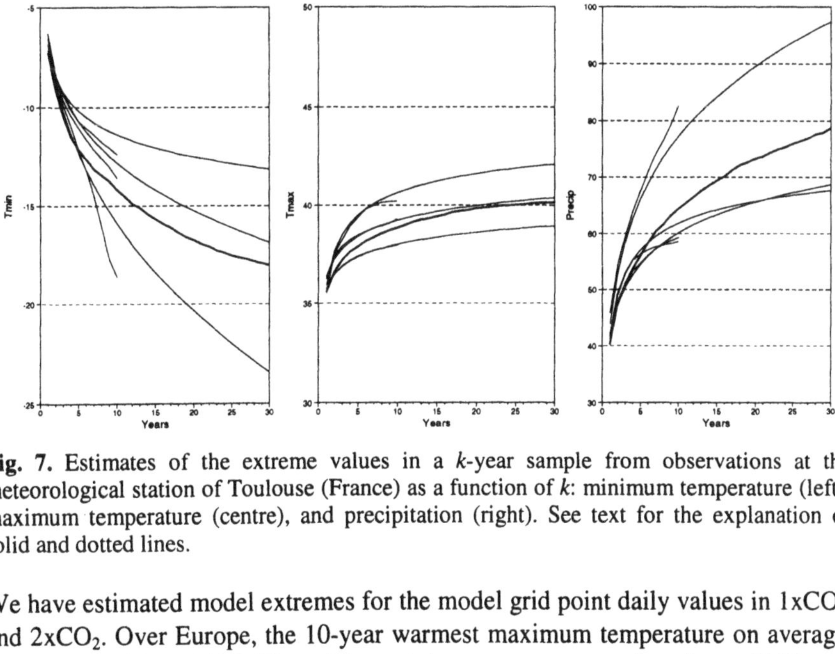

Fig. 7. Estimates of the extreme values in a *k*-year sample from observations at the meteorological station of Toulouse (France) as a function of *k*: minimum temperature (left), maximum temperature (centre), and precipitation (right). See text for the explanation of solid and dotted lines.

We have estimated model extremes for the model grid point daily values in $1xCO_2$ and $2xCO_2$. Over Europe, the 10-year warmest maximum temperature on average rises from 39°C to 44°C. The coldest minimum temperature rises from -21°C to -16°C. The largest daily value for precipitation increases from 61 mm/day to 65 mm/day. The latter increase is rather modest. In fact, the local impact may be strong but is very noisy, and therefore not significant. In contrast, the impact on temperature (Fig. 8) shows a clear pattern, indicating that whatever the grid point, the 10-year extreme is warmer in the $2xCO_2$ simulation. Over some parts of eastern Europe, the expected 10-year maximum is warmer by more than 10 K.

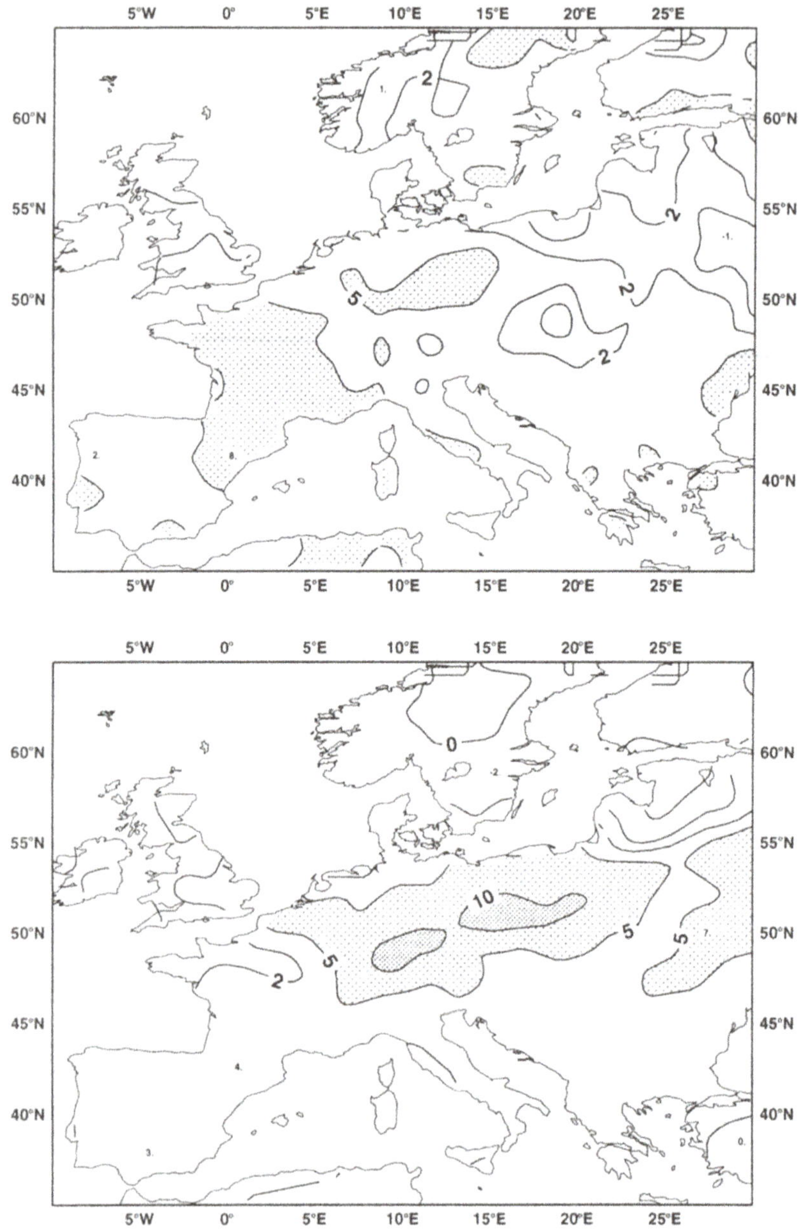

Fig. 8. $2xCO_2$-$1xCO_2$ differences in the 10-year coldest daily minimum temperature (top), and the warmest maximum temperature (bottom). Contours 0, 2, 5 and 10 K, shading above 5 K.

7. Summary and conclusion

We have analyzed the impact of a CO_2 doubling on daily minimum and maximum temperature and daily precipitation with particular attention being paid to the variability at all time-scales. Two 10-year simulations with a global climate model at high spatial resolution over Europe have been used.

The atmosphere undergoes a constant radiative forcing corresponding to a doubling of the carbon dioxide concentration, and a month-to-month varying forcing from the sea surface temperature calculated by a coupled scenario simulation at lower resolution.

The annual impact over Europe is a warming by 2 K; the mean impact on precipitation is weak. The impact on the seasonal cycle is a slight increase in the amplitude of the temperature, and a larger increase for precipitation, as precipitation increases in winter and decreases in summer. The impact on minimum temperature is generally greater than the impact on maximum temperature. The greenhouse effect impacts on the long-wave radiative fluxes which dominate during the night.

The intra-seasonal variability of temperature (daily minimum as well as maximum) decreases for both seasons. For precipitation, the only impact is obtained in winter, with an increase in the variability. The redness of the spectra of the three variables is reduced, i.e. the persistence of the phenomena is less in the 2xCO_2 simulated climate than in the 1xCO_2 one. The number of dry days is reduced in winter but increases in summer by a similar amount. The number of intense precipitation events (at the daily time-scale) is increased by 30% in winter and unchanged in summer.

It is difficult, with only 10 years of information, to identify a statistically significant impact on the interannual variability. We obtained an increase in the variability of daily maximum temperatures. The variability of minimum temperature increased only in summer. The variability of precipitation increased only in winter.

The impact on the extremes, i.e. the maximum over 10 years of the daily maximum temperature and daily precipitation and the minimum of the daily minimum temperature exhibits very large local values, due to the weak statistical stability of an extreme in a series. The geographical distribution of the impact on precipitation is chaotic. Minimum and maximum temperatures show a coherent impact, leading on average to a 5 K warming of the two extremes over Europe.

The above results rely upon two conditions: that the 10-year samples are long enough to ensure statistical stability, and that the model produces the same mechanisms as in reality to react against the radiative forcing. The first condition is not met for the interannual variability and the extremes. For the annual cycle and intraseasonal variability, t- and F-tests show that we have a stable response at the European scale; but the local features are generally insignificant. Moreover a significant response implies a response to the SST and radiative forcing, and we cannot exclude an impact of the natural decadal variability of the ocean. For the second condition, we have no definite answer. The mean climate of the model is similar but not identical to the observed one. We have attempted in this paper to check partially that the model variability was not pure numerical noise. The ARPEGE-IFS model is used daily for operational forecasting and it reproduces the weather systems with accuracy when it is maintained close to the observations. In free climate mode, the intra-seasonal variability in mid-latitudes is reasonable, even though the "blocking events" produced by the model are too zonal. However the systematic errors in the variability are generally greater than the impact of the CO_2 doubling.

The features we have discussed are not a description of what will occur in a few decades. Other simulations with other models and other SSTs will probably show different impacts. However some of our results can be also found in other simulations. We present here a coherent scenario of what could occur (irrespective of a modification of the aerosol concentration) if the greenhouse effect is enhanced. These results could then be used together with results from other models to take into account the uncertainties in the evaluation of the $2xCO_2$ impact in various environmental sciences where the climate variables are input parameters. Thus a better knowledge of the long term consequences of human activities would be available. Several national and international scientific programs are devoted to this purpose.

Acknowledgements. The authors are indebted to their colleagues from the Hadley Centre for providing the SST data from the scenario simulation, in particular to R. Jones. The help of P. Marquet in performing the ARPEGE simulations and of S. Jourdain in providing the daily series from the French climatological network must be acknowledged here, as well as the useful comments on the manuscript by D. Stephenson. This work was partly supported by the Commission of European Union (contracts ENV4-CT95-0184 and ENV4-CT97-0485), and by the French Department for Environment (ECLAT program).

8. References

Barrow EM, Hulme M (1996) Changing probabilities of daily temperature extremes in the UK related to future global warming and changes in climate variability. Clim Res 6:21–31

Courtier P, Naughton M (1994) A pole problem in the reduced gaussian grid. Q J Roy Meteorol Soc 120:1389–1407

Courtier P, Geleyn JF (1988) A global numerical weather prediction model with variable resolution: Application to the shallow water equations. Q J Roy Meteorol Soc 114:1321–1346

Déqué M, Dreveton C, Braun A, Cariolle D (1994) The ARPEGE/IFS atmosphere model: a contribution to the French community climate modelling. Clim Dyn 10:249–266

Déqué M, Marquet P, Jones RG (1998) Simulation of climate change over Europe using a global variable resolution General Circulation Model. Clim Dyn, in press

Déqué M, Piedelievre JP (1995) High resolution climate simulation over Europe. Climate Dyn 11:321–339

Fowler AM, Hennessy KJ (1995) Potential impacts of global warming on the frequency and magnitude of heavy precipitation. Natural Hazards 11:283–303

Goyette S, Laprise JPR (1996) Numerical investigation with a physically based regional interpolator for off-line downscaling of GCMs: FIZR. J Clim 9:3464–3495

Gregory JM, Mitchell JFB (1995) Simulation of daily variability of surface temperature and precipitation over Europe in the current and $2xCO_2$ climates using the UKMO Climate Model. Q J Roy Meteorol Soc 121:1451–1476

Hulme M, Conway D, Jones PD, Jiang T, Barrow EM, Turney C (1995) Construction of a 1961–90 European climatology for climate modelling and impacts applications. Int J Climatol 15:1333–1363

IPCC (1995) Climate change 1995. The science of climate change. Contribution of Working Group I to the second assessment report of the IPCC. Cambridge University Press, Cambridge, pp 572

Johns TC, Carnell RE, Crossley JF, Gregory JM, Mitchell JFB, Senior CA, Tett SFB, Wood RA (1997) The second Hadley Centre coupled ocean–atmosphere GCM: model description, spin up and validation. Climate Dyn 13:103–134

Jones RG, Murphy JM, Noguer M, Keen B (1997) Simulation of climate change over Europe using a nested Regional Climate Model. Part II: comparison of driving and regional model responses to a doubling of carbon dioxide. Q J Roy Meteorol Soc 123:265–292

Lynch P (1997) The Dolph Chebyshev window: a simple optimal filter. Mon Weath Rev 125:665–660

Mahfouf JF, Cariolle D, Royer JF, Geleyn JF, Timbal B (1994) Response of the Météo-France climate model to changes in CO_2 and sea surface temperature. Climate Dyn 9:345–362

Marinucci MR, Giorgi F (1992) A $2xCO_2$ climate change scenario over Europe generated using a Limited Area Model. Part I: present day climate simulation. J Geophys Res 97:9989–10009

Mitchell JFB, Johns TC, Gregory Jm, Tett SFB (1995) Climate response to increasing levels of greenhouse gases and sulphate aerosols. Nature (London) 376:501–504

Murphy JM, Mitchell JFB (1995) Transient response of the Hadley centre coupled ocean atmosphere to increasing carbon dioxide. Part II: spatial and temporal structure of the response. J Clim 8:57–80

Resnick SI (1987) Extreme values, regular variation and point process. Springer-Verlag, pp 320

Stephenson DB, Held IM (1993) GCM response of northern winter stationary waves and storm tracks to increasing amounts of carbon dioxide. J Clim 6:1859–1870

Timbal B, Mahfouf JF, Royer JF, Cubasch U, Murphy JM (1997) Comparison between doubled CO_2 time-slice and coupled experiments. J Clim 10:1463–1469

Wild M, Ohmura A, Gilgen H, Roeckner E (1995) Regional climate simulation with a high resolution GCM: surface radiative fluxes. Climate Dyn 11:469–486

Precipitation and snow cover variability in the French Alps

Eric Martin and Yves Durand

Abstract. The distribution of winter precipitation as analysed by the meteorological analysis system SAFRAN is validated using data from two test sites. This system, applied to the French Alps, shows that the frequency of high precipitation events is not necessarily linked to mean precipitation. Using a downscaling procedure, the system was run with General Circulation Model (GCM) outputs. The corresponding snow cover is derived with the snow model CROCUS. The results are very sensitive to the quality of the GCM run. The analyses of two future climate scenarios show that drastic changes in precipitation distribution may occur in the future.

1. Introduction

In the Alpine region, as in other temperate mountain ranges, the snow cover is strongly related to the evolution of the meteorological conditions throughout the winter. Previous studies (Martin et al, 1994) have shown that temperature is the main factor determining snow cover duration. The spatial variability is due to latitudinal effects (temperature) and orographic or Foehn effects (decreased precipitation in the east part of the French Alps). However some aspects of snow-related problems are at least as sensitive to extreme values as to mean climatic variables. In terms of variability, interannual variation is crucial for water resources as well as for winter tourism. Daily variability and extreme precipitation should be considered for floods or avalanches in winter. In the case of climate change, possible shifts in the spatial distribution of precipitation amount and distribution should also be considered.

In previous studies, Martin et al (1994, 1997) simulated the present mean snow cover of the French Alps and its sensitivity to various scenarios of climatic change. This paper addresses more precisely the question of snow cover variability and high precipitation frequency. It focuses first on the spatial analysis of high daily precipitation. The performances of the meteorological analysis system SAFRAN are tested against data from two sites and results at the scale of the French Alps are presented. In a second step, using a downscaling procedure for General Circulation Model (GCM) outputs, several reconstructions of precipitation distribution and the variability of snow cover for both the present and a future, changed climate are discussed.

2. Precipitation analyses

Adequate tools are required to assess the spatial variation and variability of precipitation at the scale of the French Alps. Observations cannot be directly used, mainly because of local effects and altitudinal gradients. Durand et al (1993) developed the meteorological analysis system SAFRAN for analyses at the scale of the French Alps. The main advantage of this system is that comparisons at a given altitude are allowed. It also calculates averages over small regions (about 1000 km²) to avoid very local effects. Martin et al (1994) used SAFRAN and the snow model CROCUS (Brun et al, 1989, 1992) to calculate a snow climatology for the French Alps. To do this, the Alps were divided in 23 regions (Figure 1) considered as homogeneous from a meteorological point of view. SAFRAN is designed to provide CROCUS with all the meteorological inputs necessary (air temperature, wind speed, humidity, cloudiness, incoming long-wave and short-wave radiation, precipitation) with a 300m vertical discretisation and an hourly time step. The system uses an optimal interpolation system to derive the main variables (temperature, wind speed, humidity, cloudiness). Incoming short-wave and long-wave radiation are derived from an atmospheric radiative model (Ritter and Geleyn, 1992). Precipitation is derived from climatologically analysed fields (Bénichou and Le Breton 1987) and observations with a daily time step. The hourly disaggregation is based on humidity, cloudiness and synoptic observations.

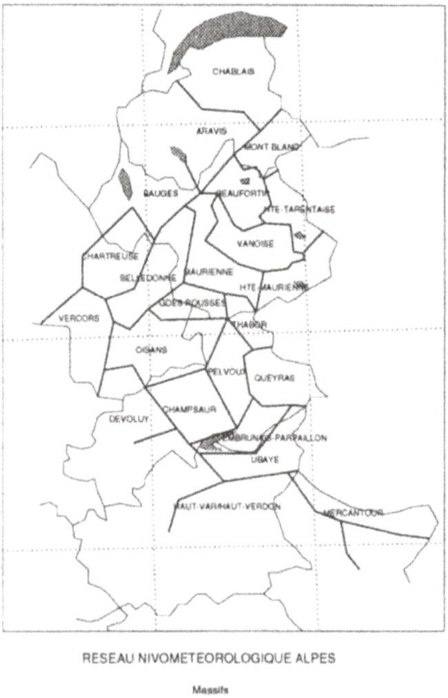

Fig. 1. Schematic map of the French Alps with the 23 massifs used in this study

The SAFRAN analyses were compared with observations at two sites not used in the system (Col de Porte, 1350m and Col du Lac Blanc, 2700m; Durand et al, 1993) for a short period (133 days). The results showed a strong underestimation at the Col de Porte site for total precipitation (270 mm in the analyses, 346 mm observed). The results were better for the second test site.

Since this study, winter precipitation at the Col de Porte has been checked and digitised for the last 36 years. Observations from the climatological network have also been added as input to the system. New tests made on 10 winters have shown that the results are improved by adding these data and that they are now very close to the observations (mean winter total precipitation analysed by SAFRAN: 514 mm, observation: 525 mm). As precipitation variability is also of great interest for climate studies, precipitation histograms (also based on 10 winters) were compared to the observational ones from the Col de Porte. Figure 2 compares the distribution of daily total precipitation observed in winter (December – February) to SAFRAN analyses. Two periods are considered for the observations. The first is the total period : 1960–96, the second is the study period 1981–91. For the total period (1960–96) the percentage of dry days is 50%, but for 1981–91 it increases to 55%. In the latter period, the percentage of days with total precipitation greater than 20 mm day $^{-1}$ is higher (10.6% compared to 9.7%), but the number of days with very high precipitation (> 40 mm day^{-1}) is stable (1.8% compared to 1.7%).

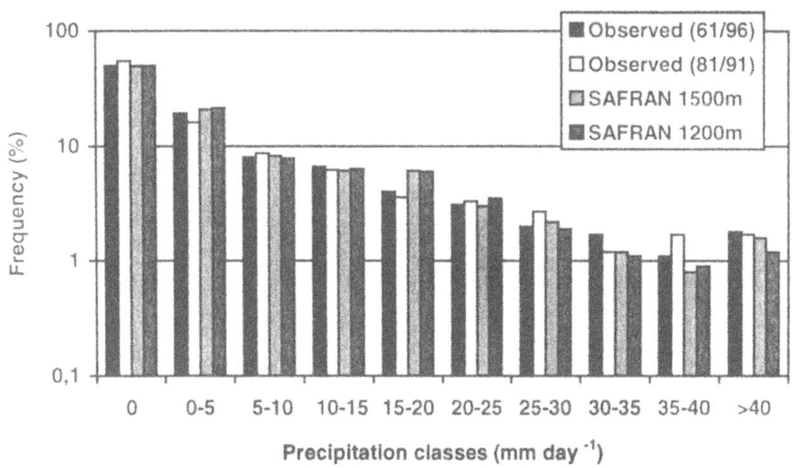

Fig. 2. Distribution of daily total precipitation in winter (December–February) at the Col de Porte Laboratory

The distribution of SAFRAN analyses at 1200 and 1500m are in good agreement with the observed distribution. Days with precipitation are underestimated (50%, observed 55%), while the frequency of low precipitation (<5 mm day^{-1}) is too high. Days with precipitation higher than 20 mm are underestimated (8.8% at 1500 m, 8.6% at 1200 m, observed 10.6%). However the analyses of very high precipitation are quite good (1.6% at 1500 m, 1.2% at 1200 m, observed: 1.7%).

The underestimation of dry days is inherent to the analysis system and the validation method: SAFRAN analyses precipitation for a given region, whose surface is about 1000 km^2. If one of the observations used as input is higher than 0, the result is likely to be higher than 0. When compared to local observations, the underestimation of dry days and the overestimation of days with low precipitation (0–5 mm day^{-1}) is normal. In contrast, the results are satisfactory for high precipitation. Thunderstorms (which are probably badly analysed by SAFRAN) are not treated here as they are very rare in winter.

Figure 3 compares the distribution of daily snowfalls (snow water equivalent in mm) with the SAFRAN analyses at 1200 and 1500m. At both elevations SAFRAN underestimates the days without snowfall and overestimates the days with snowfall lower than 5 mm day^{-1}. The analyses for 1200 m underestimate heavy snowfall, with the 1500 m analyses being better. As the elevation of the Col de Porte laboratory is 1350 m, we can conclude that SAFRAN slightly underestimates heavy snowfall for this site.

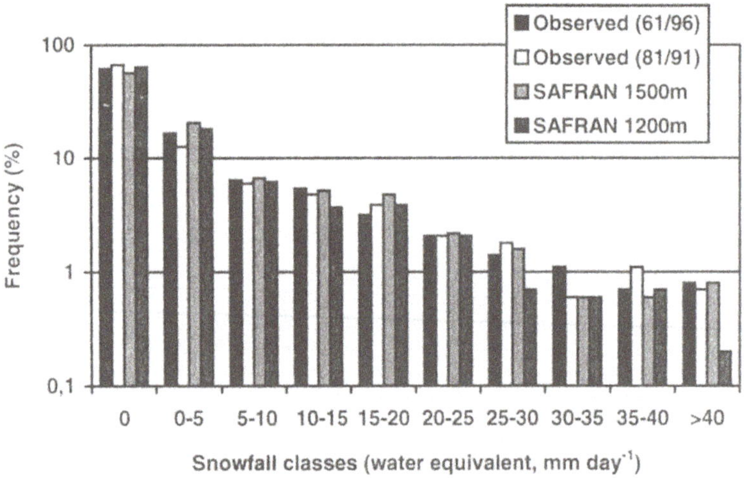

Fig. 3. Distribution of daily snowfall in winter (December–February) at the Col de Porte Laboratory

To assess the annual cycle of precipitation, comparisons were also performed at the synoptic station of Bourg-Saint-Maurice, situated at 868 m a.s.l. in the Vanoise massif. Although this observation site is used in the analysis system, its relative weight is small because many other observations are available from this region of the Alps. In winter (December–February), the percentage of days with high precipitation is increased (3.3% of days with precipitation higher than 20 mm day^{-1} for the entire year, 5.6 for winter, Figure 4). This annual variability is well reproduced in the SAFRAN analyses. Because of its situation and its lower elevation, high precipitation is less frequent at Bourg-Saint-Maurice than at the Col de Porte, situated on the western border of the Alps. With the exception of the discrimination between dry days and days with low precipitation, we consider that the distribution of precipitation calculated by SAFRAN is of good quality for this site.

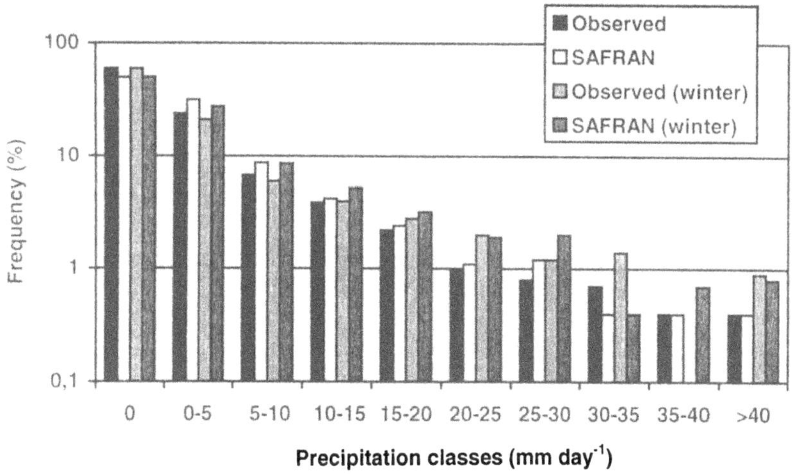

Fig. 4. Distribution of daily total precipitation at Bourg-Saint-Maurice in winter and for the entire year

3. Distribution of high precipitation at the scale of the French Alps

The results of the SAFRAN analyses at 1500 m for high precipitation events in winter for the period 1981–91 is given in Figure 5. The percentage of days with total precipitation higher than 20 mm day^{-1} varies from 1.9% to 10.3%. This very high spatial variability is consistent with the distribution of total cumulated precipitation. The frequency of high daily precipitation is greatest in the north-west part where precipitation is greatest. It decreases regularly towards the south-east because of the masking effect of the Alps. It is very low in the south-central part (1.9% for the massif Embrunais-Parpaillon) but it increases again in the extreme south: Mercantour and Haut-Var–Haut-Verdon. This is due to heavy precipitation associated with the Mediterranean low pressure areas. Heavy precipitation (also due to Mediterranean storms) is also encountered near the Italian border from Mercantour to Haute-Maurienne, but because of the lack of observations from Italy they cannot be analysed adequately by SAFRAN.

The results for precipitation >40 mm day^{-1} emphasise the very high precipitation events occurring in the south. The corresponding percentages (1.4 and 1.7) are equivalent to those encountered in the north, despite lower total precipitation (winter precipitation : 497 mm for Mont Blanc, 282 mm for the Mercantour). The spatial distribution for winter snowfall is nearly identical while values are lower (from 1 to 6.9% for snowfall >20 mm day^{-1} in water equivalent, 0 to 1.2% for snowfall >40 mm day^{-1}, not shown). On an annual basis, the differences between massifs are less pronounced than in winter. The frequency of days with total precipitation >20 mm varies from 2.4 to 7.9%. The highest reductions are located in the north-west massifs, because precipitation associated with westerly flows strongly affect this region in winter. In the south (Ubaye, Embrunais-Parpaillon), values are stable or increase slightly.

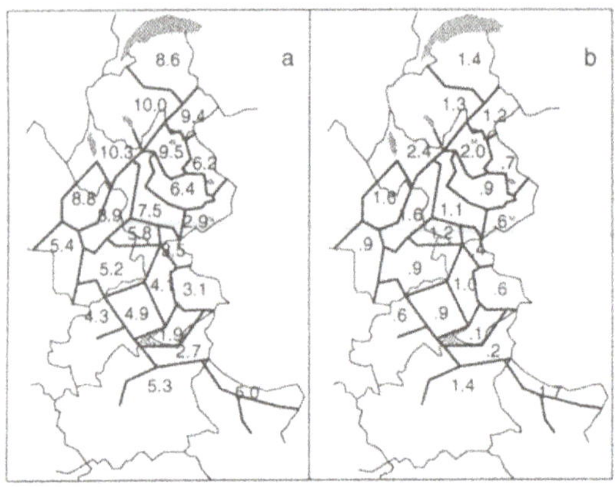

Fig. 5. Percentage of days with total precipitation higher than 20 mm day^{-1} (left) and 40 mm day^{-1} as analysed by SAFRAN for the winters of the period 1981–91

4. Precipitation distribution derived from GCM simulations (present climate)

A possible shift in precipitation distribution in the event of climate change is an important question that should be addressed. We used the downscaling method described in Martin et al (1997) to derive the precipitation distribution at 1500 m in the French Alps from various GCM runs for the present climate. The downscaling procedure is based on analogues. Each day simulated by the GCM is associated with a real, analogous day (or analogue) chosen from a 10-year database of European Centre for Medium-range Weather Forecasts (ECMWF) analyses. The selection of the analogue is done with a nearest neighbour method, the geopotential and temperature fields at 700 and 500 hPa over a small domain around the Alps are used. The distance is a Euclidean distance on a normalised variable. Once a real, analogue day is selected, SAFRAN is run with all the data available for the analogue. In a test made with real analyses of the ECMWF, Martin et al (1997) found that despite a 10% underestimation, the spatial distributions of cumulated winter and annual precipitation are well reproduced.

It is intended here to assess the capability of the analogue procedure to estimate the frequency of high precipitation and snowfalls. First of all, the dataset of the ECMWF analyses is used to test biases introduced by the analogue method itself (ECMWF in Table 1). For that purpose, the analogue must not belong to a window of +/- 10 days centred on the reference day, so as to avoid the selection of days from the same meteorological event. Three runs of the ARPEGE GCM of Météo-

France (Déqué et al, 1994) are also considered. The first two are 10-year runs with observed monthly sea surface temperatures (AMIP-type simulation). One is done with a T42 truncation, the other with a T106 truncation (T42 and T106 experiments). The last one is a 5-year run with climatological SSTs. This run (control) is the control run for the climate impact studies discussed below.

Experiment	Mont Blanc			Mercantour		
	0	>20	>40	0	>20	>40
SAFRAN analyses	48.2	9.4	1.2	66.9	5.0	1.7
ECMWF	51.1	8.0	1.0	69.1	3.8	1.1
T42	31.8	15.1	1.7	54.4	9.4	3.0
T106	44.8	10.3	1.3	72.1	5.3	1.8
Control	34.8	10.7	1.8	64.3	6.8	4.0
MPIe	43.7	11.6	0.9	71.4	7.7	3.1
HCe	53	7.1	1.3	77.2	4.6	1.1

SAFRAN, safran analyses for the period 81/91; *ECMWF*, reconstruction with the downscaling procedure using the real ECMWF analyses; *T42*, idem with an AMIP run at a T42 truncation; *T106*, idem but with a run at a T106 truncation; *Control*, idem but with a T42 run for present climate (mean SSTs); *MPIe*, idem but with 2xCO_2 scenario using SST anomalies from the MPI; *HCe*, idem but with 2xCO_2 scenario using SST anomalies from the HC.

Table 1. Frequency (%) of daily total precipitation equal to 0 mm or higher than 20 mm and 40 mm day^{-1} for the Mont Blanc and Mercantour massifs at 1500m.

Experiment	Mont Blanc			Mercantour		
	0	>20	>40	0	>20	>40
SAFRAN analyses	55.9	6.1	0.6	74.8	2.9	0.6
ECMWF	59.3	5.8	0.2	76.8	2.4	0.4
T42	47.7	2.9	0.6	69.1	5.1	0.2
T106	60.1	5.6	0.4	82.7	2.0	0.2
Control	49.7	7.2	0.7	74.3	3.8	0.9
MPIe	64.7	5.3	0.0	82.5	2.9	0.0
HCe	71.0	4.2	0.7	85.6	2.9	0.0

Table 2. Idem as table 1 but for the frequency of daily snowfalls (%)

Table 1 compares various winter total precipitation distributions at 1500 m for the Mont Blanc massif (north) and the Mercantour massif (south). Table 2 is the same for snowfall. In both massifs, the analogue procedure tends to overestimate days without precipitation and underestimate the highest precipitation (ECMWF). This bias is common to all analogue procedures, which tend to underestimate extreme events. The two AMIP runs (T42 and T106) show very different results for days without precipitation and snowfall. Although the atmospheric circulation is too zonal in both runs, the westerly and southerly winds are stronger in the T42 run. The consequence is an overestimation of precipitation in both massifs. In the T106 run, (Martin et al. 1997) snowfalls are underestimated because the climate of this run is too warm (bias 1.5°C at 700 hPa when compared to the ECMWF analyses): snowfall amounts and days with snowfall are underestimated. High precipitation frequency is quite well reproduced for Mont Blanc, but is underestimated in the

Mercantour. This is probably due to the poor simulation of the local Mediterranean circulation by the model. In the last run (control), which is only five years long, extreme precipitation is surprisingly high in the Mercantour. It seems that the winter atmospheric circulation is very strong (the result for the entire year is very similar to the SAFRAN analyses).

5. Snow cover characteristics derived from GCM simulations (present climate)

The mean snow cover duration is a good test of the accuracy of the complete method because it is the result of several snowfall and melting periods. Figure 6 compares the mean snow cover duration at 1500 m (days year^{-1}) deduced from the T42 and T106 runs (by SAFRAN/CROCUS and the analogue procedure) with the reference SAFRAN/CROCUS climatology of Martin et al (1994). In the T42 run, the northern part of the Alps is quite well simulated: for instance, the west-east gradient (Chartreuse–Haute-Maurienne) is well reproduced. In the extreme south, the snow coverage is underestimated, despite good snowfall amounts. This is probably due to poor simulation of the daily variability in the weather types by the model. The snow cover duration for the T106 runs is underestimated by at least 40 days because the climate is too warm. In terms of interannual variability, the distribution of the maximum snow depth (9 winters) is quite good in the Mont Blanc massif (about 1 m between the maximum and the minimum, Figure 7) but it is underestimated in both runs for the Mercantour.

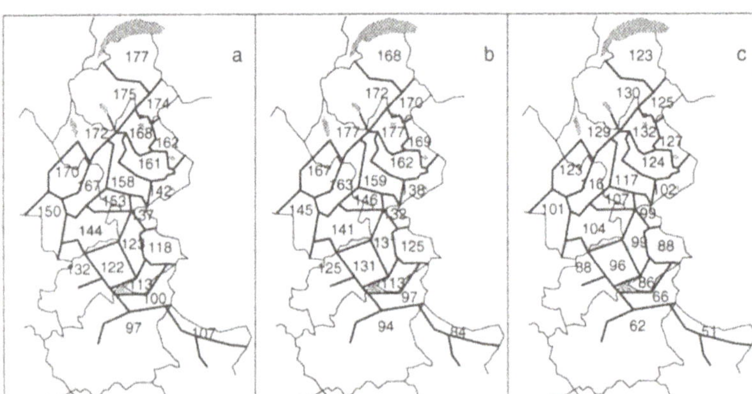

Fig. 6. Mean snow cover duration at 1500 m for a) the reference SAFRAN/CROCUS climatology b) the T42 run and c) the T106 run

Several problems are encountered using GCMs outputs when comparing the results calculated by SAFRAN and CROCUS. The most limiting factor is that the atmospheric circulation is too zonal. Southerly and south-easterly winds are very rare, which causes difficulties for the estimation of precipitation in the southern

Alps. Increasing the resolution improves the results in relation to this problem. The second problem is the **temperature bias, which is a problem for snow modelling.**

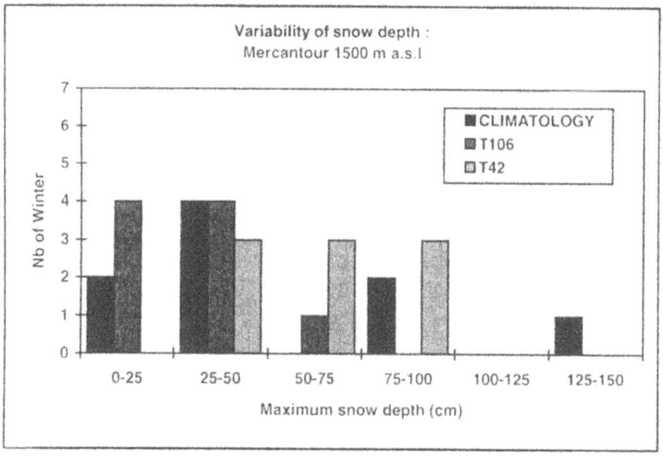

Fig. 7. Variability of the maximum snow depth for the Mercantour massif

6. Precipitation in climate change scenarios

The results of the two $2xCO_2$ time slice experiments used by Martin et al (1997) were also investigated. We focus here on precipitation distribution changes, as the question of snow cover was treated by Martin et al. (1997). Mean sea ice extent and prescribed SST changes, obtained from transient experiments performed by coupled atmosphere–ocean GCMs, were used as boundary conditions for ARPEGE/Climat. The two models that were considered are the ECHAM1/LSG developed at the Max Planck Institut für Meteorologie (MPI) (Cubasch et al. 1992) and the model developed at the Hadley Centre (HC) of the United Kingdom Meteorological Office coupled with a global ocean model (Murphy, 1994 and Murphy and Mitchell, 1994). The perturbations of the SSTs are the average of the 10-year period corresponding to a doubled atmospheric CO_2. The main differences in the SST forcing are located in the northern hemisphere, between 30°N and 60°N. The global annual mean of SST changes is +0.96 for MPI and +1.14 for HC.

Two five-year runs (called in the following MPIe for the experiment with MPI SSTs and HCe for the experiment with HC SSTs) were performed (Timbal et al., 1995). The control was a five-year run with the observed monthly mean of SSTs. The global warming of the surface air temperature in the experiments was 1.6 for MPIe and 1.9 for HCe.

In both CO_2 experiments, temperatures increase over the alpine region (at 700hPa : +1.3°C for the MPIe, +2.0°C for the HCe). Figure 8 shows the frequency of days with total precipitation higher than 40 mm day^{-1} at 1500 m. In the control run, its spatial distribution is coherent with the SAFRAN analyses for the 1981–91 period, but the values are generally higher in the western part and lower in the east.

However, the location of the maxima (north-west and extreme south) are well reproduced. In the MPIe, a slight enhancement of the atmospheric circulation can be observed in the GCM outputs. The consequence is an increased frequency of high total precipitation in some massifs: Chartreuse and Bauges (northwest), Devoluy and Ubaye (south). The frequency of precipitation higher than 20 mm day^{-1} and cumulated precipitation increases in parallel.

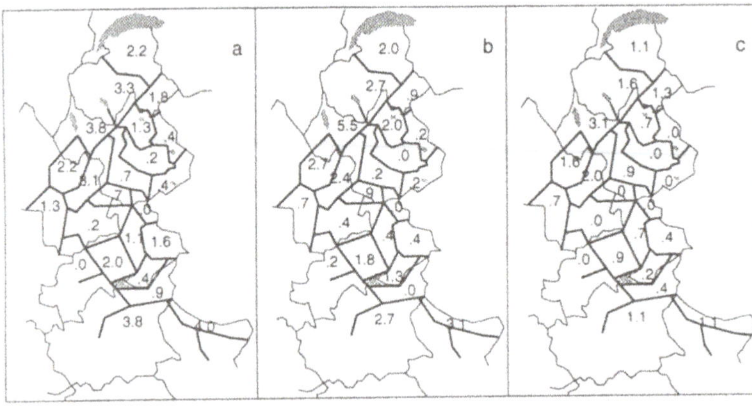

Fig. 8. Frequency of daily total precipitation >40 mm in winter (1500 m) for a) the control run, b) the MPIe, and c) the HCe

High snowfall amounts (>40mm day^{-1}) are underestimated in the control run (only five massifs with a frequency > 0). We preferred to use the threshold 20 mm day^{-1} (Figure 9). At 1500 m, despite the temperature increase, the frequency of snowfall higher than 20 mm day^{-1} increases in half of the massifs, mostly situated in the west and central part of the Alps, but cumulated snowfall increases in only one massif (Champsaur).

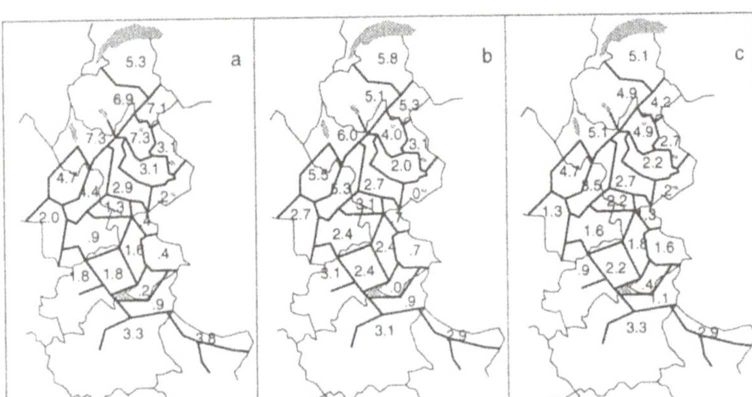

Fig. 9. Frequency of daily snowfalls >20 mm in winter (1500 m) for a) the control run, b) the MPIe, and c) the HCe

In the HCe, the atmospheric circulation is weakened. The result is a strong decrease in the frequency of high precipitation (Figure 8). The frequency at 1500 m of winter snowfall >20 mm day^{-1} remains stable in the southern part of the Alps, despite a substantial decrease in the cumulated snowfall amounts (concurrently, the number of days without snowfall increases).

7. Conclusion

The performance of the analysis system SAFRAN was tested against data from two test sites for precipitation amounts and distributions. The frequency of high daily precipitation is adequately simulated, but the system does not discriminate well between dry days and days with low precipitation because of differing analysis and validation processes (analysis at the massif scale, validation at local scale). At the scale of the French Alps, the SAFRAN analyses show that the frequency of high precipitation is not necessarily linked to cumulated precipitation amounts: the extreme south is an example of a relatively dry massif which is subject to high precipitation events.

The reconstruction of the precipitation distribution by downscaling of GCM outputs is difficult. The main problem comes from the quality of the GCM run (temperature biases, circulation that is too zonal). The run with the highest resolution (T106) is the best for precipitation variability. However, because the temperatures are too warm, snow cover and snowfalls are badly simulated. Despite the problem with temperature, it is encouraging to see that with an increase in resolution, GCMs are able to simulate the atmospheric circulation in an improved manner. For the simulation of climate change, the analysis of two runs (MPIe and HCe) showed different results: in the MPIe precipitation amounts and high precipitation frequency increase in the south whereas in the HCe, both decrease drastically. The frequency of snowfall higher than 20 mm day^{-1} decreases in almost all cases (except for Champsaur in the MPIe), but it remains stable in the south for the HCe.

Large uncertainties persist over high precipitation. The quality of the GCM run is a key factor. Important points are temperature biases or zonality of the atmospheric circulation. Another crucial parameter is the atmospheric circulation variability: lee cyclogenesis in the Mediterranean sea is associated with high snowfall events in the southern Alps several times per winter. Usually, GCMs underestimate this important phenomenon. The downscaling method should also be checked; in particular, comparisons with other downscaling methods (e.g. nested models) should be conducted. The use of several long-term runs from various GCMs could be of great interest. Another important point to be taken into account is the decadal scale variability. The observations of snow cover in the region show that groups of years with high snow cover alternate with groups of years with low snow cover. It would be interesting to verify the ability of GCMs to reproduce these trends. Finally, for improved accuracy, the data set of reference days for the downscaling procedure should be increased.

Acknowledgements. This research was partly supported by the EC Environment and Climate Research programme (contract ENV4-CT95-0184, Climatology and Natural Hazards).

8. References

Bénichou P, Le Breton O (1987) Prise en compte de la topographie pour la cartographie des champs pluviométriques statistiques. La Météorologie VII(19) Oct 87, pp 23–34

Brun E, Martin E, Simon V, Gendre C, Coléou C (1989) An energy and mass model of snow cover for operational avalanche forecasting. J Glaciol 35:333–342

Brun E, David P, Sudul M, Brunot G (1992) A numerical model to simulate snow-cover stratigraphy for operational avalanche forecasting. J Glaciol 38:13–22

Cubasch U, Hasselman K, Höck H, Maier-Reimer E, Mikolajewicz U, Santer BD, Sausen R (1992) Time dependent greenhouse warming computations with a coupled ocean–atmosphere model. Clim Dyn 8:55–69

Deque M, Dreveton C, Braun A, Cariolle D (1994) The ARPEGE/IFS atmosphere model: a contribution to the French community climate modelling. Clim Dyn 10:249–266

Durand Y, Brun E, Mérindol L, Guyomarc'h G, Lesaffre B, Martin E (1993) A meteorological estimation of relevant parameters for snow models. Ann Glaciol 18:65–71

Martin E, Brun E, Durand Y (1994) Sensitivity of the French Alps snow cover to the variation of climatic variables. Ann Geophysicae 12:469–477

Martin E, Timbal B, Brun E (1997) Downscaling of general circulation model outputs: simulation of the snow climatology of the French Alps and sensitivity to climate change. Clim Dyn 13:45–56

Murphy JM (1994) Transient response of the Hadley-Centre coupled ocean–atmosphere model to increase of carbon dioxide. Part I : control climate and flux adjustment. J Clim 8:36–56

Murphy JM, Mitchell JFB (1994) Transient response of the Hadley-Centre coupled ocean–atmosphere model to increase of carbon dioxide. Part II : spatial and temporal structure of response. J Clim 8:57–80

Ritter B, Geleyn F (1992) A comprehensive radiation scheme for numerical weather prediction models with potential applications in climate simulations. Mon Weather Rev 120:303–325

Timbal B, Mahfouf JF, Royer JF, Cariolle D (1995) Sensitivity to prescribed changes in sea surface temperature and sea-ice in doubled carbon dioxide experiments. Clim Dyn 12:1–20

Influence of forest cover in the Eastern United States on regional climate

Robert L. Walko, Roger A. Pielke, Sr. Pier Luigi Vidale, Joe Eastman and Lixin Lu

Abstract. Using the climate version of the Regional Atmospheric Modeling System (CLIMRAMS), we show the influence of forest cover change in the eastern United States on regional climate. By estimating what the natural landscape was prior to European settlement we can assess using CLIMRAMS what the climate would be for a given year using current observed weather at the lateral boundaries. By replacing this landscape with a USGS evaluation of current landscape, we can demonstrate the importance of landscape in determining climate in the region, and contrast this change with existing scenarios of other anthropogenic causes of climate change.

1. Introduction

The influence that climate and climate change have on forests and on the biosphere in general is an important physical relation in the whole earth ecosystem. We present results from an atmospheric simulation model that show that weather and climate conditions are directly influenced by vegetation properties and can show strong sensitivity to landuse change. Many of these atmospheric conditions, such as air temperature and precipitation amount, have a strong influence on vegetation. Consequently, to understand how vegetation might evolve, one must also understand how vegetation influences climate. We make this point because we feel that it is important to view atmospheric and biospheric phenomena as an integrated system.

We first present results that demonstrate that changes in vegetation can cause changes in atmospheric conditions that have a direct impact on vegetation. We then describe an interactive coupling between an atmospheric and a vegetation model that captures the influence of each component upon the other and simulates the evolution of climate and vegetation as a combined system.

2. Simulation Results

The Regional Atmospheric Modeling System (RAMS) employed in this study is a general purpose atmospheric simulation model. In addition to predicting air velocity, temperature, pressure, and water vapour content, as well as liquid and ice precipitation and radiative transfer, RAMS contains a sub-model called LEAF-2 which solves energy and water conservation equations for vegetation and multiple soil layers, and evaluates conductive, turbulent, and radiative exchanges between them and the atmosphere. LEAF-2 contains multiple "patches" or sub-grid areas of surface grid cells, each characterized by a specific landuse type (forest, cropland, etc.) so that a grid cell may be covered by the same fractional area of each landuse type that occurs in reality. Vegetation characteristics in LEAF-2 are normally held constant in time for short simulations or vary seasonally according to a prescribed function in longer simulations, with no response to specific weather phenomena. Thus, vegetation exerts a one-way influence on atmospheric phenomena with no feedback. The vegetation characteristics that are needed for prediction of atmospheric and soil variables in RAMS are height, roughness height, albedo, leaf area index, transmissivity (to radiative transfer), a root profile, and several parameters that control stomatal resistance. Transpiration rate is evaluated from these parameters and from soil moisture and air temperature, humidity, and radiation.

We present results of simulations conducted for south Florida in the summer. This is a wet season and convective storms with moderate to heavy precipitation are an almost daily occurrence. Many of the storms are associated with sea breeze fronts that develop in the daytime on both the east and west coasts. Depending on prevailing winds, one of the sea breeze fronts often penetrates inland and crosses most of the Florida Peninsula. The strongest precipitation tends to occur where this front collides with the front developing from the opposite coast. The simulations were initialised from an atmospheric sounding in Florida taken at 1200 UTC on 1 July 1973, a case that has been studied previously (Pielke and Mahrer 1978). Two-way interactive grid nesting was used in which a finer mesh with 10 km spacing covering a 420 km by 500 km area was centred within a coarser mesh with 40 km spacing covering a 1680 km by 1920 km area. Both grids were centred over the Florida Peninsula. Because these grids are too coarse to resolve convective vertical motion, convection was parameterised using a modified Kuo scheme (Tremback 1990). Ambient winds were light easterly, which supported westward propagation of the East Coast sea breeze front.

Two simulations of this case were carried out under identical conditions except that one was run with vegetation classes observed for 1993 and the other with vegetation observed in 1900. Figure 1 shows the vegetation coverage for these two times. During the 20th century, considerable areas of forest were cleared and converted to agricultural, industrial, or residential use. In addition, many swampy areas were drained. These changes affect surface fluxes, most importantly of sensible and latent heat, which in turn influence the strength of the sea breeze and the convective stability of the atmosphere.

Figure 2 shows convective precipitation accumulated over the 12-hour simulation, which encompasses the daytime convective period. With all other model simulation parameters being identical, the alteration in landuse between 1900 and 1993 has resulted in a significant redistribution of precipitation, as well as an overall reduction in the area-averaged precipitation amount. Although these results were obtained for meteorological conditions on a single day, the comparison simulations have since been carried out with continual meteorological observations used to nudge the lateral boundaries of the coarse grid for the entire months of July and August 1973. This much larger statistical sample, representing more average summertime conditions, has resulted in qualitatively the same trends between 1900 and 1993 but the results are even more pronounced. It should be noted that the two-month simulation time is still short compared with response times of vegetation and soil processes, so the simulation represents transient rather than equilibrium climates.

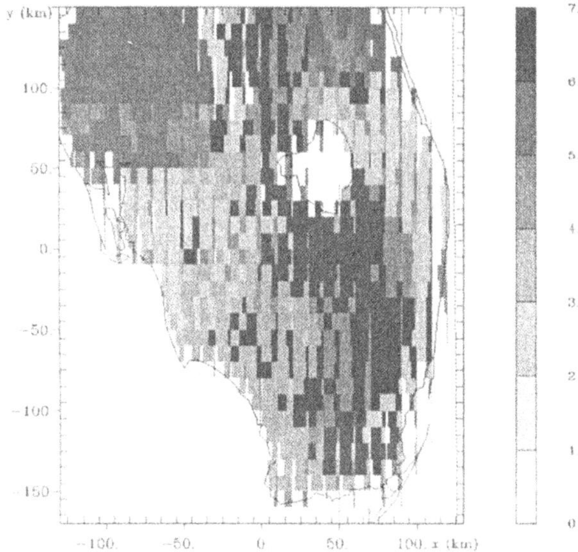

Fig. 1a. Landuse map of the southern Florida Peninsula for 1900 as represented in the RAMS simulations. Surface grid cells are squares centred on the inward-pointing tick marks. Sub-grid patch areas within each grid cell are represented as a subdivisions in the east-west (x) direction, resulting in the small rectangular areas. For clarity, only the southern part of the Florida Peninsula, a subset of the fine grid domain, is shown. Landuse class numbers are (1) water, (2) crops, (3) needleleaf evergreens, (4) miscellaneous classes, (5) marsh, (6) mixed woodland, and (7) swampy grassland.

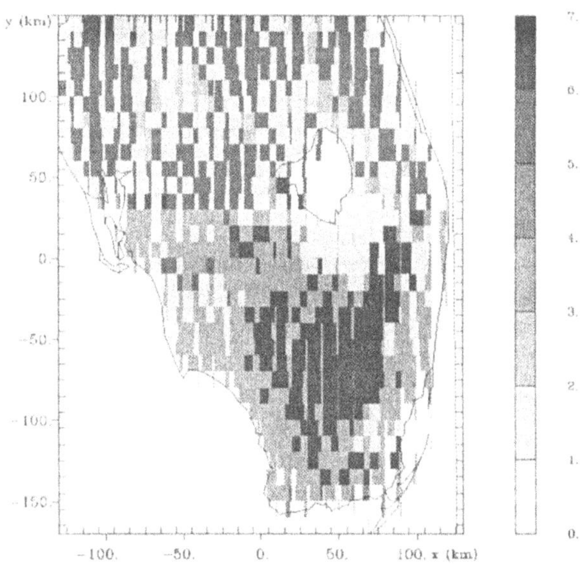

Fig. 1b. Same as Figure 1a but for 1993 landuse data.

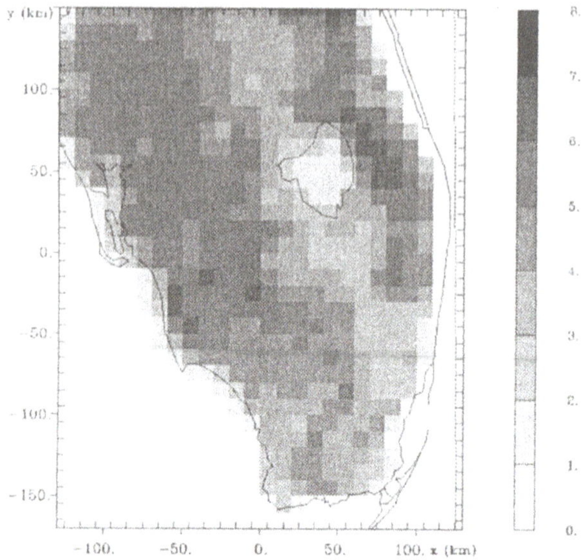

Fig. 2a. Accumulated convective precipitation (mm) for simulations conducted with 1900 landuse.

RAMS/LEAF-2 has been applied in numerous other studies where landuse characteristics play a major role in determining atmospheric response. Recently, Vidale et al. (1998) conducted atmospheric simulations for the Boreas Experiment, with vegetation parameters and initial soil moisture specified from detailed observations. Simulations were carried out for several days. Atmospheric temperature and surface fluxes evolved in good agreement with surface and aircraft observations. Figure 3, from Vidale et al. (1998), shows an example of sensible and latent heat fluxes computed by the simulation and measured by an aircraft along a particular transect across the domain. Very strong horizontal gradients are evident and result from the inhomogeneity of land surface characteristics. Such variations in fluxes induce local and mesoscale solenoidal circulations which not only influence plant microclimates but can also act as triggers for moist convection and precipitation in preferred areas, with obvious strong feedbacks on vegetation.

A pronounced example of convective response to landuse change was demonstrated in simulations by Pielke et al. (1997) for Georgia in the southeastern USA for meteorological conditions on 26 July 1987. Current vegetation patterns, in which much of the original forest has been cleared in patchy areas, resulted in the significant development of convective storms and precipitation, while the homogeneous forest representative of natural conditions that existed a century or more ago resulted in only weak convective storms and precipitation.

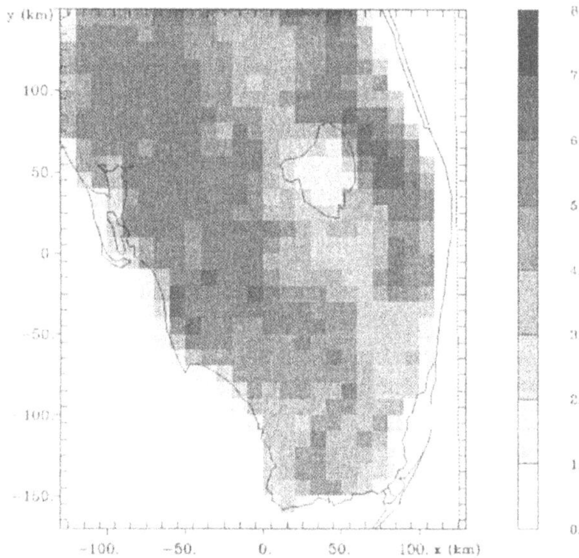

Fig. 2b. Accumulated convective precipitation (mm) for simulations conducted with 1993 landuse.

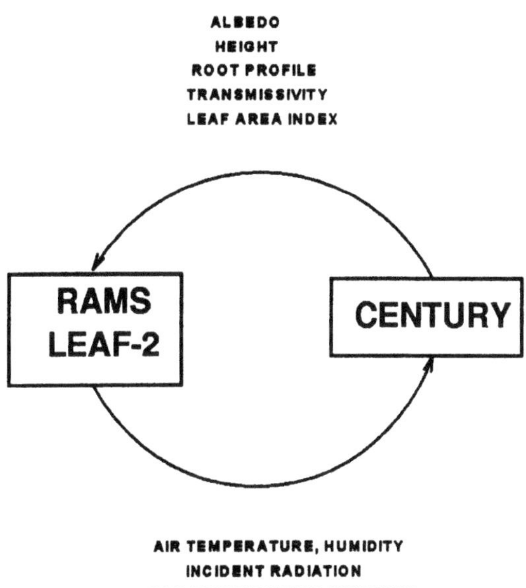

Fig. 3. Physical parameters communicated between coupled models RAMS/LEAF-2 and CENTURY.

3. Coupled Atmospheric and Ecosystem Models

Recognising that the atmosphere and vegetation are mutually interdependent, we have coupled RAMS/LEAF-2, which specialises in the effect of vegetation on atmospheric conditions, with CENTURY, which describes plant growth and evolution under the influence of weather and climate conditions. In the coupled system, RAMS/LEAF-2 provides atmospheric and soil conditions required by CENTURY to describe the plant environment, while CENTURY provides vegetation characteristics of direct importance to the atmosphere that develop in response to plant life-cycles and evolution (Figure 3). Early simulation results show that the response of the combined system are realistic. The coupled models are expected to be a valuable tool in providing more accurate estimates of climatic response to anthropogenic landuse change.

4. Conclusions

The simulation results described above are evidence that landuse change, such as deforestation, can cause significant changes in atmospheric conditions that in turn influence vegetation growth and evolution. From this, we argue that the atmosphere and biosphere should be viewed as a combined system with many internal interactive processes rather than as separate systems where a focus is placed on one-way interaction, depending on one's particular interest or point of view. In pursuing the goal of modelling these interactions and simulating climate and ecosystem change, we have coupled the models RAMS/LEAF-2 and CENTURY, which provides a representation of the interactive processes between the atmosphere and biosphere.

Acknowledgements. We would like to acknowledge Martin Beniston, the Institute of Geography, University of Fribourg, and the Swiss National Science Foundation, who provided support for the first author to attend the 1997 Workshop on Global Change Research in Wengen, Switzerland. This research was supported by United States Geological Survey Contracts #14-08-0001-A0929 and #1434-94-A-01275, NSF Grant No. ATM-9306754, EPA Grant #R824993-01-0, and NASA Grant No. NAG-5-2302.

5. References

Pielke RA, Lee TJ, Copeland JH, Eastman JL, Ziegler CL, Finley CA (1997) Use of USGS-provided data to improve weather anclimate simulations. Ecol Appl 7:No. 1

Pielke RA, Mahrer Y (1978) Verification analysis of the University of Virginia three-dimensional mesoscale model prediction over south Florida for July 1, 1973. Month Wea Rev 106:1568–1589

Tremback CJ (1990) Numerical simulation of a mesoscale convective complex: Model development and numerical results. Ph.D. Thesis, Colorado State University, Dept. Atmospheric Science, Fort Collins, CO 80523

Vidale PL, Pielke RA, Barr A, Steyaert LT (1998) Case study modeling of turbulent and mesoscale fluxes over the BOREAS region. J Geophys Res, in press

Extremes of moisture availability reconstructed from tree rings for recent millennia in the Great Basin of Western North America

Malcolm K. Hughes and Gary Funkhouser

Abstract. The western USA, from the eastern slopes of the Rocky Mountains to the Pacific Ocean, is rich in tree-ring archives of climate variability. We make particular reference to tree rings from the long-lived conifers of the high mountains of the Great Basin. These are among the few tree-ring records in the world likely to yield reliable information on not only interannual and interdecadal time scales, but also century scale change, since thousand-year and older trees growing at low density in open stands are relatively common in this region. These records have yielded well-verified reconstructions of precipitation which contain evidence on all time scales of extremes more intense or more persistent than those known for the twentieth century. The implications of these findings for forests are discussed.

1. Introduction

The western USA, from the eastern slopes of the Rocky Mountains to the Pacific Ocean, is rich in two very different, but related, resources. First, economically, socially and ecologically significant forests cover a substantial portion of a region in which questions of sustainability of resource use become ever more pressing as human population and economic activity grow. Most of the region may be classified as semi-arid, and it is this feature that is in large part responsible for the presence of the second resource. This is the remarkable natural archive of climate variability, especially in moisture availability to plants, contained in the annual rings of the trees of mid- and high elevations in the region's mountains.

Since Andrew Ellicott Douglass's pioneering work in Arizona in the first decade of this century, much work has been published which describes the development and use of the techniques of dendrochronology to assign precise and accurate calendar dates to individual growth layers (rings), the extraction of information on past climate and other environmental features from tree rings in the American West, and the physiological, ecological and climatological bases for these activities. Hughes and Graumlich (1996) provide a convenient entry point to this extensive literature, particularly relevant to the present article. In particular, they summarise work by several authors in which tree-ring chronologies of a thousand years length and more from sites in the American West have been used as natural archives of past climate. Here we will focus on the climate information contained in a group of such tree-ring archives from locations in and around the Great Basin, with a particular emphasis on moisture availability. The reasons for doing this arise from two recent reports. In the first, Dettinger et al (in press) show that, at both ENSO (El Niño –Southern Oscillation) and interdecadal time scales, the total amount of winter precipitation delivered to the cordillera of western North America has been remarkably stationary during the last 115 years, and perhaps back to the early eighteenth century. They also found evidence that spatio-temporal patterns of variation in precipitation amounts across the region also showed considerable consistency back to 1710. In the second report, Hughes and Graumlich (1996) report further evidence from a moisture-sensitive tree-ring chronology for the occurrence of intense multidecadal moisture deficit in the region near Mono Lake, California between approximately AD 900 and AD 1350. This corroborates the earlier work of Stine (1994), based on geomorphological and other evidence of low stands in the level of Mono Lake and other water bodies. If both sets of findings are sound, it would become clear that records covering the centuries before AD 1710 may be valuable in providing an adequate description of natural climate variability, at least in the interior American West.

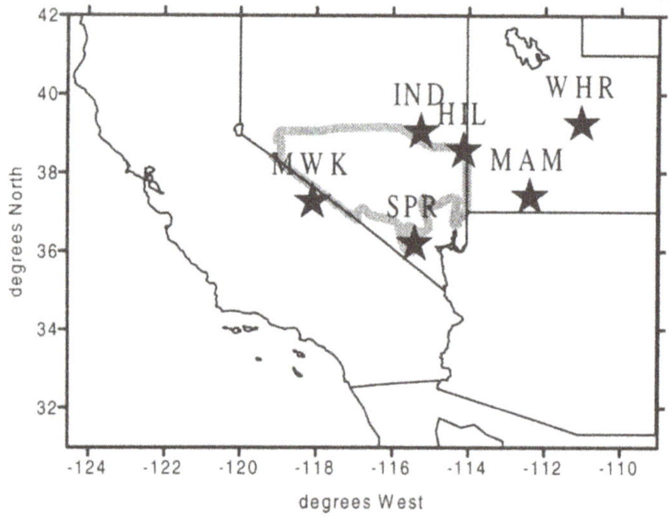

Fig. 1. Locations of the sites from which the six lower forest border *Pinus longaeva* chronologies were taken, indicated by a star and a three-letter code. The grey line indicates the boundary of Nevada Climate Division 3. MWK marks the location of the Methuselah Walk chronology.

The question is also raised of how the variability of recent centuries differs from that in earlier centuries, and of what that might mean for a better understanding of the causes of that variability. Thus, it is important to confirm that the indications from the Stine (1994) and Hughes and Graumlich (1996) work of a greater incidence of intense, persistent moisture deficit before approximately AD 1500 are indeed accurate. This is the objective of the work reported here, in which the precipitation record derived from the Methuselah Walk *Pinus longaeva* tree-ring chronology (Hughes and Graumlich, 1996), is compared to that derived from a set of several such records from sites in and close to the Great Basin. These data are available largely as a result of the dedicated efforts and foresight of the late Donald A. Graybill (Graybill, 1987).

2. Tree-ring records of variability in moisture availability on interannual to century time scales

The records used here are taken from six locations at the lower elevational limit of *Pinus longaeva* distribution in and close to the Great Basin (Table 1, Figure 1). Not only is each chronology 1700 years or more long, but they all share a number of other characteristics that render them particularly suitable for the reconstruction of interannual to century-scale variability in moisture availability. The region is semi-arid, and so plant growth is very likely to be limited primarily by moisture availability, particularly at a species' lower elevational limit. Detailed analyses have indeed established that this is the case for *P. longaeva* (Fritts, 1969). Great care was taken in the selection of sites and of individual trees to ensure a strong moisture signal, whilst minimising problems of excessive sensitivity resulting in too high a fraction of missing rings (LaMarche, 1982). The sampled stands are characteristically very open, the large distances between trees and lack of undergrowth leading to minimal interaction between trees and but a slight influence of other forest phenomena such as fire on the tree-ring record. Cook et al. (1995) have demonstrated that not only is the maximum wavelength of the recoverable climatic information signal limited by the length of the individual series making up a chronology, but that all but the most conservative detrendings of the individual ring width series will result in a serious degradation of the low-frequency signal. Their results indicate that the mean segment length in each chronology should be maximised, for example by removing unnecessary short segments, and that the absolute minimum detrending dictated by knowledge of non-climatic trends should be applied. If sufficient samples are available for a given chronology, it may be necessary to exclude those showing strong, obviously non-climatic trend. This approach has been taken in the development of the six chronologies used here. In every case mean segment length is greater than 650 years (Table 1), and age-related trend was removed by fitting either a straight line of zero or negative slope, or a negative exponential curve.

Site	Site code	State	Lat.N	Lon.W	Elev. (m)	Dates (- indicates BC start date)	Number of segments (segments >250 years)	Mean segment length (years)	Loading on PC1
Indian Garden	IND	NV	39 05	115 26	2900	-3261 - 1980	60 (59)	653	0.43
Mammoth Creek	MAM	UT	37 39	112 40	2590	-759 - 1989	49 (42)	740	0.56
Methuselah Walk	MWK	CA	37 26	118 10	2805	-6000 - 1996	295 (283)	748	0.47
Spring Mountains	SPR	NV	36 19	115 42	3000	218 - 1984	45 (43)	838	0.60
Wild Horse Ridge	WHR	UT	39 25	111 04	2590	286 - 1985	37 (35)	755	0.54
Hill 10842	HIL	NV	38 56	114 14	3050	-201 - 1984	45 (44)	1040	0.58

Table 1. Location and length of the Great Basin lower forest border bristlecone pine chronologies.

Recent recalibration of the long Methuselah Walk chronology, extended to AD 1996, confirms that such chronologies contain a record of interannual to decadal scale variability in moisture availability (adjusted $R^2 = 0.4$, $F=42.5$, $p<0.0001$, $R^2_{prediction} = 0.37$) (Figure 2). $R^2_{prediction}$ gives 'some indication of the predictive capability of the regression model' (Montgomery and Peck (1992), p.177). The instrumental time series used was the Nevada Division 3 record for 1932–1994 for July–June precipitation, taken to be representative of a large part of the Great Basin (Figure 1). The July–June year was chosen on the basis of LaMarche's (1974) finding that such lower forest border trees in the White Mountains (the location of Methuselah Walk) respond mainly to variation in previous summer/fall and current late spring precipitation. Hughes and Graumlich (1996) point out some weaknesses of an earlier calibration, differing only in its end date of AD 1979. In particular, it seems that although the tree rings effectively capture decadal fluctuations in precipitation, they fail to capture some dry years, particularly those where each of the months February through May were 1 – 3 °C cooler than the well predicted dry years.

They suggest that this problem may result because the 'effective moisture deficit experienced by *Pinus longaeva* is greater in years of early thaw, resulting in particularly restricted ring growth'. The new, extended calibration does, however, confirm that the tree rings capture multi-year to decadal variations, for example, both the above average period in the late 1970s and early 1980s, and the much discussed drought of the late 1980s and early 1990s (Figure 2).

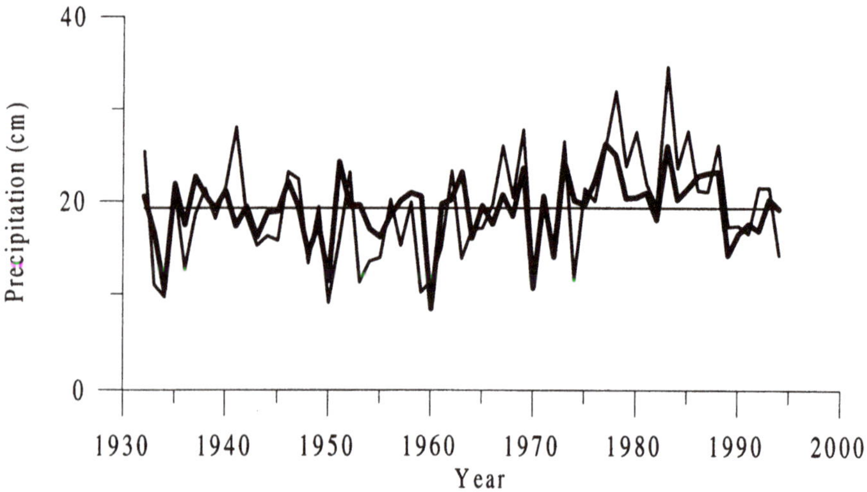

Fig. 2. Comparison of actual Nevada Division 3 July–June precipitation (fine line) with that reconstructed from the Methuselah Walk tree-ring chronology (thick line) AD 1932–1994.

3. A new reconstruction for the Great Basin

Given that lower forest border *Pinus longaeva* ring-width chronologies resemble white noise, with interannual variation having much greater amplitude than that on longer time scales, it is reasonable to view multidecade and century scale features with scepticism. This led Hughes and Graumlich (1996) to search for independent evidence for the major extended moisture deficits reconstructed for the period AD 900 to 1400. They found this confirmation in Graumlich's (1993) tree-ring based reconstruction of precipitation in the neighbouring Sierra Nevada, and in Stine's (1994) reconstruction of Mono Lake levels.

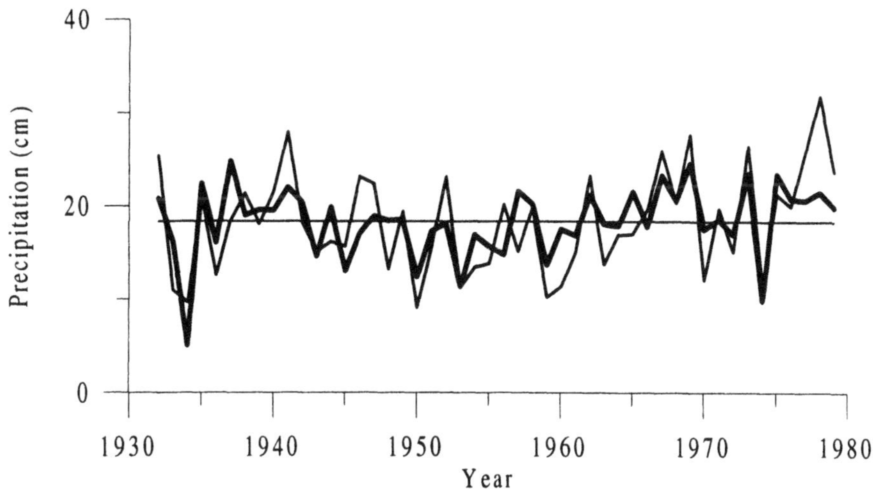

Fig. 3. Comparison of actual Nevada Division 3 July–June precipitation with that reconstructed from principal component 1 (PC1) of the six Great Basin lower forest border chronologies AD 1932–1979.

We decided to conduct a further test by using the network of chronologies mentioned above (Figure 1 and Table 1). As all six are strongly intercorrelated, we will use the first principal component (PC1) as representative of the set. PC1 accounts for 53% of the total variance, and the loadings for the chronologies are broadly similar in size (Table 1). This indicates that PC1 is not dominated by the Methuselah Walk chronology. PC1 calibrates even more strongly on Nevada Division 3 July–June precipitation than does the Methuselah Walk chronology alone (Figure 3), adjusted $R^2 = 0.48$, $F=45.23$, $p<0.0001$, $R^2_{prediction} = 0.47$. As with the calibration of the Methuselah Walk chronology on divisional precipitation, PC1 captures, at least in the calibration period, much of the interannual to interdecadal variability of precipitation. A reconstruction based on PC1 of the 5 chronologies other than Methuselah Walk has slightly less favourable calibration and cross-validation statistics (adjusted $R^2 = 0.4$, $F=33.1$, $p<0.0001$, $R^2_{prediction} = 0.37$). When the three reconstructions of Nevada Division 3 July–June precipitation are compared (Figure 4), it is clear that the same pattern of increased incidence of multidecadal precipitation deficits between AD 400 and AD 1400 as compared with the last 500 years is evident in Methuselah and the reconstruction based on PC1 of all six Great Basin chronologies. There is slightly less similarity between the reconstructions based on Methuselah and on PC1 of the other five Great Basin chronologies, and Methuselah departs most clearly from both PC1 reconstructions in the eighteenth century. The PC1 reconstructions provide some confirmation of the findings of LaMarche (1974), Stine (1994) and Hughes and Graumlich (1996).

Few of the driest non-overlapping 50-, 20- and 10-year periods in the 1762-year long reconstruction based on PC1 of the six Great Basin chronologies fall in the twentieth century, or even in the last 500 years (Table 2). The droughts that have caused much concern in the twentieth century, such as that of the late 1940s and early 1950s, are much less intense than those of the last 1600 years. For example, there are 56 non-overlapping 5 year periods reconstructed as being drier than the driest in the twentieth century up to 1979 (1949–1953, mean = 15.25). For example, the most severe 5 year period, 1919–23, had a mean precipitation of 11.83 cm. A similar comparison was made using the reconstruction based on the Methuselah Walk chronology for the period 6000 BC to AD 1994, focusing on the recent multi-year drought of the late 1980s and early 1990s. The trees record 227 drier non-overlapping 4-year periods since 6000 BC (2.85 per century) and of these only six have occurred since AD 1700: 1833–6, 1841–4, 1779–82, 1946–9 and 1855–8 in order of increasing dryness.

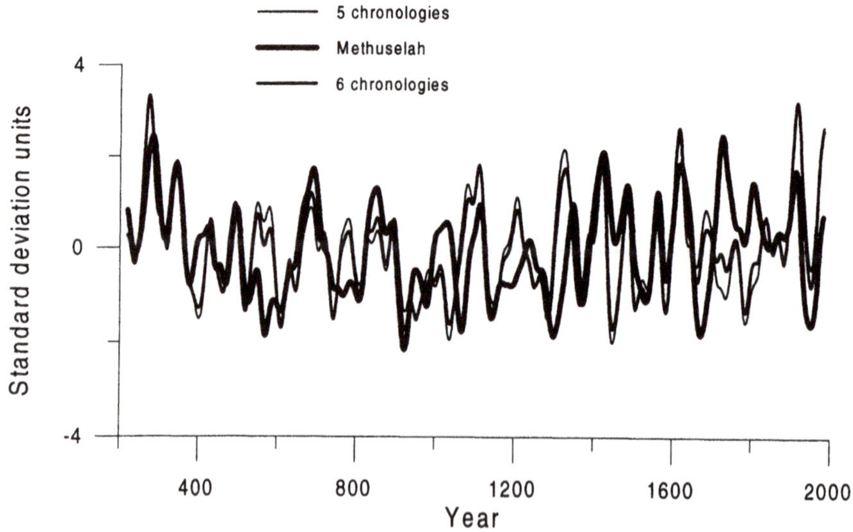

Fig. 4. Comparison of Nevada Division 3 July–June precipitation reconstructed from) the Methuselah Walk *Pinus longaeva* chronology, PC1 of the six Great Basin lower forest border chronologies, and PC1 of the same set without Methuselah Walk. All have been smoothed using a 50-year low-pass Gaussian filter.

	Rank	50 yr periods	total precip (cm)	20 yr periods	total precip (cm)	10 yr periods	total precip (cm)
	1	917-966	16.72	731-749	15.95	917-926	14.55
	2	1494-1543	16.90	941-960	16.00	1496-1505	14.84
	3	368-418	16.93	1430-1449	16.15	789-798	15.03
	4	1425-1474	17.11	1359-1378	16.19	1537-1546	15.06
	5	1242-1291	17.12	916-935	16.21	1575-1584	15.25
	6	1020-1069	17.13	1490-1509	16.22	560-569	15.32
	7	601-650	17.19	1275-1294	16.27	1622-1631	15.41
	8	1125-1174	17.23	1572-1591	16.38	1146-1155	15.42
	9	732-781	17.27	384-403	16.39	602-611	15.46
	10	1333-1382	17.40	1028-1047	16.42	510-519	15.46
Number of periods in 20th century		0		0		0	
Number of periods ending after AD1500		1		2		4	
Driest period in 20th century (amount in cm.)		1914-1963 (18.0)		1921-1960 (16.87)		1949-1958 (16.22)	
Rank of driest period in 20th century		20		21		33	

Table 2. Ten driest non-overlapping periods of 50, 20 and 10 years length in Nevada Division 3 precipitation reconstructed using PC1 of six lower forest border stripbark chronologies.

4. Discussion

The results presented here support the conclusion that there was a greater incidence of intense persistent moisture deficits after AD 400 and before approximately AD 1500, at least in the Great Basin of North America. This pattern is most evident in the reconstructions based on the Methuselah Walk chronology and on PC1 of six Great Basin chronologies including Methuselah Walk. The possibility that the pattern of intense persistent droughts between AD 900 and 1350 was stronger in the western Great Basin than the east will be investigated in future work, using a more spatially extensive array of similar chronologies. It is evident that, on these time scales, the twentieth century has experienced relatively few intense and persistent droughts, as also reported by Graumlich (1993) for the neighbouring Sierra Nevada. In the case of one-year extreme droughts, Hughes and Brown (1992) and Hughes et al (1996) find the twentieth century to be relatively drought-free, based on the incidence of very small rings in *Sequoiadendron gigantea* of the Sierra Nevada, California over the last 3000 years.

Thus, in considering a reasonable scenario for natural variability in moisture availability for tree growth in this region, we suggest that it would be prudent to assume a higher incidence of drought, on time scales from interannual to centennial, than would be expected on the basis of the instrumental record or the last 300 years. Hughes and Brown (1992) produced evidence for periods in time in which the incidence of extreme one-year droughts such as that of 1977 (experienced in the Sierra Nevada, but not in the Great Basin) was three times greater than in the instrumental record. It is conceivable that the occurrence of as many as 12 such droughts in a century would have significant ecological impact on that region's forests, for example, by increasing fire frequency to the extent that fuel condition rather than fuel type is important, and perhaps by providing a stimulus to forest insect populations and pathogens. We have shown, in this and prior work (Hughes and Graumlich, 1996), that decadal and multidecadal extreme moisture deficits were experienced more frequently before AD 1500 than since. Swetnam and Betancourt (in press) describe a range of ecological effects of climate variability in the American Southwest, including 'regionally synchronised fires, insect outbreaks, and pulses in tree demography (births and deaths)'. They portray these effects as being manifest 'across scales, from annual to decadal, and from local to mesoscale'. It would be reasonable to assume that the ecological effects of climatic extremes would increase in intensity, and change in character, as their length increases from decadal to centennial.

We need to consider carefully the implications of findings such as those reported here for the use of concepts such as 'natural range of variability' in the management of forest resources. The question could be posed in the terms 'how long a period of record is needed?'. It would probably be more productive to consider what range of relevant climate variability might be expected for the existing and expected boundary conditions of the climate system. Rather than using natural archives such as dendrochronological records in an actuarial sense to calculate what might happen on the basis of what has happened, they should be used to cast light on the operation of the climate system, so that estimates of impending climate variability might be derived with a sound physical basis. In this context, the major result reported here, the increased frequency of very persistent moisture deficits before AD 1500, raises questions about the operation of the climate system. Could such extreme, persistent, mesoscale events occur by chance within global circulation patterns as observed in the modern period, or would they require some global-scale modification of the circulation? There is some evidence that the latter may be the case. First, Stine (1994) reported low stands of lakes in southern South America at the same times as those in the American West. He postulated a contraction of both circumpolar vortices as a possible explanation for this. This would, presumably, result from a diminution in the equator–pole thermal gradients. Second, there is a broad similarity between the multi-century pattern over the last 1600 years reconstructed here for the Great Basin, and the broad patterns of accumulation measured on the high-elevation ice-caps at Quelccaya in Peru, and Dunde in Tibet (Thompson, 1996). This could be mere coincidence, or it may represent a feature of variability in the hydrological cycle on a global scale. This suggestion should be treated with great caution. Rather than use generic terms such as, for example, 'Medieval Warm Period' or 'Little Ice Age', into which an unsubstantiated meaning of global synchroneity at decadal to multidecadal time scales may be read (Hughes and Diaz, 1994; Bradley and Jones, 1992), we prefer to present our results for inclusion in global-scale quantitative analyses designed to detect temporal and spatial pattern objectively, as for example in the work of Mann et al. (in press). The existing network of high-resolution proxy climate records is capable of yielding useful information on these issues for the last three or four centuries, but insufficient data are available for longer periods.

Acknowledgments. The work reported here is largely based on materials and preliminary analyses developed by the late Donald A. Graybill. Funding has been provided by the U.S. National Oceanographic and Atmospheric Administration's Paleoclimatology program, grant number NA66GP0311.

5. References

Bradley RS, Jones PR (eds) (1992) Climate since AD 1500. Routledge, London

Cook ER, Briffa KR, Meko DM, Graybill DA, Funkhouser G (1995) The segment length curse in long tree-ring chronology development for palaeoclimatic studies. The Holocene 5:229–235

Dettinger MD, Cayan DR, Diaz HF, Meko DM (in press) North–South precipitation patterns in western North America on interannual-to-decadal time scales. J Climate

Fritts HC (1969) Bristlecone pine in the White Mountains of California: growth and ring-width characteristics: The University of Arizona Press, Tucson

Graumlich LJ (1993) A 1000-year record of temperature and precipitation in the Sierra Nevada: Quaternary Res 39:249–255

Graybill DA (1987) A network of high elevation conifers in the western U.S. for detection of tree-ring growth response to increasing atmospheric carbon dioxide Proceedings International Symposium on Ecological Aspects of Tree-Ring Analysis, pp 463–474

Hughes MK, Diaz HF (1994) Was there a Medieval Warm Period, and if so, where and when? Climatic Change 26:109–142

Hughes MK, Graumlich LJ (1996) Multimillennial dendroclimatic records from western North America In: Bradley RS, Jones PD, Jouzel J (eds) Climatic variations and forcing mechanisms of the last 2000 years: Springer Verlag, Berlin, pp 109–124

Hughes MK, Touchan R, Brown P (1996) A multimillennial network of giant sequoia chronologies for dendroclimatology. In: Dean J, Meko D, Swetnam TW (eds) Tree Rings, Environment and Humanity, Proceedings of the International Conference, Tucson, Arizona, 1994 May 1994, Radiocarbon, Tucson, pp 225–234

Hughes MK, Brown PM (1992) Drought frequency in central California since 101 BC recorded in giant sequoia tree rings: Clim Dyn 6:161–167

LaMarche VC Jr (1974) Paleoclimatic inferences from long tree-ring records. Science 183:1043–1088

LaMarche VC Jr (1982) Sampling strategies. In: Hughes MK, Kelly PM, Pilcher J, LaMarche VC (eds) Climate from tree rings: Cambridge University Press, Cambridge pp 2–6

Mann ME, Bradley RS Hughes MK (in press) Global-scale temperature patterns and climate forcing over the past six centuries. Nature (London)

Montgomery DC, Peck E (1992) An introduction to linear regression analysis. Wiley, New York

Stine S (1994) Extreme and persistent drought in California and Patagonia during mediaeval time. Nature (London) 269:546–549

Swetnam TW, Betancourt J (in press) Mesoscale disturbance and ecological response to decadal climatic variability in the American Southwest. J Clim

Thompson LG (1996) Climatic changes for the 2000 years inferred from ice-core evidence in tropical ice cores. In: Jones PD, Bradley R, Jouzel J (eds) Climatic variations and forcing mechanisms of the last 2000. NATO Advanced Research Series I, 41: 281–295

Predictive models of tree-growth: preliminary results in the French Alps

Lucien Tessier, Thierry Keller, Joel Guiot, Jean-Louis Edouard and Frédéric Guibal

Abstract. The analysis of tree-ring–climate relationships provides models (response functions) of tree-growth calibrated on the inter-annual variability of climate. Output of GCMs can be used as inputs of these models in order to evaluate the change in radial growth induced by climatic change. A spatio-temporal approach applied to a large data set of ring-width chronologies and meteorological data allows the response of trees to be evaluated for different populations of various species in numerous habitats.
Such a study was carried out, firstly on ten populations in south-eastern France, then on populations at high-altitude sites. The species involved were *Larix decidua*, Mill., *Pinus sylvestris*, L., *Abies alba*, Mill., and *Picea abies* Karst. The calibration of tree ring to climate relationships was based on the monthly values of precipitation and temperature provided by meteorological stations more or less distant from the tree sites. Outputs of GCMs were obtained from the ARPEGE model of Météo-France with a simulation on a large grid (2°79 in latitude and 3° in longitude) for the hypothesis of a CO_2 doubling.
Results show that climatic change can induce either an increase or a decrease in the mean radial growth. For most of the tree-populations, no significant change was apparent. The results suffer from insufficient data related to the spatial representation of climate (i.e. stations that are too far from the tree populations, only two grid points available from the GCMs, etc.).
In a new research project supported by the European Union, this basic methodology is being proposed, with some improvements in the statistical techniques. The data set is being extended to the whole of south-western Europe, with three target areas centred on the Alps, the Pyrenees, the Mediterranean area, and southern Italy. The spatial coverage of both tree rings and meteorological data will be improved. The spatial variability of climate will be more precisely taken into account through the availability of GCM simulations on finer grids (40–50 km),.

1. Introduction

"Tree-rings, from the past to the future" was the title of a recent international workshop dealing with dendrochronology (Ohta et al. 1995). From the first pioneering works (Douglass 1936) to the explosion of the subject that has accompanied the development of possibilities brought by intensive computer calculations (Fritts 1962, 1976), climate has been in the centre of the concepts involved in tree-ring analysis. Until now, the contribution of dendrochronology to climate change research has focused on the reconstruction of climate history and environmental changes, mainly during the last 2000 years, with an annual resolution. The book "Climate since AD 1500" (Bradley and Jones 1995) provides a good state-of-the-art account of such reconstructions.

The emergence of "global change research" attracted the attention of scientists to the simulation of tree-growth responses to the climate changes projected by Atmospheric General Circulation Models (AGCMs) (IPCC 1990; Pan and Raynal 1995). As stated by Hughes (1995), global change research requires different approaches to those adopted by most dendrochronologists in recent decades. Two methodological challenges are involved in dendroclimatology: 1) low frequency analysis of climatic variations; 2) the identification and extraction of new climate information from tree rings. However the main challenge remains in linking future climate changes, as predicted by AGCMs, to induced impacts on tree growth. Such approaches need firstly to link the AGCM outputs with a model providing estimated radial tree growth from climatic parameters. The second objective is move from tree growth to forest productivity, but there are great uncertainties over the identification of these relationships.

We present here the coupling of an AGCM simulation with statistical models of tree-ring–climate relationships for 24 tree populations in the French Alps. The results of this study will be discussed in relation to a strategy to generalise the approach over a larger area (south-western Europe), thereby including more spatial climatic and ecological variability and allowing an approach to the estimation of forest productivity.

2. Global Methodology

Trees growing in temperate zones and particularly those in mountain regions contain a primary, climate-dependent signal (Tessier et al. 1997). Variation in annual tree-ring widths from one year to the next is strongly linked to inter-annual climate variability.

The establishment of a relation between radial growth variations and climatic variables is achieved in two steps:

1. Standardisation of tree-ring series to produce a site chronology: after crossdating and standardisation of basic chronologies (one chronology for each core), a master indexed chronology is built for each population. The average of the basic chronologies expresses the high-frequency signal common to all the trees, and is strongly linked to climate.
2. Calibration of the site chronology with instrumentally recorded climatic data.

The basic method used here for standardisation is the commonly used methodology in dendroclimatic reconstruction (Fig 1). Complete descriptions of these mathematical and statistical processes are given by Fritts (1976) and Cook & Kairiukstis (1990). The calibration step involves Artificial Neural Networks (ANN). This new statistical process enables the introduction of non-linearity in the tree-ring–climate relationships (Caudill and Butler 1992; Keller et al. 1997). Once a master chronology of standardised indices is produced for each population, the tree-ring–climate relationship is established over the period with instrumental climatic data. The robustness of the relationship is validated using bootstrap methods (Guiot et al. 1995).

The next step deals with the use of the calibrated relationships to derive annual radial growth from the climatic perturbation provided by the AGCM outputs. Predicted growth is then compared to actual observed growth.

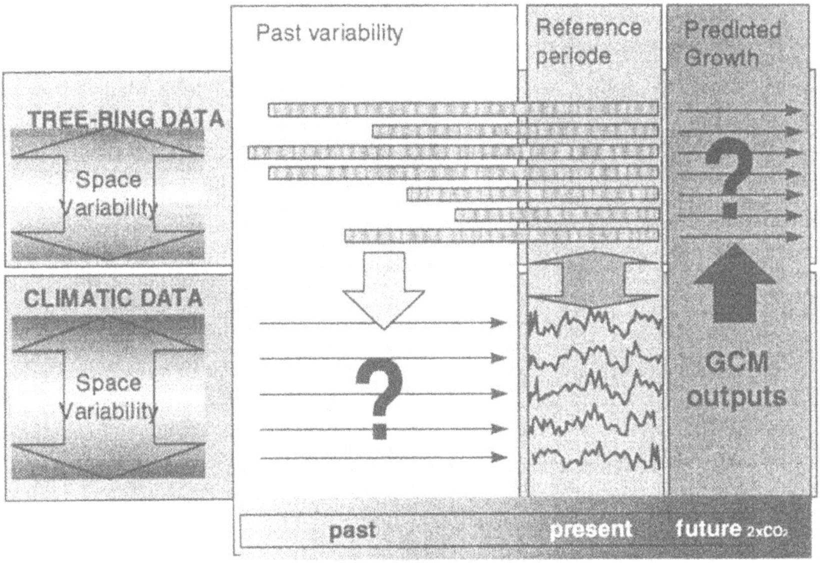

Fig. 1. Basic methodology of dendroclimatology. Tree-ring–climate relationships can be established for the reference period. Past climate can be then reconstructed from tree-rings and future growth evolution can be estimated from AGCM outputs

3. Data and Processing

3.1 Tree-ring data

Twenty-four populations were used for this study (Fig. 2, Table 1). They cover several bioclimatic zones from the strict Mediterranean area to the highest mountain forest belt of the northern Alps. Five species were studied:

- *Larix decidua* Mill.
- *Pinus sylvestris* L.
- *Picea abies* Karst.
- *Abies alba* Mill.
- *Pinus uncinata* Ram.

The partition of the region into six zones is based on the ecological classification proposed by Ozenda (1981). At each site, a minimum of ten dominant or co-dominant trees (three cores per tree) was sampled.

It is only for *Larix decidua* that the diversity of biotopes occupied by that species (the highest mountain forest belt) is represented from the northern internal Alps (Zones A, B, C) to the southern Alps (D, E). *Pinus sylvestris* forests are mainly represented in their Mediterranean biotopes (Zone F), from the Mediterranean (ETOI) to the supra-Mediterranean (BVAU, VENT); only MAP1 is in the coniferous montane forests. The three other species are only represented by one (*Pinus uncinata*), and two (*Abies alba, Picea abies*) populations.

Site	*Zone*	*Code*	*Species*	*Latitude N*	*longitude E*	*Altitude*	*nb trees*	*nb cores*
Plan de l'Aiguille	A	120B	*Picea abies*	45.90	6.86	1800 m	20	49
Blaitière		BLAI	*Larix d.*	45.91	6.88	1900 m	20	46
Maurienne		MAA1	*Abies alba*	45.28	6.83	1600 m	10	29
Maurienne		MAL3	*Larix d.*	45.25	6.40	1650 m	6	10
Maurienne		MAL1	*Larix d.*	45.20	6.73	1400 m	10	30
Maurienne	B	MAL2	*Larix d.*	45.20	6.74	1800 m	10	30
Maurienne		MAP1	*P. sylvestris*	45.20	6.73	1400 m	10	29
Orgère		ORGB	*Larix d.*	45.22	6.66	2130 m	20	51
Orgère		ORGH	*Larix d.*	45.22	6.67	2300m	20	62
Achard	C	ACHA	*P. uncinata*	45.12	5.90	2000 m	20	39
La Barrière		TAIL	*Picea abies*	45.07	5.90	1800 m	13	35
Chardonnet		CHAR	*Larix d.*	45.03	6.55	2100 m	18	46
Oriol	D	ORIO	*Larix d.*	44.80	6.60	2150 m	17	42
Morgon		MORG	*Larix d.*	44.50	6.42	1930 m	18	47
Saint-Ours		SO	*Larix d.*	44.48	6.80	1940 m	14	42
Roche-la-Croix		RLCS	*Larix d.*	44.46	6.80	2250 m	12	36
Crête des Mélèzes		CDM	*Larix d.*	44.35	6.25	1630 m	15	43
Forêt de la Blanche	E	FDLB	*Larix d.*	44.35	6.43	1780 m	15	45
Beauvezer		MI40	*Larix d.*	44.12	6.63	2150 m	11	21
Beauvezer		SF70	*Abies alba*	44.13	6.65	1750 m	13	33
Merveilles		MERV	*Larix d.*	44.09	7.44	2150 m	30	88
Bois de Vaucluse		BVAU	*P. sylvestris*	44.40	5.58	1300 m	20	60
Ventoux	F	VENT	*P. sylvestris*	44.18	5.26	1000 m	10	28
Etoile		ETOI	*P. sylvestris*	43.41	5.51	750 m	15	38

Table 1. Description and location of population involved in this study. The division in 6 zones (A, B, C, D, E, F) proposed is inspired form ecological sub-division by Ozenda (1981).

A : Northern intra alpine zone (continental influences
B : transition intra alpine zone (continental and mediterranean influences)
C : external alpine zone (oceanic influences)
D : central intra alpine zone (continental influences)
E : Mediterranean Alps (Mediterranean and influences)
F : Mediterranean zone (Eu and supra mediterranean zones)

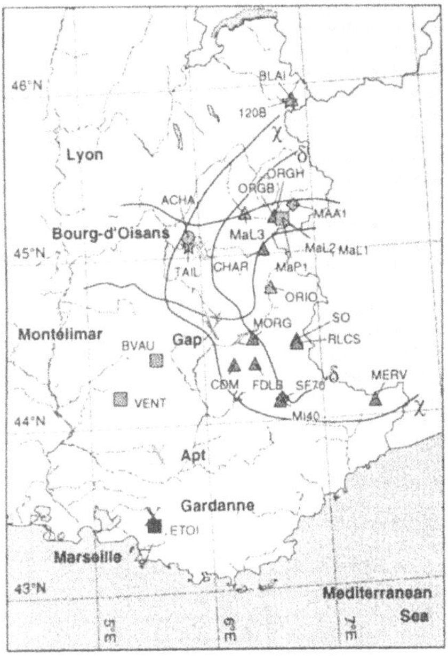

Figure 2. Location of the populations and the meteorological stations. The colors of sites represent each set composed by meteorological station(s) and associated populations. The division lines are extracted from ecological sub-division proposed by Ozenda (1981).

Northern-Southern Alps transition area :

Northern limit of the transition zone

Southern limit of the transition zone

Continentality indices :

A : limit between the external pre-alpine zone and the intermediate zone

B : limit between intermediate and intra-alpine zone

Meteorological stations: Temperature, Precipitation

Tree species: Larix decidua, Picea abies, Abies alba, Pinus uncinata, Pinus sylvestris

3.2 Climatic data

Calibration of tree-ring–climate relationships involves each indexed master chronology as the dependant variable. Predictors (regressors) are the monthly mean temperature and the monthly precipitation for each biological year, i.e. from October t-1 to September t (t is the current year of the annual ring). For statistical reasons, the meteorological data involved in the calibration (Fig. 2 Tab. 2) are selected in order to provide sufficiently long series in relation to the number (24) of climatic variables being considered in the response functions. The meteorological stations that have been used are approximately representative of the variation in climate from the northern to the southern Alps. Climatic outputs are given by the ARPEGE AGCM of Météo-France, in which the atmospheric model is coupled with an oceanic surface model. The grid resolution (2°79 latitude and 3° longitude) provides only four grid points for the whole area (Tab. 2). The two nearest points have been used.

3.3 Simulation of climate change and its impact on tree-growth

The ARPEGE model for both monthly precipitation and temperature and for each grid point simulates the CO_2 doubling effect on climate. The simulated anomalies for the closest grid point is added to each climatic series used in the calibration processes (Fig. 3). For temperature, the warming simulated is from about 2° to 3°K, with an important peak for late winter and early spring (about 5°K). Precipitation has a complex pattern; only spring appears to have a homogeneous increase.

The simulated climatic data are then introduced as predictors in each tree-ring–climate relationship to obtain, for each population, the estimated ring width after a CO_2 doubling. The same operation is repeated using the mean monthly values of the calibration period (1901–1980) as predictors.

station	var.	Location&Altitude	Grid-point	Populations
Lyon	TM	4.83e-45.72n - 300m	4.00e-46.05n	120B, BLAI, MAA1, MAL1, MAL2, MAL3, MAPI, ACHA, TAIL, ORIO
Le Bourg d'Oisans	PR	6.03e-45.05n - 730m		
Gap	TM	6.08e-44.56n - 735m	3.75e-43.25n	OrgH, OrgB, CHAR, MORG, SO, RLCS, CDM, FDLB, SF70, MI40, MERV
	PR			
Montélimar	TM	4.81e-44.57n - 75m	3.75e-43.25n	BVAU, VENT
Apt	PR	5.23e-43.87n - 234m		
Marseille	TM	5.38e-43.30n - 75m	3.75e-43.25n	ETOI
Gardanne	PR	5.47e-43.45n - 283m		

Table 2. Location of meteorological stations, corresponding grid-points and populations involved

4. Results

Figure 4 enables the comparison of ring widths predicted by the model with:

- Observed mean values measured for the period 1901–1980.
- Estimated mean values obtained from the mean climate of the same period.

Taking into account the standard deviation of the observed tree-ring width values, there is good agreement between the measured and simulated values of ring width for 1901–1980, which indicates the robustness and efficiency of the growth model. The climatic change simulated by the AGCM affects only six populations: five of them with a positive effect and one with a negative effect. The variations observed for other populations are statistically insignificant. Even for the populations connected to meteorological data provided by Lyon station, where an increase of about 5°K is simulated for February (Fig. 3), only one population (MAA1) reacts. Before any interpretation of these results, it is necessary to keep in mind that the radial growth predictive model does not take into account the potential direct effect of the CO_2 doubling on photosynthesis and water-use efficiency (Larcher 1980; Mooney et al. 1991; Reynolds et al. 1992). Also, the calibration applied on standardised series only takes into account the inter-annual variability of ring width and climate. In addition, any positive or negative trend effect (Becker et al., 1994; Bert, 1992) induced by a correlative trend in temperature and precipitation cannot be taken into account.

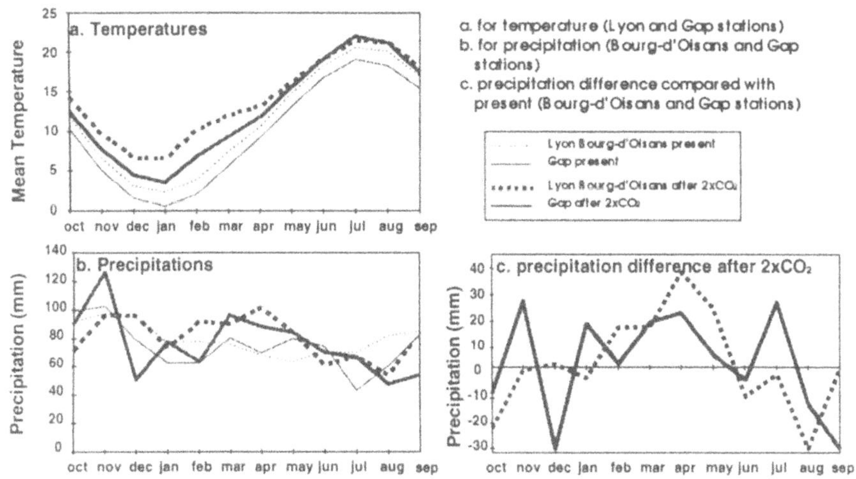

Fig. 3. Climatic perturbation simulated by an atmospheric CO_2 concentration doubling

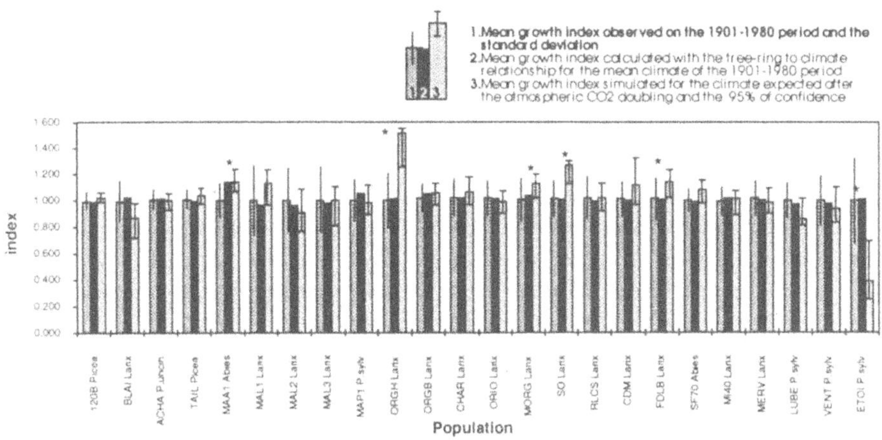

Fig. 4. Effect on radial growth of an atmospheric CO_2 concentration doubling (North-South gradient) estimated with Neural Network

5. Discussion

The statistics which measure the link between climatic factors (precipitation and temperature) and radial tree-growth cannot confirm that a cause–effect phenomenon exists. To reach such a conclusion, coherency has to be analysed on the basis of our ecological knowledge about species and biotopes.

For *Larix decidua*, five populations from among the 15 that were analysed react to the simulated climate. For these five populations, better growth is simulated. Differences in the degree of reaction can be only analysed in relation to the altitudes of the sites (Tab.1). The highest (2300m) population (ORGH) reacts the most strongly to the simulated climate change. However, FDLB (one of the lowest sites) also reacts positively while RLCS at 2250 m altitude shows no significant change. More precise ecological characterisation is necessary to identify the factors (altitude, topography, exposition, soil characteristics, genetics, etc.) involved in the ability of trees to react to climate change. Moreover, the simulated climate, with only four grid points, does not take into account the spatial diversity of climate very well, in an area experiencing oceanic (zone B), continental (zone A, C, D) and Mediterranean influences (E).

Nevertheless, all the *Larix decidua* populations have been sampled in the sub-alpine forest belt near the timberline. In such an environment, temperature can be considered as the most limiting factor both by extending the tree activity period and accelerating growth processes. At higher altitudes, *L. decidua* completes its growth in less than four months (June–September, Ladefoged, 1970). Favourable temperatures at the beginning or end of the growth season can extend the growth period by about 15 to 30 days (Tranquillini 1964; Larcher 1980). The ring width directly results from the length of this growing season, through cambial activity and the phenological phases of leader development from spring to autumn. The generally positive reaction of *L. decidua* populations to the simulated climate change can be explained by such a mechanism.

The *Pinus sylvestris* population from l'Etoile (ETOI) constitutes the southernmost population from all studied sites. At 750m altitude, this population is included in the supra-Mediterranean zone. The negative growth reaction to the simulated climate has to be seen in terms of the behaviour of this species in such an environment. Previous studies (Tessier 1984), performed on a large data set gathered in a region extending from Marseille to Gap, showed that in the most critical situations in terms of water stress (whatever its origin may be, climatic or edaphic), *P. sylvestris* reacts by a severe reduction in growth rate and almost total summer dormancy. This behaviour can be related to stomatal regulation (Aussenac & Granier, 1978). In the hypothesis concerning the impacts of temperature increases, evapotranspiration would soon lead to depletion of water reserves. Without any compensating increase in precipitation, it is plausible that increased summer dryness will produce a reduction of radial growth.

Table 3 (interpretation) summarises the interpretation of the positive and negative impact of climate change on the two most sensitive populations. The interpretation takes into account the site location, the supposed limiting factor of tree growth, the effect of the climatic perturbation on ecophysiological processes according to the nature of the limiting factor and the consequences on tree growth: wider or narrower annual tree rings.

For the Alps, all reactive populations are located in the more continental zones (B, D, E). These zones are characterized by stronger climatic variations than peripheral zones. The harsher climatic conditions in winter may explain such a reaction. The ETOI *Pinus sylvestris* population also experiences harsh climatic conditions (summer drought). These results indicate that it is only the populations that are submitted to critical life conditions that are sensitive to the simulated climatic perturbation.

	Etoi (Pinus sylvestris)	OrgH (Larix decidua)
Location	Southern extension limit of this species in France	Altitudinal limit of the species (2300 m)
Limiting Factor	**Hydric** Impact on : Availability of water reserves	**Thermic** Impact on : Duration of vegetation period Metabolism intensity
Effect of the Climatic Perturbation	**Evapo-transpiration increase without compensatory precipitation** ➢ Maximal water stress	**Summer temperature increase** ➢ Longer vegetation period ➢ Better metabolic balance
Effect on Tree Growth	**Earlier and more frequent growth stop in Summer** ➢ Narrower annual tree ring	**Longer growth period** ➢ Wider annual tree ring

Table 3. Interpretation of the impacts on radial tree growth of the climatic perturbation induced by an atmospheric CO_2 concentration doubling

6. Conclusion and outlook

To be coherent, an analysis of the spatial response of trees needs to cover a larger data set for each species in order to cover a larger biotope typology. Only *Larix decidua* is represented here by a sufficiently large number of populations to cover this spatial diversity, which is why this must remain a preliminary study. However, the results indicate positive (*L decidua*) and negative (*Pinus sylvestris*) impact of climatic change on tree growth.

The results are consistent with what is known about tree ecophysiology and the distribution of the two species. The first results obtained on a reduced data set (Keller et al. 1997) are confirmed by this larger data set. It is also confirmed that the complexity of the combination of climate variability and stand variability and the individual reactions of different species make the estimation of the impact of climate change on tree growth and ecosystems very difficult. Although the results are significant, several points should be considered in order take into account climate and tree-ring variability and tree behaviour.

The attempt made here involves only two grid-points for the climatic simulation. Future studies will use the most accurate simulation available from the ARPEGE model with a grid of 40 to 50 km. Such a scale is more compatible with the climatic, topographic and ecological variations in mountains. The new grid density will probably enable precipitation to be taken into account, which is very sensitive to spatial variability.

As with the simulated climatic data set, the meteorological and tree-ring data sets need to be extended and their densities increased in order to represent the entire range of climate–site combinations for the different species. This extended data set should cover as many regional climatic zones as possible. Particular attention needs to be paid to the boundaries of both climatic zones and species distributions. It is at these boundaries that the most important changes can be expected. Such improvements need a global sampling strategy and a global methodological approach of the two data sets (climate and tree rings).

It is also anticipated that there will be progress in dealing with the calibration of tree-ring–climate relationships through the better use of synthetic parameters (i.e. ETP, ETR, water balances, etc.) than through basic monthly parameters such as temperature and precipitation. Such a development will enable the introduction of ecophysiology into the response-function models. One has also to keep in mind that the results about tree-growth are not immediately transferable to forest productivity. The next challenge will be to move from predicted tree-growth to predictions about forest production.

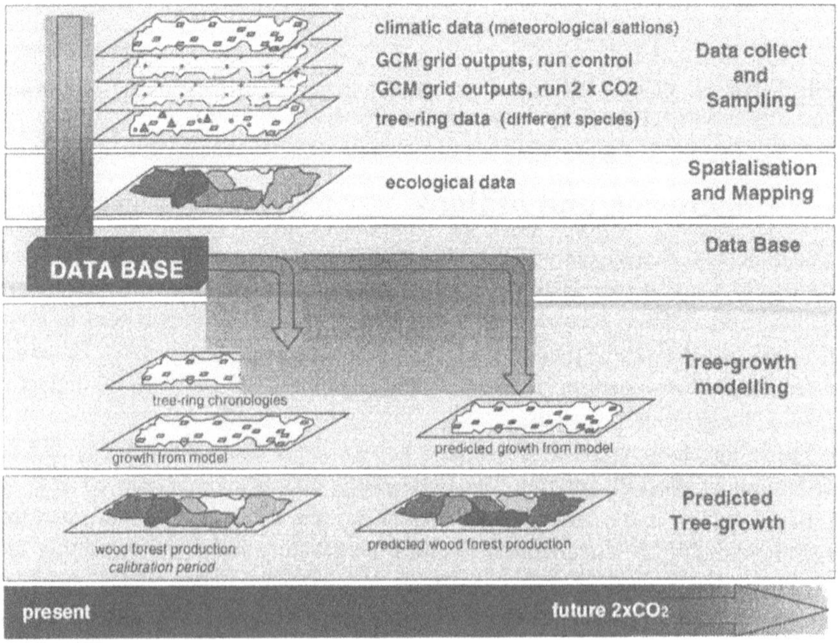

Fig. 5. Global strategy to predict future forest production from tree rings

As a conclusion, we present (Fig. 5) the global strategy for such an approach. This is the general strategy developed for the European project FORMAT (Forest Modeling Assessment from Tree-rings). The first objective, similar to that described above, is to obtain the growth responses of tree populations to climate change as predicted by a 2xCO_2 scenario for three target areas, covering the climatic and ecological variability of south-western Europe (Fig. 6). The second objective will be to map tree growth in order to assess the impact of climate change on forest production. We hope to present the results of that study in the meeting planned for 2001.

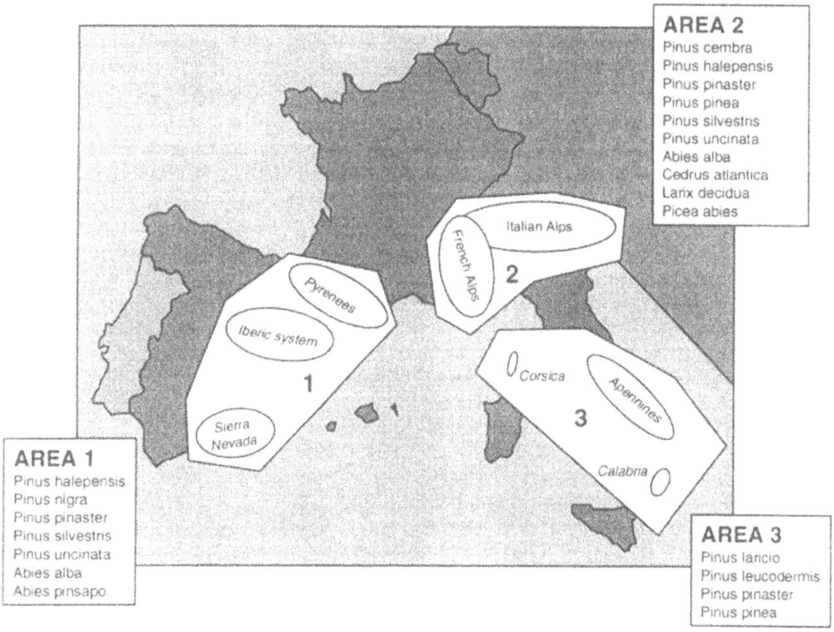

Fig. 6. Target area for future studies on the estimation of forest production

7. References

Aussenac, G and Granier, A (1978) Quelques résultats de cinétique journalière du potentiel de sève chez les arbres forestiers. *Ann. Sci. For 35(1), 19-32*

Becker, M., Bert, G.D., Bouchon, J., Dupouey, J.L., Picard, J.F. & Ulrich, E. (1994) Long term changes in Forest Productivity : the Dendroecological Approach. *in «Forest decline and air pollution effects in the French mountains». G. Landmann & M. Bonneau (eds), Springer Verlag, 1-12.*

Bert, D. (1992) Silver fir (*Abies alba* Mill.) shows an increasing long term trend in the Jura mountains. *Lundqua Report "Tree rings and Environment", Proceedings of the International Dendrochronological Symposium, Ystadt, South Sweden, 3-9 September 1990, 34, 27-29.*

Bradley R. S. and Jones Ph. (1995) Climate Since AD 1500. *Bradley and Jones eds. Routledge Publishing Company, London, 706p.*

Caudill, M. & Butler, C. (1992) Understanding Neural Networks, vol. 1. *A Bradford Book, The MIT Press Cambridge, Massachusetts, 309p.*

Cook and Kairiukstis, (1990) Methods of dendrochronology. Applications in the envirronmental sciences. *Kluiwer Academic Press. Dortrecht 394p.*

Douglass, A.E. (1936) Climatic Cycles and tree growth, Vol. III, A study of cycles. *Carnegie Inst. Washington pub., 289p.*

Fritts, H.C. (1962) An approach to dendroclimatology, screening by means of multiple regression techniques. *J. Geophys. Res.*, 67, 1413-1420.

Fritts, H.C. (1976) Tree-rings and Climate. *Academic Press, London, New-York, San Francisco, 576p.*

Guiot, J., Keller, T., Tessier, L. (1995) Relational databases in dendroclimatology and new non-linear methods to analyse the tree response to climate and pollution. *In Ohta, S., Fujii, T., Okada, N., Hughes, M.K., and Eckstein, D., eds., Tree Rings: From the Past to the Future. Proceedings of the International Workshop on Asian and Pacific Dendrochronology. Forestry and Forest Products Research Institute Scientific Meeting Report 1: 17-23.*

Hughes, M.K. (1995) Tree rings and the challenge of global change research. *In Ohta, S., Fujii, T., Okada, N., Hughes, M.K., and Eckstein, D., eds., Tree Rings: From the Past to the Future. Proceedings of the International Workshop on Asian and Pacific Dendrochronology. Forestry and Forest Products Research Institute Scientific Meeting Report 1: 1-7.*

IPCC (1990) Climate Change. The IPCC Scientific Assessment, J. T. Houghton, G. J. Jenkins & J. J. Ephraums (eds) Cambridge University Press.

Keller, T., Guiot, J., Tessier, L. (1997) Climatic effect of atmospheric doubling on radial growth in south eastern France. *Journal of Biogeography, 24, 857-864.*

Ladefoged, C. W. (1970) The periodicity of wood formation. *Dan. Biol. Skr., 7 (n°3).*

Larcher, W. (1980) Physiological Plant Ecology. *2nd edition, Springer Verlag.*

Mooney, H.A., Drake, B.G., Luxmoore, R.J., Oechel, W.C. & Pitelka, L.F. (1991) Predicting ecosystem responses to elevated CO_2 concentrations. *Bioscience 41, 2, 96-104.*

Ohta, S., Fujii, T., Okada, N., Hughes, M.K. and Eckstein, D. (1995) *Tree Rings from the Past to the Future*, Proceedings of the International Workshop on Asian and Pacific Dendrochronology. Published by Forestry and Forest Products Research Institute. 286p.

Ozenda (1981) Végétation des Alpes sud-occidentales. *Editions du CNRS, 258p.*

Pan, Y. & Raynal, D.J. (1995) Predicting growth of plantation conifers in the Adirondack mountains in response to climate change. *Can. J. For. Res.* 25, 48-56.

Reynolds, J.F., Hilbert, D.W., Chen, J., Harley, P.C., Kemp, P.R. & Leadley, P.W. (1992) Modeling to response of plants and ecosystems to elevated CO_2 and climate change. *Carbon Dioxide Research Division, United States Department of Energy.*

Tessier, L. (1984) Dendroclimatologie et écologie de *Pinus silvestris* L. et *Quercus pubescens* Willd. dans le sud-est de la France. *Thèse de Doctorat ès Sciences Université d'Aix-Marseille III, Faculté de St-Jérôme, Marseille.*

Tessier, L., Guibal, F., Schweingruber, F.H. (1997) Research strategies in dendroecology and dendroclimatology in mountain environments. *Climatic Change, 36, 499-517.*

Tranquillini, W. (1964) The physiology of plants at high altitudes. *Annual Review of Plant Physiology, 15, 345-362*

Documenting the effects of recent climate change at treeline in the Canadian Rockies

Brian H. Luckman and Trudy A. Kavanagh

Abstract. Dendrochronology and tree-ring densitometry are used to reconstruct summer temperatures and treeline dynamics in the Columbia Icefield area from 1600 AD to the present. Detailed studies at three sites, less than 10 km. apart, show different responses to regional climate over this interval. At a north-facing site, *Abies lasiocarpa* have maintained a population by vegetative regeneration with little change in the treeline ecotone over the last 400 years. In contrast, at a warmer, south-facing site, the *Picea engelmannii* dominated treeline shows catastrophic dieback during the late 1600s. Extensive upslope migration of treeline by seedling establishment occurred during the 20th Century. At an adjacent valley-floor site, tree clumps established during the 18th–19th centuries but exhibit no subsequent population expansion beyond their borders. These results suggest that the response of treeline ecotones to climate change varies with both local site conditions and the response of individual species. These data can provide important inputs to simulation models and to resource managers who wish to understand the effect of climate change on ecosystem dynamics.

1. Introduction

Ecotones are important indicators of environmental change because they represent a physical manifestation of the range limits of species and are therefore sensitive to changes in the prevailing environmental conditions. Investigations have targeted upper and lower treeline sites because of the biological records they provide (e.g. tree rings, pollen data) and the physical evidence of current and former shifts of the ecotone. In alpine areas, the close juxtaposition of treelines to physical systems (e.g. glaciers, ice-core and varve records), which also provide excellent paleo-environmental proxies, enhances the possibility for developing a more detailed picture of response than is possible in many other environments.

This paper provides some detailed examples of changes in the treeline ecotone in the Canadian Rockies. Although the studies reported in this paper are not complete and form a limited geographical sample, they provide insights into the nature and complexity of change in this environment and some indication of the direction of future changes that may occur in response to regional, anthropogenically enhanced climate change. The focus is on biological indicators of change. The paper summarises the history and nature of past climate variability at the Columbia Icefield and examines the response of the forest to those variations.

2. The Columbia Icefield Area of the Canadian Rockies

The Columbia Icefield is the largest ice mass within the Canadian Rockies and lies astride the continental divide in Banff and Jasper National Parks (Figure 1). The area adjacent to the eastern outlet glaciers of the Icefield (Athabasca, Dome and Saskatchewan) is readily accessible by road and has been the focus of many glacier and paleoenvironmental studies (e.g., Heusser 1956; Beaudoin and King 1990; Luckman 1986, 1988). In addition to the availability of physical records of environmental change, this region is free from major natural and anthropogenic disturbances to the vegetation. There is no evidence of human occupation prior to the first European visits in the late 1890s, and the area has been protected as a National Park for most of the present century. There have been no logging, agricultural or animal husbandry activities in the region and direct human impact has been confined to a small area immediately adjacent to the road. In addition, there is little obvious evidence of fire at these isolated high elevation sites. Therefore, it is an ideal area in which to investigate the response of vegetation to natural climate variation.

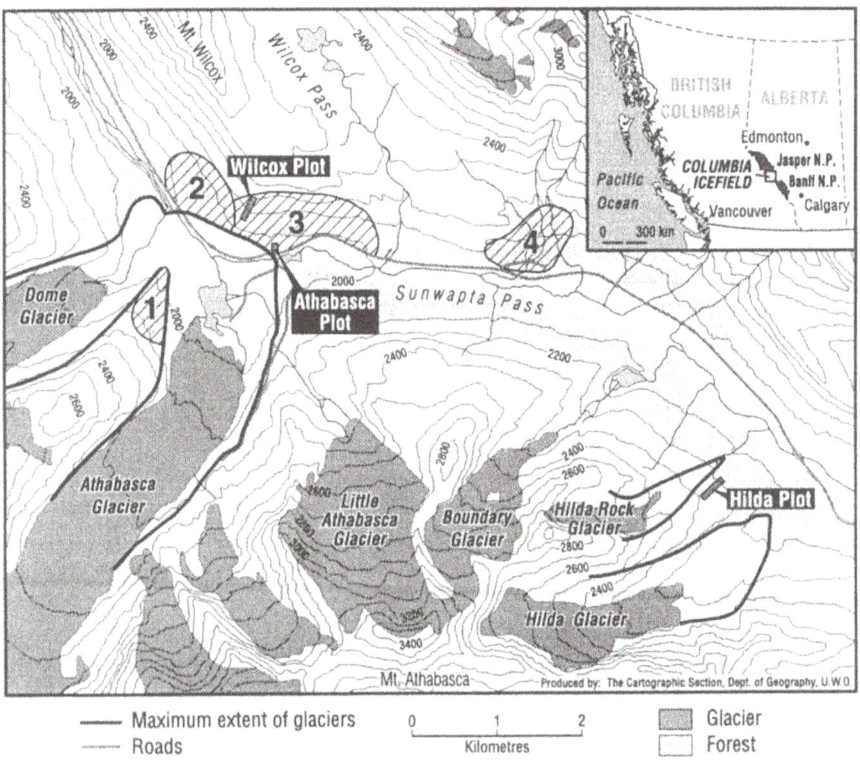

Fig. 1. Study sites at the Columbia Icefield. The numbered chronology development sites are: 1=Athadome; 2=Ancient Forest; 3=Icefields (primary snag collection area); 4=Sunwapta Pass (living-tree densitometric chronology site sampled by F. Schweingruber in 1984). Athabasca and Dome Glaciers reached their maximum Holocene extents in 1843/44 and 1846 respectively (Luckman 1988). The inset diagram shows the location of the Columbia Icefield in western Canada.

Several previous studies relating to treeline dynamics have been carried out at the Columbia Icefield using different lines of evidence. Beaudoin (1986) reconstructed Holocene treeline fluctuations in the Wilcox and Sunwapta Passes using pollen ratio data and showed that the treeline was higher than present during the early Holocene. Studies elsewhere in the Canadian Rockies indicate that the treeline had fallen to present elevations by ca. 4000 years B.P. (Luckman and Kearney 1986, Vance *et al.* 1995). The presence of snags and coarse woody debris above the present treeline provides physical evidence of former higher treelines in this area (Luckman 1986, 1990, 1994). Recent seedling establishment above the present treeline indicates that the treeline is currently advancing upslope (Kearney 1982).

3. Records of Environmental Change Over the Last Millennium

3.1 The Instrumental Climate Record

As in most mountain areas, meteorological data are extremely sparse for this region. There are short seasonal records from a climate station near the Athabasca Glacier, but the only long climate records are from valley floor sites some 100 to 150 km distant from the Icefield (Janz and Storr 1977; Luckman and Seed 1995; Luckman 1997). A regional temperature anomaly series was developed from four stations within the Rockies and the Rocky Mountain Trench (Banff, 1888–1994; Donald/Golden 1891–1990; Jasper 1916–1994; and Valemount 1916–1989) to provide an overview of recent climatic history (Luckman *et al.* 1997).

Figure 2 summarises the annual and seasonal temperature trends for the central Canadian Rockies expressed as anomalies from the 1961–1990 mean. The annual temperature series shows an average rise of 1.43°C per century but can be characterised as a general warming trend which peaked in the late 1930s and 1940s, a cooler interval in the 1940s and early 1950s, variable conditions through the 1960s and 1970s, and the warmest period of record during the 1980s and 1990s. This annual record largely reflects the influence of the winter season (January–March) which shows both the greatest range (over 12°C) and steepest increases in temperatures over this interval (ca. 3.2°C per 100 years). The spring (April–June) and summer (July–September) seasons show a broadly similar trend (ca. 1.3°C per 100 years) and pattern, but the autumn (October–December) record has no significant trend. Figure 2 also shows an alternative summer season (June–August), which is more appropriate for vegetation studies. This shows cool summers through the first part of the record with increasingly warmer summers until the early 1940s, a return to cooler summers through the 1940s and 1950s, followed by a highly variable period in the 1960s. The warmest summers occurred during the early 1970s, and the last two decades have been somewhat cooler and stable with no long-term trend.

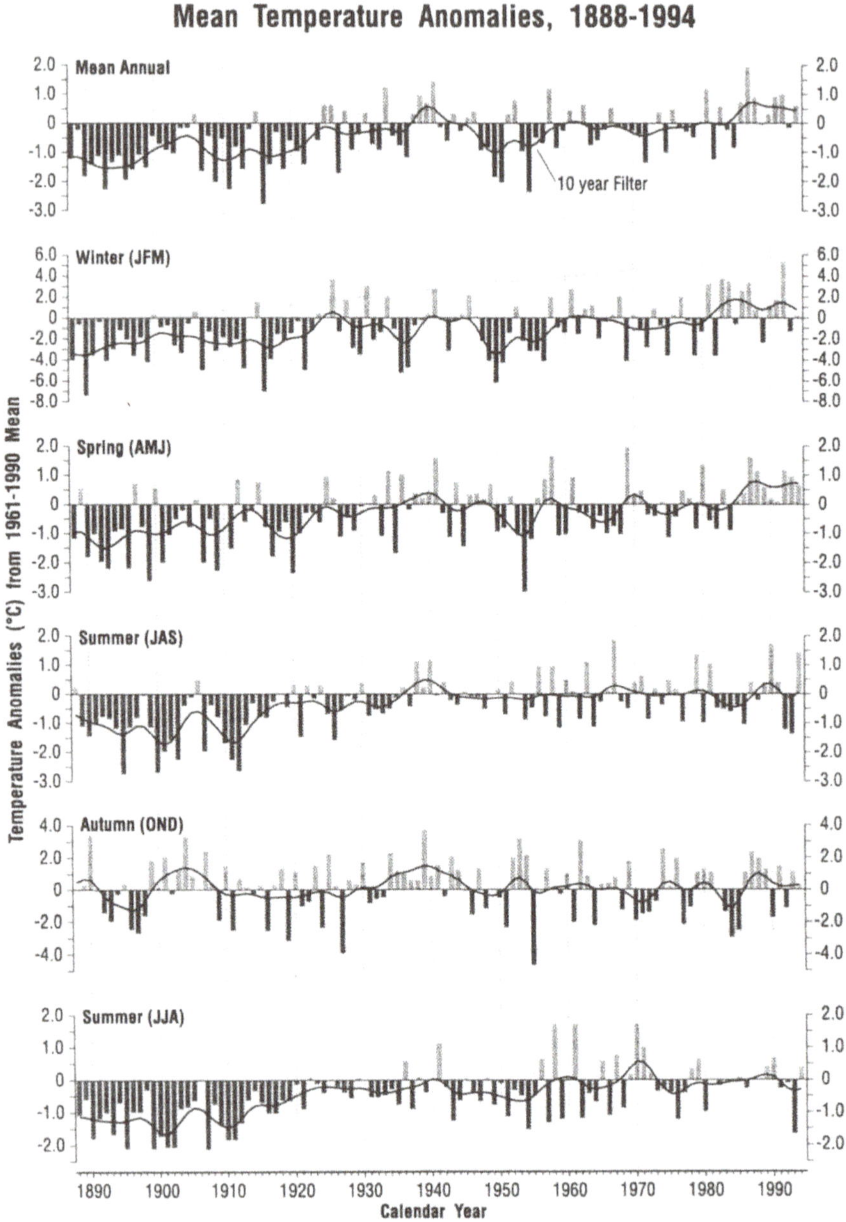

Fig. 2. Regional temperature anomalies for the central Canadian Rockies, 1888-1994. The anomalies are calculated with reference to the 1961-1990 mean. Note the differences of the temperature scales of the seasonal series.

3.2 Proxy Temperature and Glacier Records

Early dendrochronological studies at the Columbia Icefield (Hamilton 1987; Luckman *et al.* 1984) focused on developing ring-width series for a stand of old *Picea engelmannii* (Parry) Engelm. trees adjacent to the moraines of the Athabasca Glacier (Area 2, Figure 1) and from snag material lying on the surface near the treeline (Area 3, Figure 1). Ring-width series from these and adjacent sites were used to develop a master chronology extending back to 1073 AD (Luckman 1993). Tree-ring densitometry was performed on selected snag samples (*Picea engelmannii* and *Abies lasiocarpa* (Hook.) Nutt.) at the Swiss Federal Institute for Forest, Snow and Landscape Research, WSL/FNP, Birmensdorf. These results were combined with an earlier densitometric chronology from living *Picea engelmannii* (Schweingruber 1988) at an adjacent site in Sunwapta Pass (Area 4, Figure 1) to produce a composite maximum density chronology over the same interval. Luckman *et al.* (1997) used these chronologies to develop a summer temperature reconstruction for the Columbia Icefield area. Regional April–August temperature anomalies were reconstructed using multiple regression with the maximum density and ring-width of the growth year and prior year as predictors (Figure 3). Correlation of this reconstruction with the regional temperature record (Figure 4) is 0.65 for the 1890–1982 interval.

The reconstruction in Figure 3 spans the period of the "Little Ice Age" and is the longest densitometrically-based summer temperature record from North America. Almost all the reconstructed temperature anomalies are negative indicating that summer temperatures for most of the last 900 years at the Icefield were considerably lower than those experienced during recent decades. The mean temperature anomaly for 1101–1900 was 0.71°C below the 1961–1990 reference period and 0.33°C below the 1891–1990 mean of the instrumental record. The only part of the reconstructed record with decadal positive temperature anomalies is the late 11th century. Since this part of the reconstruction is based on fewer than five samples from young trees, it is possible that these results may be unduly influenced by juvenile effects and should be interpreted more cautiously than the later, better replicated sections of the reconstruction. Significantly cooler summer intervals are reconstructed for the 1200 to 1300s, the mid to late 1400s, the late 1600s, and the 19th century, which was the most extended cold period in the record. The average temperature anomaly from 1781–1900 was -1.04°C, and summer temperatures during the first half of the 19th century were particularly severe.

Corroboration of this temperature reconstruction is provided by regional and local glacier records, which show significant periods of glacier advance in the late 1200s and early 1300s, the early 1700s, and during the 1800s (Figure 3; Luckman *et al.* 1997). Glaciers generally reached their Little Ice Age maxima in the mid 19th century (Luckman 1996). The reconstruction shows the strong warming from the late 19th to the mid 20th centuries, which is also seen in the instrumental record. Warmer intervals are reconstructed for the late 11th century, between 1350 and 1440, and parts of the 18th and 20th centuries. The 1961 to 1990 reference period is clearly warmer than any equivalent-length period over the last 800 years. This summer temperature reconstruction supports paleoenvironmental data that indicate fluctuating climatic conditions during the "Little Ice Age" rather than a long period of sustained cold of several centuries duration (Bradley and Jones 1993, 1995; Briffa *et al.* 1990, 1992).

4. Treeline Responses to Climate Change

Treeline is the ecotone between the forest and tundra biomes where trees are growing at their autecological limit and where a change in environmental conditions may result in a change in treeline position (Brubaker 1986, 1988; Fritts 1976; Rochefort *et al.* 1994; Stevens and Fox 1991). The treeline position reflects the dynamic balance between recruitment and mortality in this ecotone. The physical position of the treeline can advance due to seedling establishment beyond the present treeline, or due to a change in growth form from krummholz to erect stems (Kullman 1987; Lavoie and Payette 1992). Treeline retreat results when adult trees die and there is no seedling establishment (Kullman 1987; Kullman and Engelmark 1991; Lloyd and Graumlich 1997). In a situation where there is limited mortality and *in situ* seedling establishment, the population may be maintained with no change in the treeline position.

Tree populations may lag shifts in climate by decades to centuries (Davis 1986). In the absence of disturbance, population change is slowed by long juvenile or adult periods and plasticity in growth and reproduction. Where climate conditions reduce tree growth, and thus delay sexual maturation, the resulting lengthy juvenile period slows population change, particularly treeline advance. Similarly, a long adult period (which characterises most treeline species) also reduces the rate of population change, such as treeline retreat, because adult trees are far more tolerant of harsh climatic conditions than seedlings. Tolerance of harsh conditions is achieved through plasticity of growth and reproduction; trees reduce overall growth or assume the krummholz growth form, thus mitigating the harsh climatic conditions. Sexual reproduction decreases during periods of severe conditions (Kullman 1997) and those species capable of layering reproduce asexually and thus maintain the population. Factors that speed population change include high seed production, wide dispersal of seed and high rates of seedling survival. During periods of benign climate, those species that are capable of rapid production and dispersal of large quantities of seed can establish large numbers of seedlings that will ensure a rapid increase in population and thus treeline advance.

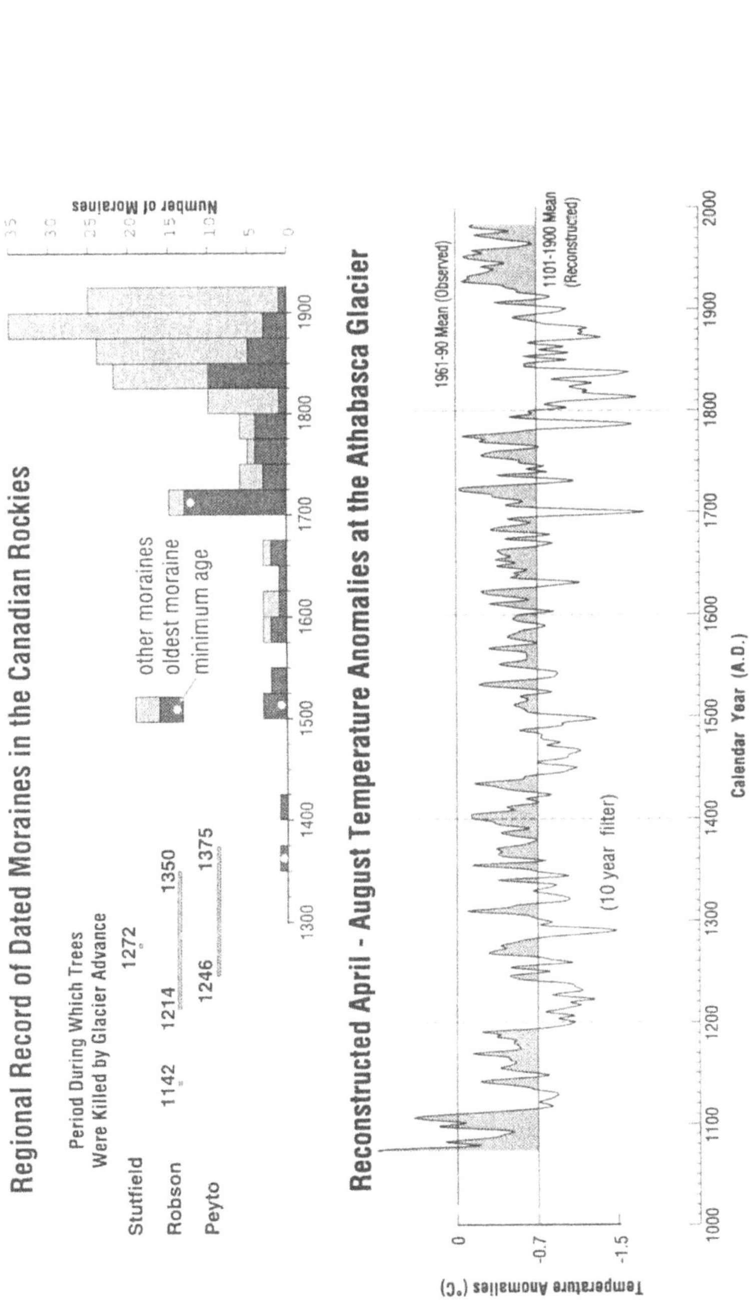

Fig. 3. Comparison of the April-August (summer) temperature reconstruction for the Columbia Icefield area with the regional record of dated Little Ice Age moraines in the Canadian Rockies. The moraine record is a composite of dated moraines at about 60 sites (for details see Luckman 1996). The temperature reconstruction shown has been smoothed with a 10 year gaussian filter (after Luckman et al. 1997).

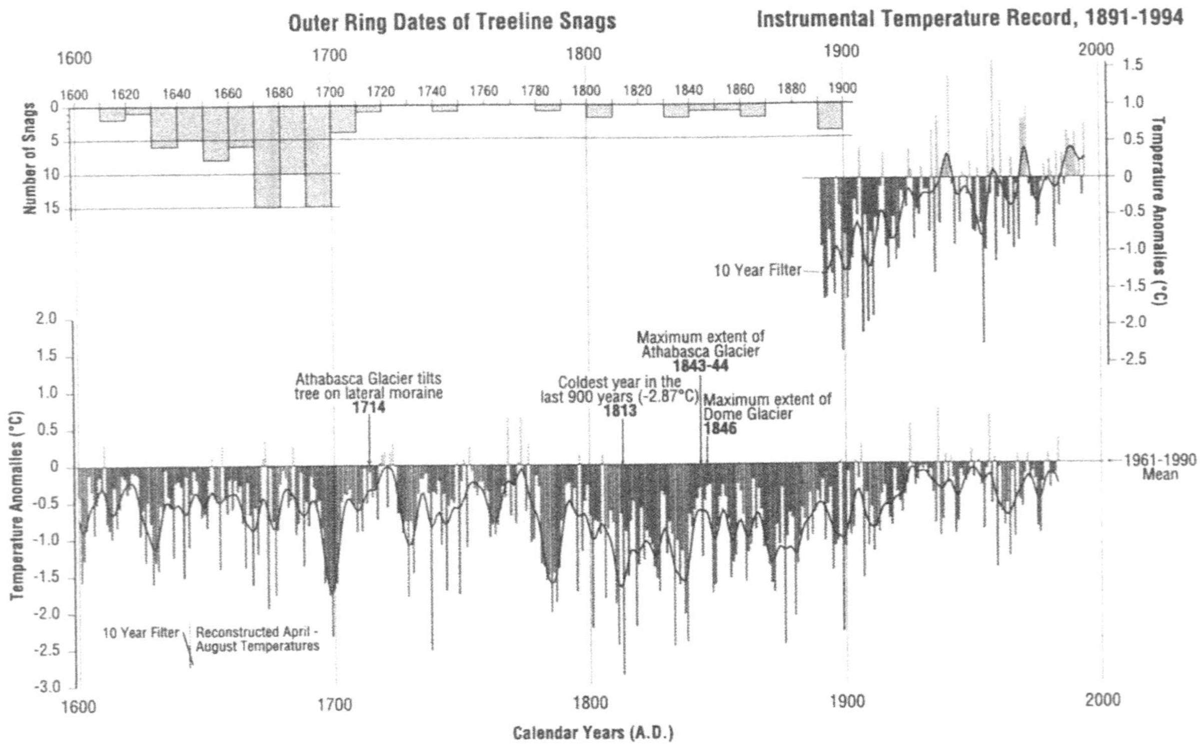

Fig. 4. Environmental history at the Columbia Icefield 1600 to 1994. The outer ring dates of 87 snags lying on the surface at the Wilcox site are grouped in 10 year intervals (see Table 1). These outer ring dates provide minimum estimates for the death dates of trees that formerly grew at or above present treeline. The dating of local glacier advances is indicated on the record of reconstructed April to August temperature anomalies. The regional instrumental climate record of temperature anomalies is also shown. (Redrawn with revisions from Luckman et al. 1997).

Population changes at treeline may also result from a variety of natural and anthropogenic causes including snow avalanche (Johnson 1987), fire (Arsenault and Payette 1992; Billings 1969), wind (Veblen *et al.* 1989), outbreaks of insects (Veblen *et al.* 1991) or disease, grazing (Earle 1993) or logging. The sites in this study were specifically selected to eliminate such confounding effects, thereby focusing the investigation on the response of treeline populations to variations in climate.

4.1 Study Sites

The preliminary results presented here are part of a project to reconstruct the dynamics of the treeline ecotone at the Columbia Icefield and examine vegetation response to local and regional climate variability over the past 400 years. Three sites were examined in detail: two from a south-facing slope below Mount Wilcox, and a third on a north-facing slope on the northeast flank of Mount Athabasca (see Figure 1). The south-facing slope was selected because large snags (>40 cm diameter) lying on the surface within and above the presently advancing treeline indicate that the treeline was formerly higher at this site. Dendrochronological studies show that many of these trees died in or before the late 17th and early 18th centuries (Luckman 1986, 1994). In addition, historic photographs (Schäffer 1911) indicate considerable changes in tree cover since the early 1900s. During the Little Ice Age, the Athabasca Glacier advanced to the base of this slope, where an ice-damaged tree precisely dates its maximum Neoglacial extent to 1843–44 A.D. (Luckman 1988). Two sites were studied in detail on this slope: one at mid-slope (Wilcox) and the other on the valley floor (Athabasca), immediately adjacent to the Little Ice Age moraine of the Athabasca Glacier. The north-facing (Hilda) site is located 8 km southeast of the south-facing site. It is on an interfluve above and between two tributary valleys containing small glaciers but the site is not directly exposed to their microclimatic influence. This site has a more gentle slope and is at a higher elevation than the Wilcox site. There are no snags above the treeline at this site.

5. Treeline Dynamics at the Columbia Icefield

Although treeline shifts involve changes in the balance between mortality and recruitment at a site, the record of these processes is not preserved in equal detail over time, nor can it be addressed at the same spatial scales of sampling. The discussion of ongoing studies of treeline dynamics in this area is divided into two parts: first, a general review of changes over the last 900 years based on studies of the snag population; and secondly, a more detailed appraisal of changes over the last 300 years from studies of living trees.

5.1 Mortality and Recruitment Over the Last Millennium

Between 1981–83 and in 1991 extensive sampling of standing and fallen snags was carried out over a large area on the Mount Wilcox slope to develop a long tree-ring chronology for this site (Area 3, Figure 1). Over 120 snags were sampled and 78% of these were successfully crossdated. Table 1 summarises the temporal distribution of inner and outer ring dates for dated snags from this slope. Results for the period after 1600 are shown in more detail in Figure 4.

Most sampled snags contained piths, but because cross-sections were cut at variable distances above the rootstock (where present), these pith dates provide only approximate estimates for the date of tree establishment. Few of the sampled snags retained bark and their irregular weathered surfaces indicate the loss of an unknown number of outer rings. Although the ring-series preserved in these snags do not provide as accurate a record of recruitment and mortality as can be obtained from living or recently dead trees, they can be used to give a broad outline of treeline fluctuations at the site over the last 900 years. Based on the inner ring dates of the snags, the primary phase of colonisation above the present treeline was between 1300 and 1600 (Table 1). Over 64% of the snags had inner dates between 1345 and 1532. Few snags began growth after the mid 1660s and no living trees sampled in the Wilcox plot (see below) predates 1800.

Period	Innermost Ring	Outermost Ring
1070–1099	2	0
1100–1199	6	0
1200–1299	8	2
1300–1399	28	2
1400–1499	30	3
1500–1599	22	11
1600–1699	5	68
1700–1799	5	7
1800–1899	0	12
1900–1980	0	1
Total	106	106

Note: These data have been revised from those shown in Luckman 1994, Fig. 2 and include 12 dated snags from the Wilcox sample plot

Table 1. Inner and outer ring dates from snags at the Wilcox site, Columbia Icefield, Alberta

The distribution of outer ring dates in the snag population shows a marked mortality event in the late 17th century (Figure 4). Seventy percent of the outer ring dates occur between 1645 and 1705. Only a small sample of snags was identified to species, but there are approximately equal numbers of *Picea engelmannii* and *Abies lasiocarpa,* with a broad age range from 100 to almost 500 years in both species. Therefore, this mortality event affected two different species with a wide range of ages, rather than a single age cohort. The concentration of dates in the late 1600s is remarkable. The temperature reconstruction shown in Figures 3 and 4 indicates that the years 1696 to 1701 were the most sustained period of cold summers in the whole record (6 years with an average temperature anomaly of -1.75°C, and no year with an anomaly less than -1.54°C). This 1690s "cold snap" was the most severe sequence of summers at this site since the late 1400s. Cool periods of similar length did occur in the 19th century, but there was no significant tree population present on the slope to respond at that time (see below). Several glaciers also constructed their outermost Little Ice Age moraines in the early 1700s (Figure 3) during the subsequent slightly warmer interval. At this time, the Athabasca Glacier was close to its maximum Holocene position and tilted a tree at its western trimline in 1714 (Heusser 1956; Luckman 1988). This combination of tree death and glacier data provides strong circumstantial evidence that the 1690s "cold snap" was real and probably resulted in considerable tree mortality, which produced a regression of the treeline in the late 17th to early 18th century. Kullman (1997) has provided empirical evidence for a similar event from a long-term treeline study in Sweden that supports this conclusion. He documented the effect of one extremely cold year (1966), as well as continued cooling, on the defoliation and death of many *Picea abies* trees in subsequent years. As a result, the treeline moved downslope about 50 m.

Detailed investigations at the Athabasca site (see below) yielded eight snags from within the 14 sampled tree clumps; no coarse woody debris was encountered between the clumps. Although it is possible that some material may have been removed from the surface of this area by visitors early this century, there was no evidence of fire, cutting, or removal of material from within the sampled tree clumps. Five of the eight snags were crossdated. This limited data set indicates that these trees established between 1700 and 1790 and died between 1856 and 1879. The narrow range of death dates from relatively young trees of varying ages (66 and 160 years) suggest that limited mortality may also have occurred during the cold periods of the late 19th century.

There are no snags above present treeline in the Hilda plot suggesting that there has been no recent downslope shift of treeline position. Coarse woody debris was recovered from within the forest and the existing tree clumps, but was not found between these clumps. This wood was generally poorly preserved and only 12 (20%) of these samples were crossdated. Nine snags had outer ring dates between 1860 and 1930. There is no evidence of a significant mortality event preserved at this site.

The snag records from these three sites indicate quite different histories. At the Wilcox site, the snag record provides evidence of a major mortality event and recession of the treeline during the late 17th century. The limited snag data from the Athabasca site suggest some mortality of a subsequent generation during colder conditions in the mid to late 19th century. The treeline at the Hilda site appears to have been stable throughout this interval and lacks any evidence of significant mortality events.

5.2 Treeline Dynamics in the Last 200 Years

Large sampling plots (8250 m^2) were established at both the Hilda and Wilcox sites. These plots extend 275 m upslope from the subalpine forest into the alpine tundra zone (Figure 5). The south-facing Wilcox site has a steep (32°) slope extending from 2050 m to 2195 m elevation. The continuous subalpine forest limit occurs at about 2060 m and many large snags are found in the upper portions of this plot. The Hilda site, 8 km southeast, has a lower mean slope (17°). The base of the plot is similar in elevation (2190 m) to the top of the Wilcox plot and extends upslope to 2280 m. The tree limit occurs at about 2240 m. Within these two plots, the position of all living and dead trees (snags) were mapped to the nearest 0.1 m and species, height, age and mode of establishment (seed or layering/clonal) were recorded for each living stem. In this study, a tree is defined as an individual with a height greater than 1.4 m; all smaller stems were defined as seedlings. Seedling density was tallied in 5 m x 5 m quadrats. In both plots, a subsample of quadrats was mapped to show the position of seedlings (1057 at Wilcox, 1130 at Hilda) to the nearest 0.1 m and species, height, age and mode of establishment recorded. Internode lengths were measured for nine years on a subset of these seedlings (449 at Wilcox and 265 at Hilda).

The Athabasca plot is located at about 1975 m elevation, close to the base of the Mount Wilcox slope, immediately adjacent to the Little Ice Age moraine of the Athabasca Glacier. Historic photographs indicate the presence of ice within 50 m of the site in the early 1900s (Luckman 1988; Schäffer 1911). Since that time, the glacier has retreated approximately 1.5 km across the valley. In 1995, Parks Canada and Brewster Transport built a new visitor facility at a site on a small knoll just beyond the terminal moraine of the Athabasca, which included several unmodified tree clumps. A detailed inventory was carried out at the proposed construction site (ca. 5000 m^2) and a salvage dendrochronology project was undertaken immediately prior to site clearance (Kavanagh and Luckman 1995). Sampling involved basal sectioning of all stems (and some rootstocks) in 14 clumps. Information from these cross-sections allowed a detailed investigation of the chronological development of these clumps. Within each clump, the position of all trees and seedlings was mapped to the nearest 0.1 m, and species, height, age and mode of establishment were recorded. Few seedlings were present in the areas between tree clumps.

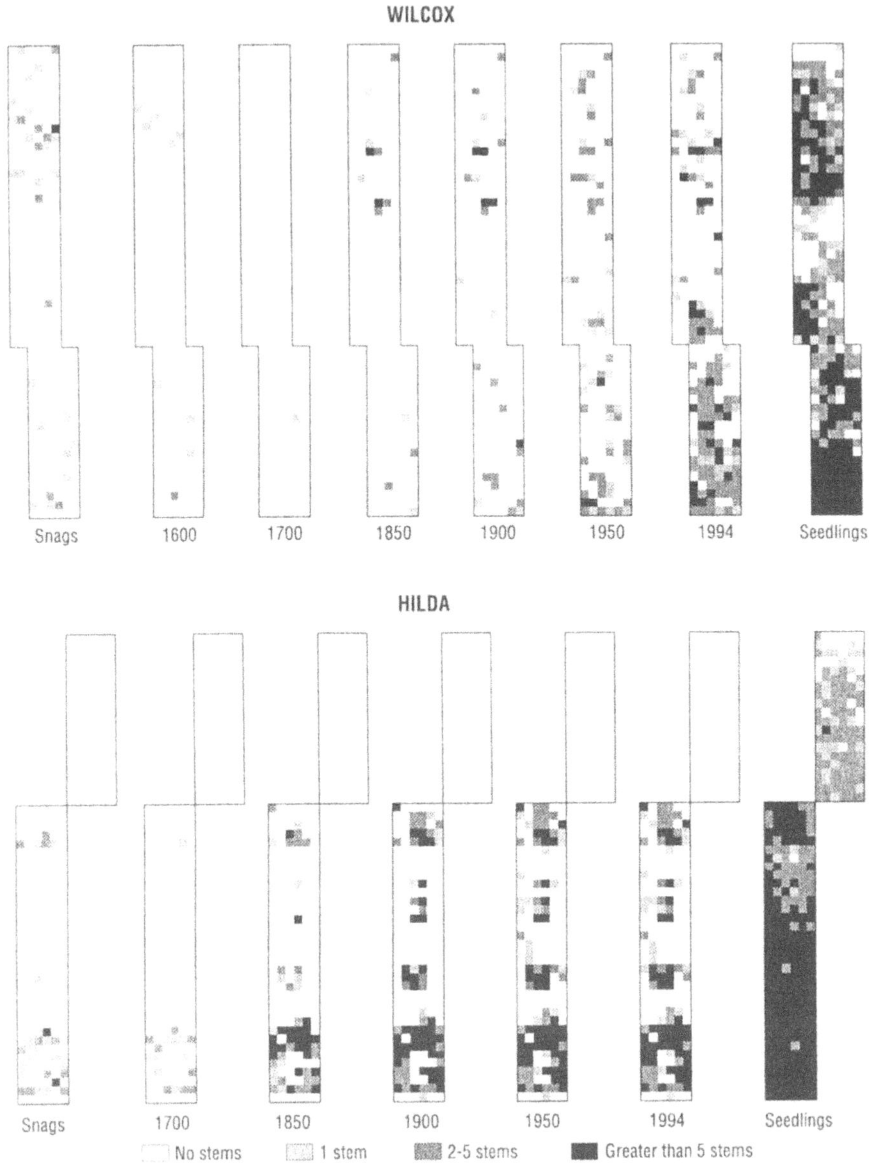

Fig. 5a. Spatial distribution of snags, trees and seedlings for selected years since 1600 for Wilcox and Hilda sites. The snag data indicate the number of individuals present in each 5m x 5m quadrat in 1994. The tree and seedling data indicate the number of living stems in each quadrat in the year shown. Each plot is 30m wide by 275m long.

Forest Composition. Table 2 summarises the forest structure at the three sites. *Abies lasiocarpa* dominates the vegetation in the Hilda plot (>90%), whereas *Picea engelmannii* comprises more than 50% of the tree population in the Wilcox plot and more than 85% of the trees in the Athabasca plot. Tree establishment is overwhelmingly the result of clonal expansion at Hilda, whereas establishment by seed dominates at the Wilcox plot. More than 60% of the trees in the Athabasca plot established clonally.

Species	Wilcox (South-facing) Seed	Clonal	Hilda (North-facing) Seed	Clonal	Athabasca (Valley-floor) Seed	Clonal
Trees						
Abies lasiocarpa	163	61	90	705	9	8
Picea englemannii	276	52	31	5	36	68
Pinus albicaulis	5	--	--	--	--	--
Pinus contorta	--	--	--	--	--	--
Total	444	113	121	710	45	76
	(557)		(831)		(121)	
Seedlings						
Abies lasiocarpa	3636	99	5416	2541	1	15*
Picea engelmannii	1854	13	327	3	--	150*
Pinus albicaulis	--	--	28	--	--	--
Pinus contorta	1	--	--	--	--	--
Total	5490	112	5761	2544	1	165*
	(5602)		(9305)		(166)	

*estimated values

Table 2. Tree and seedling structure in three study plots at the Columbia Icefield, Alberta (by mode of establishment)

Tree density, determined for 5 m x 5 m quadrats, is shown in the 1994 panels of Figure 5. In both the Hilda and Wilcox plots, tree density decreases with increasing elevation. Although tree stems are found throughout the Wilcox plot, no trees occur above 2240 m elevation at Hilda. Several tree clumps are clearly visible in the lower part of the Hilda plot on Figure 5b. All the trees at the Athabasca plot (Figure 5c) were located in tree clumps.

Seedling structure is similar to forest structure in the two large plots (Table 2). Although the majority of seedlings are *A. lasiocarpa* in both plots, 3058 (84%) of the 3636 *A. lasiocarpa* seedlings at Wilcox are located in the lowest 5 m and *P. engelmannii* dominates the remainder of the plot. At both sites, the majority of seedlings established from seed during the late 1960s and early 1970s. There are very few seedlings present in the Athabasca plot and, with the exception of one seed-established seedling located between two clumps, all seedlings are the result of layered branches in the tree clumps. Figure 5 also shows seedling density at the three sites. As expected, seedling density decreases with increasing distance upslope from the forest edge at the base of the plot at both the Wilcox and Hilda sites.

Forest response to climate change. The major elements of the April–August temperature reconstruction (Figure 4) include a rapid cooling in the late 1600s, variable conditions in the 1700s and very cold conditions throughout the 1800s. The instrumental record (inset, Figure 4) and reconstruction show that spring and summer temperatures have been warming since the early 1900s and the last 30 years have been the warmest on record. Forest vegetation in the three study sites at the Columbia Icefield has responded quite differently to these changes. Figure 5(a–c) shows the spatial pattern of tree establishment at the three sites.

Wilcox Plot. Figure 5a shows the density of all snags and coarse woody debris mapped during the detailed plot study. Many snags were present in the upper areas of the plot. Only a limited sample of snags (12) were successfully crossdated, but these indicate that of the ten trees living within the plot in 1600, nine have outer dates prior to 1700. The late 17th century cooling event probably killed most of the trees within the plot at that time. By 1850, some establishment had occurred, mostly in the upper portions of the plot. Establishment continued at low rates and by 1950 trees were present in low numbers throughout the plot. The majority of trees established during the late 1960s and early 1970s. Seedling density in 1994 was high throughout the plot except for the middle and upper areas where steep slopes with exposed bedrock and coarse rock materials provide an unfavourable substrate for seedling establishment.

Hilda Plot. In 1700, trees were established in the forest at the base of the plot and in a tree clump in the middle of the plot (Figure 5b). There are no tree snags present in the upper areas of the plot and there is no evidence of extensive tree mortality at the end of the 17th century. Tree recruitment continued throughout the 1700s and early 1800s, and by 1850 all the tree clumps mapped in 1994 were present, with most trees established by layering. Between 1850 and 1900 establishment was limited to increasing tree density in and adjacent to tree clumps, mainly by clonal expansion. Some establishment occurred beyond tree clumps between 1900 and 1950 in the lower areas of the plot, but there is no further change in forest structure by 1994. Seedling density in 1994 is high throughout the plot, including the upper section where there have been no trees present for at least the past 400 years.

Athabasca Plot. The pattern of tree establishment in this plot is similar to the pattern at the Hilda site. Three clumps were established by 1700, and there is no evidence of extensive tree mortality following the "cold snap" of the late 17th century. No tree snags were found outside clump limits. Eight of the 14 tree clumps established by 1850 and the remaining six were all present by 1950. However, unlike the other plots, very few seedlings were present in the plot in 1995; only one seedling established from seed was found between clumps. The remaining shoots were layered stems in the tree clumps.

Although there has not been a significant change in the spatial pattern of tree establishment at the Athabasca site, there has been a significant change in growth form. Prior to 1900, no stem was taller than 1 m (Figure 6), indicating the clumps had a krummholz growth form. After 1900, the trees developed erect stems and stem height increased at a fairly constant rate for all trees. Trees in the Athabasca plot were quite short (<5 m) in comparison to trees in the two other plots (>10 m).

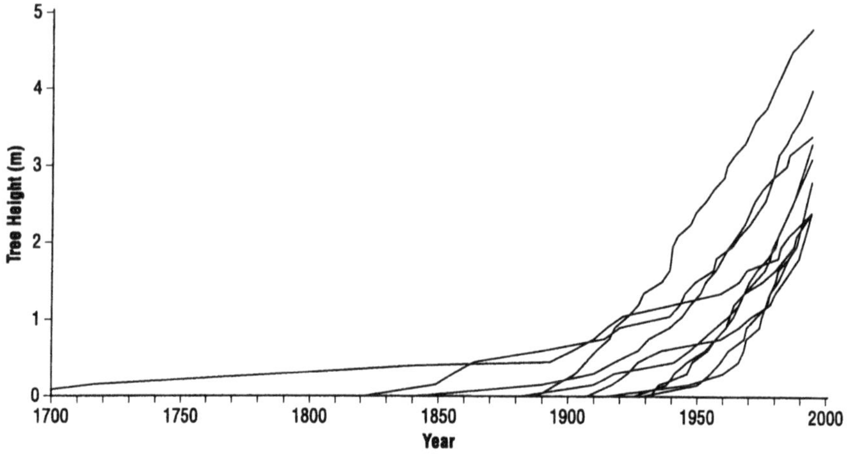

Fig. 6. Height/age relationships for the tallest tree in 10 of the 14 tree clumps sampled in the Athabasca sample plot.

Spatio-Temporal Patterns of Establishment. Figure 7 summarises the spatio-temporal pattern of tree establishment in the three plots. Few quadrats were occupied in the Wilcox plot until 1900, and the rate of spatial expansion of trees was low until 1950, after which it increased substantially. The presence of seedlings in 75% of the quadrats suggests that trees will occupy a large portion of the plot in the near future. Similar low rates of spatial expansion of the treeline occurred in the Athabasca plot until 1900, but unlike the Wilcox plot, these low rates continue to the present. The lack of seedling establishment during the 1940s, 1960s and 1970s has resulted in very limited spatial expansion of tree cover at this site. The Athabasca trees have a flagged appearance indicating that wind is an important factor at this site. Flagging was not observed at the other sites. Over 80% of trees in the Athabasca plot have broken leaders, compared to 23% and 29% in the Wilcox and Hilda plots, respectively. The flagging and lack of seedlings suggest that the Athabasca site is quite exposed and that the effect of wind may have been a significant factor limiting tree growth.

The temporal pattern of tree establishment at Hilda differs from both other plots. Tree establishment occurred at a steady rate until about 1950 and then declined. The differences in the rate of tree recruitment between the Hilda and Wilcox plots from 1900 to 1994 are clearly seen in Figure 7 and primarily reflect the higher growth rate at Wilcox. A sample of internode lengths obtained from seedlings (Figure 8) indicates a four-fold difference in growth rates between the two plots. Although there are significant differences in species composition of seedlings at the two plots (primarily *A. lasiocarpa* at Hilda and *P. engelmannii* at

Wilcox), differences in growth rates between plots are similar for both species (4.3:1 for *P. engelmannii* and 3.8:1 for *A. lasiocarpa*). The greater growth rates result in a faster conversion of "seedlings" to "trees" in the Wilcox plot, producing a much more visible change in the vegetation cover. Seedlings are present in over 90% of the quadrats at Hilda suggesting that the treeline will advance upslope at some point in the future at this site.

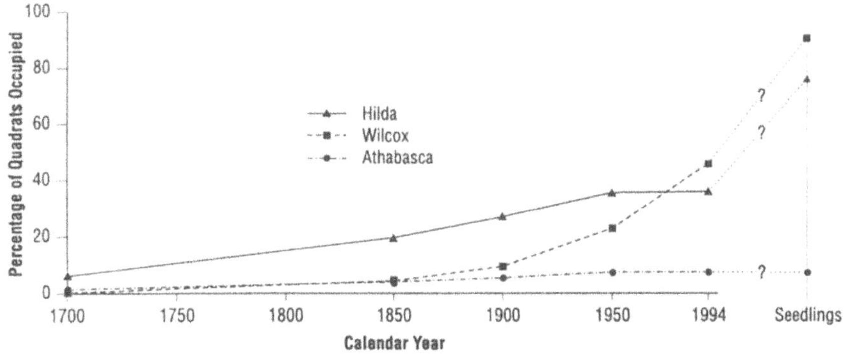

Fig. 7. Temporal evolution of colonisation at the Wilcox, Athabasca and Hilda study plots. The percentage of quadrats occupied by trees is a useful comparative measure of the changes in colonisation at these sites. The seedling data indicate probable future trends, barring a significant mortality event.

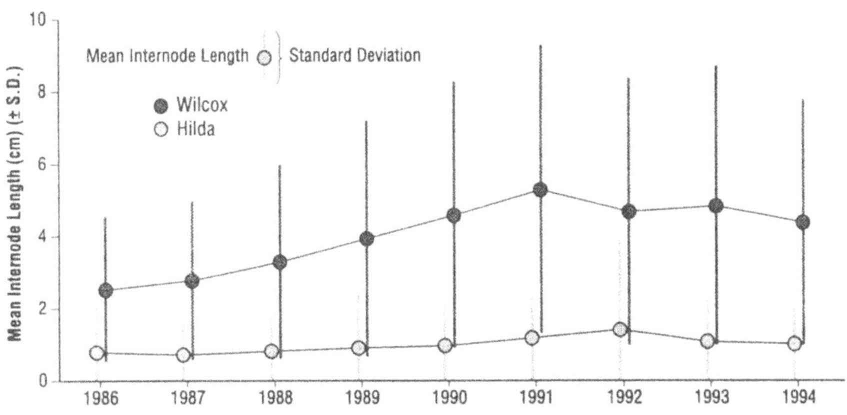

Fig. 8. Mean internode lengths for the years 1986 to 1994 for seedlings at the Wilcox (n=449) and Hilda (n=269) study plots. The bars indicate ±1 standard deviation about the mean.

6. The Relationships Between Tree Recruitment and Climate

Figure 9 shows the temporal pattern of tree establishment in detail for the three plots. The thermal index shown on the left axis is a cumulative plot of monthly June to August positive temperature anomalies in °C from 1890 to 1994. Three main periods of warmer than average summer temperatures are indicated, namely the 1930s, the late 1950s, and the 1960s and 1970s. The tree establishment curves in Figure 9 are differentiated by site, species and mode of establishment. Only the seed-established trees of both species at Wilcox appear to follow a temporal pattern that is similar to the thermal index. Smaller pulses of establishment in the early 20th century are dwarfed by the large pulse of the 1960s and 1970s. The lack of similar response in the Hilda and Athabasca plots is because the majority of trees in these plots established by layering, a process that occurs independent of temperature controls. A complicating factor is that, because of the slower growth rates at Hilda, most of the stems recruited after 1955 are still classified as seedlings. When the relationships between seedling establishment and climate are examined (Figure 10) it is clear that extensive seedling establishment took place in both plots during the warmer summers of the late 1950s and 1960s. A decrease in the rate of recruitment in the late 1960s and early 1970s is observed in both plots (particularly at Hilda) and appears to be a response to several cooler than average summers. Warmer summers in the mid-1970s again result in increased rates of seedling establishment in the late 1970s.

The results from the three sites reported above suggest that each appears to have responded quite differently to regional climate changes. The Wilcox site presents a "classical" history of strong treeline fluctuations in response to climate. The earlier higher treeline shows considerable dieback in the later 1600s which is probably associated with a period of much colder summer temperatures. This contrasts sharply with the lack of evidence for widespread mortality at the Athabasca and Hilda sites at this time. Limited establishment by seed took place at all sites between 1700 and 1900, and, although there was an increase in density at Hilda during this time, there was no real change in the position of treeline. In the 20th century, the sites have responded differently to warming. The high rates of seedling recruitment during the late 1960s and early 1970s, combined with high growth rates, have resulted in treeline advance at Wilcox. In the Athabasca and Hilda plots, clonal expansion continues to increase density in the tree clumps, but there is no real change in the treeline position. Although there are seedlings established well beyond tree limit at Hilda, they are as yet too small to result in a visible change in treeline position. The treeline will eventually advance upslope in this plot but is lagged as a result of slower growth rates. The observed differences in the response of vegetation at the three sites seem to reflect both microclimate and differences among the sites themselves.

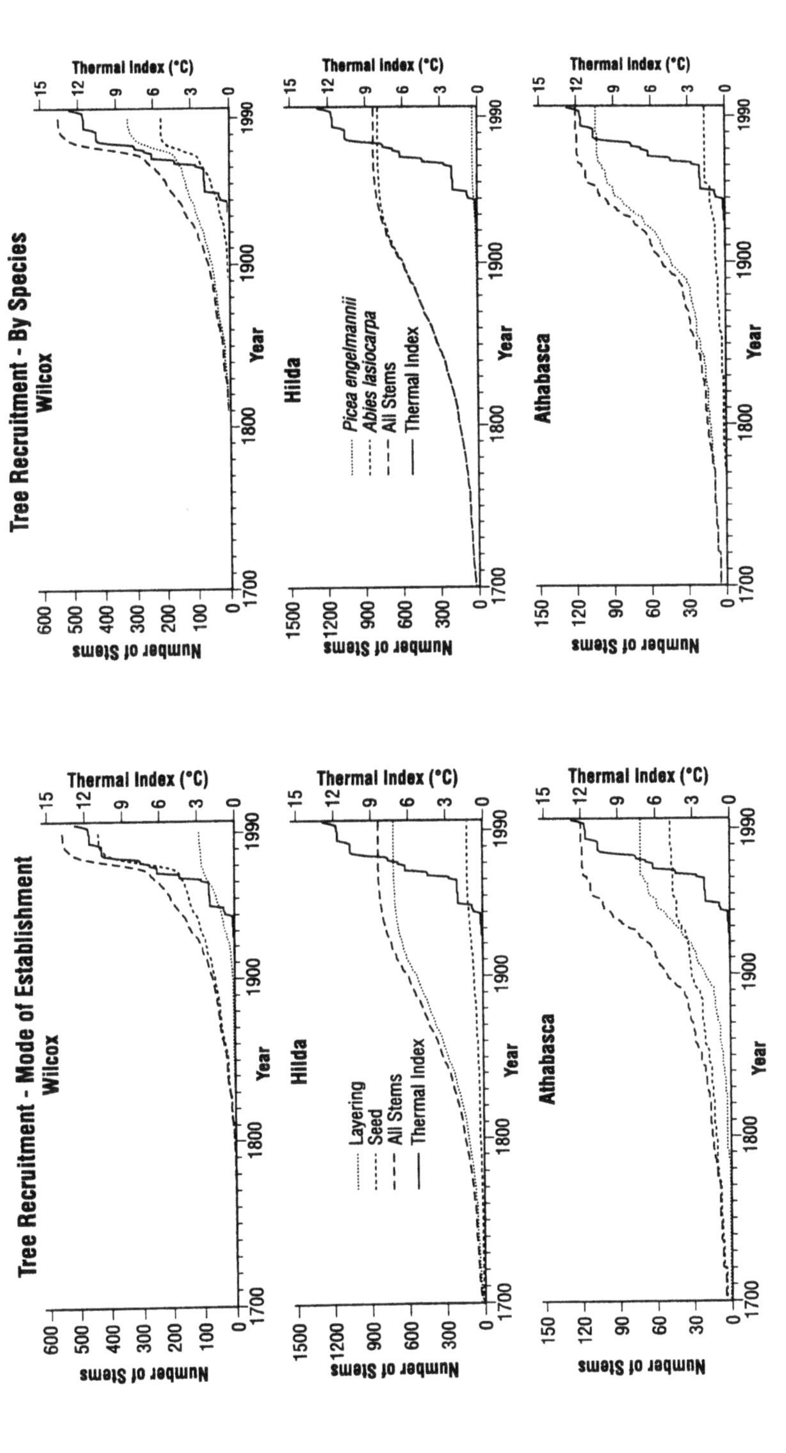

Fig. 9. Temporal pattern of tree establishment in the Wilcox, Hilda and Athabasca study plots. The population at each site is differentiated by mode of establishment (left hand diagrams) and species (right hand diagrams). The thermal index shown is a cumulative plot (1890 to 1994) of mean monthly June to August positive temperature anomalies in °C.

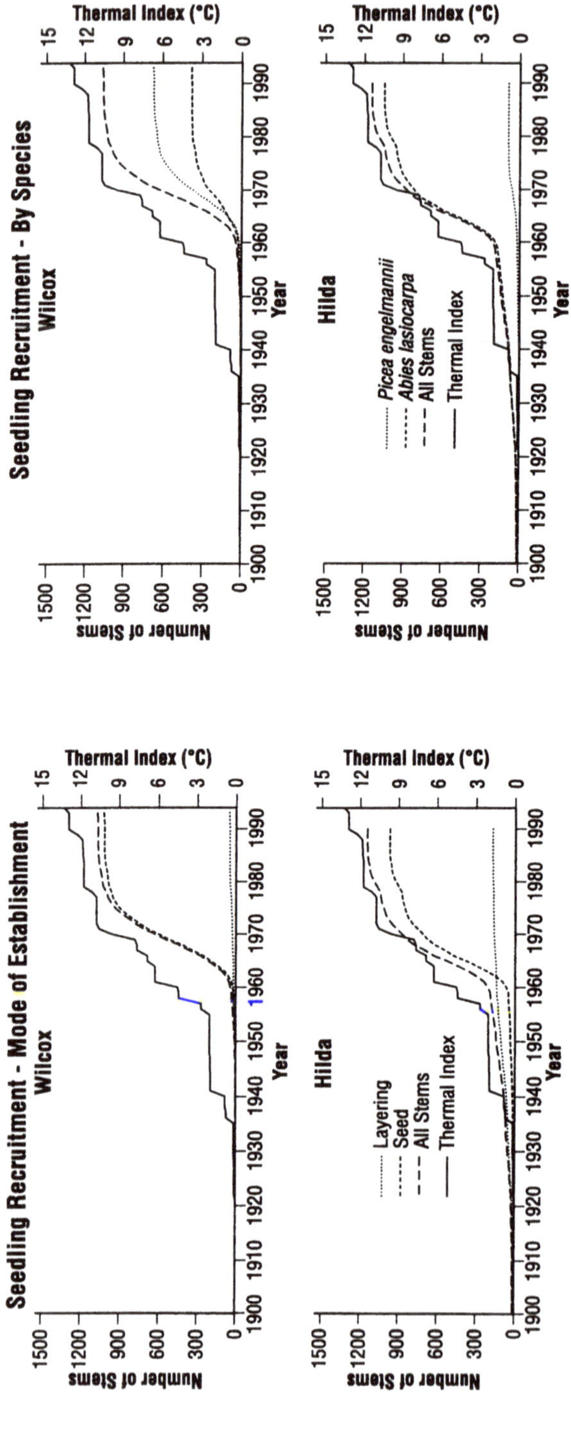

Fig. 10. Temporal pattern of seedling establishment at the Wilcox and Hilda study plots. (For explanation see Figure 9.)

Limited meteorological investigations at Wilcox and Hilda indicate that temperatures are on average about 1.5°C warmer at the Wilcox site during the summer. The Wilcox site faces south and receives higher levels of insolation. Limited field observations suggest that the slope is snow-free much earlier in the season than the Hilda site. In addition, the slope at Wilcox is steeper, better drained and therefore drier than the Hilda site. Differences in the vegetation cover support this inference; species tolerant of dry conditions, such as *Juniperus communis*, *Arctostaphylos uva-ursi* and *Shepherdia canadensis*, provide ground cover at Wilcox, whereas *Cassiope tetragona*, *C. mertensiana* and *Phyllodoce* sp., typically found in moist habitat, provide much of the ground cover at Hilda. The Hilda site is cooler and wetter resulting in lower growth rates compared to the Wilcox site. Although the Athabasca plot is at the base of the Wilcox slope, it appears that local conditions inhibit seedling establishment outside the clumps. The ground cover in this area (*Arctostaphylos uva-ursi, Shepherdia canadensis*, and *Salix* sp.) is indicative of dry soil conditions, and the combination of low soil moisture and high winds may have been sufficient to prevent recent seedling establishment.

7. Commentary

Many studies in North America describe the response of the treeline ecotone to regional variations in climate, but the topics of treeline advance and recession have not received equal attention. Although there are observations from many sites in North America that document treeline advance in response to warmer temperatures (*e.g.* Brink 1959; Franklin *et al.* 1971; Hessl and Baker 1997; Magee and Antos 1992; Payette and Filion 1985; Rochefort *et al.* 1994), the effect of cooling climate on treeline dynamics is more difficult to document and has been limited to observations of the presence of coarse woody debris above the present treeline (*e.g.* Luckman 1985). Few studies have reconstructed the details of population dynamics in the treeline ecotone (Earle 1993; Lloyd and Graumlich 1997), yet these types of studies are important because they identify the controls on processes. Soil moisture (Lloyd and Graumlich 1997) and length of growing season as controlled by snow cover (Earle 1993; Ettl and Peterson 1995a, 1995b) have also been identified as important controls on the establishment and growth of seedlings at treeline. The processes involved in tree mortality and the resulting regression of treeline are less well understood, although Kullman's (1997) long-term observations suggest extreme cooling events causing defoliation can result in tree death. Establishment and mortality patterns vary in space and time and are dependent on the individual differences in response to micro-site variations in soil properties, insolation and wind (Kienast *et al.* 1987).

The effects of global and regional climate variability are modulated by microclimatic effects at the local scale, especially in alpine regions, which are topoclimatically diverse. This topoclimatically-induced variation will be further diversified by additional local site factors, such as slope or drainage, which modify the growth environment. These physical variations are filtered through the diverse genotypical and phenotypical responses of individual tree species, which may respond in different ways to the same variations in climate. This web of interrelationships suggests that responses to climate change at treeline sites within a given region will be diverse, complex, and may occur at different time-scales. The limited examples discussed here do not provide a systematic sample of possible treeline responses within the central Canadian Rockies, although they do demonstrate the diversity of changes that can occur within a small area in response to similar macroclimatic forcing.

Although treeline environments are sensitive to climatic variability, not all treeline sites are equally sensitive. Moreover, even within the same area, the limiting climatic conditions may vary (e.g. exposure, snow cover, temperature and precipitation) between different segments of the same treeline. Any attempt to evaluate the probable response of treeline ecotones to future climate changes must recognise the influence of local site conditions in modulating those changes. Detailed paleoenvironmental case studies provide insight into the likely range of effects in any specific region. They also provide information on physical clues that can be used to assess the responsiveness of different types of treeline sites. For example, reconnaissance mapping of snags or seedling density could be used to classify the sensitivity of sites or to identify those under greatest stress. This information would be particularly useful in the design and selection of a network of long-term monitoring sites in which it may be more important to examine the range of potential changes rather than simply characterising the average condition of the ecotone.

The local heterogeneity of vegetation responses documented in this study contrasts with the results of simulation models that assume homogeneous responses over broad scales. For particular sites, this diversity of response may be much more significant in modelling the response of the ecotone to climate variability. Therefore, the data from this and other similarly detailed studies may provide important input for the parameterization of simulation models, and provide critical information to resource managers who wish to understand the effect of climate change on ecosystem dynamics.

Acknowledgements. Dendrochronological research in the Columbia Icefield area has been funded by grants to Luckman from the Natural Sciences and Engineering Research Council of Canada for many years, and has been carried out in conjunction with many former colleagues and students. We thank E. Schär and F. Schweingruber for densitometric analyses; P. Jones, K. Briffa and E. Seed for assistance in the preparation of data; E. Seed for assistance in the field; Brewster Transport for logistical support; and David Mercer (UWO) for Cartography. We also thank Parks Canada, in particular Jim Todgham from Jasper, for logistical and financial support of our research within Banff and Jasper National Parks.

8. References

Arsenault D, Payette S (1992) A postfire shift from lichen–spruce to lichen–tundra vegetation at treeline. Ecology 73:1067–1081

Beaudoin AB (1986) Using *Picea/Pinus* Ratios from the Wilcox Pass Core, Jasper National Park, Alberta, to Investigate Holocene Timberline Fluctuations. Géographie physique et Quaternaire 40:145–152

Beaudoin AB, King RH (1990) Late Quaternary Vegetation History of Wilcox Pass, Jasper National Park, Alberta. Palaeogeography, Palaeoclimatology, Palaeoecology 80:129–144

Billings WD (1969) Vegetational pattern near alpine timberline as affected by fir–snowdrift interactions. Vegetatio 19:192–207

Bradley RS, Jones PD (1993) "Little Ice Age" summer temperature variations: their nature and relevance to recent global warming trends. The Holocene 3:367–376

Bradley RS, Jones PD (eds) (1995) Climate since A.D. 1500. (2nd Edition) Routledge, London

Brink VC (1959) A directional change in the subalpine forest–heath ecotone in Garibaldi Park, British Columbia. Ecology 40:10–16

Briffa KR, Bartholin TS, Eckstein D, Jones PD, Karlén W, Schweingruber FH, Zetterberg P (1990) A 1,400-year tree-ring record of summer temperatures in Fennoscandia. Nature (London) 346:434–439

Briffa KR, Jones D, Bartholin TS, Eckstein D, Schweingruber FH, Karlén W, Zetterberg P, Eronen M (1992) Fennoscandian summers from A.D. 500: temperature changes on short and long timescales. Climate Dynamics 7:111–119

Brubaker LB (1986) Responses of tree populations to climatic change. Vegetatio 67:119–130

Brubaker LB (1988) Vegetation history and anticipating future vegetation change. In: Agee JK, Johnson DR (eds) Ecosystem Management for Parks and Wilderness. University of Washington Press, Seattle, Washington, U.S.A. pp 41–61

Davis MB (1986) Climatic instability, time lags, and community disequilibrium. In: Diamond J, Case T (eds) Community Ecology. Harper and Row, New York, U.S.A. pp 269–284

Earle CJ (1993) Forest dynamics in a forest–tundra ecotone, Medicine Bow Mountains, Wyoming. PhD dissertation, University of Washington

Ettl GJ, Peterson DL (1995a) Growth response of subalpine fir (*Abies lasiocarpa*) to climate in the Olympic Mountains, Washington, USA. Global Change Biology 1:213–230

Ettl GJ, Peterson DL (1995b) Extreme climate and variation in tree growth; individualistic response in subalpine fir (*Abies lasiocarpa*). Global Change Biology 1:231–241

Franklin JF, Moir WH, Douglas GW, Wibery C (1971) Invasions of subalpine meadows by trees in the Cascade Range, Washington and Oregon. Arct Alp Res 3:215–224

Fritts HC (1976) Tree Rings and Climate. Academic Press, London and New York

Hamilton JP (1987) Densitometric tree-ring investigations at the Columbia Icefield, Jasper National Park. M.Sc. Thesis, University of Western Ontario, London, Canada

Hessl AE, Baker WL (1997) Spruce and fir regeneration and climate in the forest–tundra ecotone of Rocky Mountain National Park, Coloroda, USA. Arct Alp Res 29:173–183

Heusser CJ (1956) Postglacial environments in the Canadian Rocky Mountains. Ecological Monographs 26:253–302

Janz B, Storr D (1977) The Climate of the Contiguous National Parks: Banff, Jasper, Kootenay, Yoho. Project Report No. 20, Applications and Consultation Division, Meteorological Applications Branch, Environment Canada, Toronto, Canada

Johnson EA (1987) The relative importance of snow avalanche disturbance and thinning on canopy plant populations. Ecology 68:45–53

Kavanagh TA, Luckman BH (1995) Tree clump development near the Interpretive Centre, Columbia Icefield, Jasper National Park, Alberta. Contract Report to Parks Canada, Jasper

Kearney MS (1982) Recent seedling establishment at timberline in Jasper National Park. Can J Bot 60:2283–2287

Kienast F, Schweingruber FH, Bräker OU, Schär E (1987) Tree-ring studies on conifers along ecological gradients and the potential of single-year analyses. Can J For Res 17:683–696

Kullman L (1987) Long-term dynamics of high-altitude populations of *Pinus sylvestris* in the Swedish Scandes. J Biogeogr 14:1–8

Kullman L (1997) Tree-limit stress and disturbance. A 25-year survey of geoecological change in the Scandes Mountains of Sweden. Geografiska Annaler 79A:139–165

Kullman L, Engelmark O (1991) Historical biogeography of *Picea abies* (L.) Karst. at its subarctic limit in northern Sweden. J Biogeogr 18:62–70

Lavoie C, Payette S (1992) Black spruce growth forms as a record of a changing winter environment at treeline, Quebec, Canada. Arct Alp Res 24:40–49

Lloyd AH, Graumlich L (1997) Holocene dynamics of treeline forests in the Sierra Nevada. Ecology 78:1199–1210

Luckman BH (1986) Reconstruction of Little Ice Age events in the Canadian Rockies. Géographie physique et Quaternaire XL:17–28

Luckman BH (1988) Dating the moraines and recession of Athabasca and Dome Glaciers, Alberta, Canada. Arct Alp Res 20:40–54

Luckman BH (1990) Mountain areas and global change – a view from the Canadian Rockies. Mountain Res Developm 10:183–195

Luckman BH (1993) Glacier fluctuations and tree-ring records for the last millennium in the Canadian Rockies. Quaternary Science Rev 16:441–450

Luckman BH (1994) Climate conditions between ca. 900–1300 A.D. in the southern Canadian Rockies. Climatic Change 26:171–182

Luckman BH (1996) Reconciling the glacial and dendrochronological records of the last millennium in the Canadian Rockies. In: Jones PD, Bradley RS, Jouzel J (eds) Climatic Variations and Forcing Mechanisms of the Last 2000 years. Springer-Verlag, Berlin, pp. 85–108

Luckman BH (1997) Developing a proxy climate record for the last 300 years in the Canadian Rockies – some problems and opportunities. Climatic Change 36:455–476

Luckman BH, Briffa KR, Jones PD, Schweingruber FH (1997) Tree-ring based reconstruction of summer temperatures at the Columbia Icefield, Alberta, Canada, A.D. 1073–1983. The Holocene 7:375–389

Luckman BH, Jozsa LA, Murphy PJ (1984) Living seven-hundred-year-old *Picea engelmannii* and *Pinus albicaulis* in the Canadian Rockies. Arct Alp Res 16:419–422

Luckman BH, Kearney MS (1986) Reconstruction of Holocene Changes in Alpine Vegetation and Climate in the Maligne Range, Jasper National Park, Alberta. Quatern Res 26:244–261

Luckman BH, Seed ED (1995) Fire–Climate Relationships and Trends in the Mountain National Parks. Final Report: Contract C2242-4-2185, Parks Canada, Ottawa

Magee TK, Antos JA (1992) Tree invasion into a mountain-top meadow in the Oregon Coast Range, USA. J Veg Sci 3:485–494

Payette S, Filion L (1985) White Spruce expansion at the tree line and recent climatic change. Can J For Res 15:241–51

Rochefort RM, Little RL, Woodward A, Peterson DL (1994) Changes in sub-alpine tree distribution in western North America: a review of climatic and other causal factors. The Holocene 4: 89–100

Schäffer MTS (1911) Old Indian Trails of the Canadian Rockies. G.P Putnam's Sons, New York; republished in Hart EJ (ed) (1980) "A Hunter of Peace", Whyte Foundation, Banff, Alberta

Schweingruber FH (1988) A new dendroclimatic network for western North America. Dendrochronologia 6:171–180

Stevens GC, Fox JF (1991) The causes of treeline. Ann Rev Ecol System 22:177–191

Vance RE, Beaudoin AB, Luckman BH (1995) The paleoecological record of 6ka BP climate in the Canadian Prairie Provinces. Géographie physique et Quaternaire 4:81–98

Veblen TT, Hadley KS, Reid MS, Rebertus AJ (1989) Blowdown and stand development in a Colorado subalpine forest. Can J For Res 10:1218–1225

Veblen TT, Hadley KS, Reid MS, Rebertus AJ (1991) Methods of detecting past spruce beetle outbreaks in Rocky Mountain subalpine forests. Can J For Res 21:242–254

Annual- *versus* decadal-scale climatic influences on tree establishment and mortality in northern Patagonia

Ricardo Villalba and Thomas T. Veblen

Abstract. Current capability to predict climate-induced vegetation change is limited by an inadequate understanding of the effects of climate variations on tree regeneration, tree mortality and disturbance regimes. Climatic controls of vegetation patterns are most evident at broad spatial and temporal scales. However, over periods ranging from a few years to several decades, the influence of climatic fluctuations on forest dynamics has been poorly recognized. This study explores the role that climatic fluctuations at annual- to decadal-scales have on *Austrocedrus chilensis* establishment and mortality at the forest–steppe border in northern Patagonia, Argentina, from 37 to 43°S latitude. Recent episodes of tree establishment and mortality are reconstructed based on age frequency distributions from living trees, and dendrochronological dating of snags, respectively.

Austrocedrus establishment at the forest–steppe ecotone appears to be episodic in relation to climatically distinct episodes. Wet-cool summers prevailing for a decade or longer facilitate tree establishment. For the past 150 years, tree establishment at the forest–steppe ecotone has involved interactions between climate and disturbance (mainly fire and grazing). The nature and intensity of these interactions have varied through time. Episodic mortality is associated with extremely dry-warm climatic events during a single summer, or more commonly during two consecutive summers. Most episodes of tree establishment and mortality are concurrent with years of above- and below-average tree growth, respectively. However, regional episodes of *Austrocedrus* establishment are related to decadal-scale climatic events, whereas mortality episodes are mainly controlled by extreme, annual-scale climatic fluctuations. The distinction between the effect of short- *versus* long-lasting climatic fluctuations on forest dynamics is crucial to properly predicting forest response to climatic changes.

1. Introduction

Climatic variability is an important exogenous factor affecting regeneration dynamics and successional processes in forest communities (Davis 1986; Prentice 1986). If climate were stationary, vegetation dynamics would consist of fluctuations mainly due to year-to-year weather variations, succession triggered by disturbances, and gap-phase regeneration (Prentice 1992). These stochastic processes can be predicted from physiological and life-history characteristics of the species and information on climate and natural disturbances. However, vegetation processes also vary spatially in response to changes in broad-scale patterns of climate and changes through time in response to long-term variations in climate (Prentice 1992).

Recent concern over the ecological effects of anthropogenic climate change has accelerated efforts to understand and quantify climate-induced vegetation change (Bartlein et al. 1986; Clark 1990; COHMAP members 1988; Davis 1989a,b; Graham et al. 1990; Melillo et al. 1996; Overpeck et al. 1990; Solomon 1986). Both modelling and paleoecological approaches have proven useful in elucidating the effects of climate change on vegetation dynamics. Current capability to predict climate-induced vegetation change is limited, however, by spatial and temporal scale problems (Neilson 1986), and inadequate understanding of the effects of climate variation on disturbance regimes (Graham et al. 1990; Overpeck et al. 1990).

Climate affects vegetation directly and indirectly. Growing-season and winter weather conditions directly determine the survival and vigour of plant species. Changes in vigour and growth rates affect relative competitive abilities among species, which in turn influence patterns of succession, mortality, and gap formation (Shugart 1984). Climate indirectly affects the frequency, magnitude, type and extent of disturbance (Overpeck et al. 1990). Climate-model results indicate that some warmer climates tend to be characterised by more frequent droughts and convective wind storms (Overpeck et al. 1990). These weather conditions are likely to increase disturbances such as catastrophic windthrows, forest fires (due to drought, wind, and enhanced natural ignition sources), and insect outbreaks (due to pre-disposition of trees to attack). Thus, climate-induced vegetation change is likely to be mediated both by the direct effects of climate variation on the performance of plants and indirectly through altered disturbance regimes (Johnson and Larsen 1991, Sirois and Payette 1991).

The interrelated systems of climate, vegetation, and physical landscape are highly dynamic on time scales ranging from hours to million years. The control of vegetation patterns by climate is most obvious at coarse spatial and temporal scales (Webb 1987), but at time scales of decades to several centuries, the effects of climatic variation on vegetation dynamics have been less obvious. This is often due to the difficulty of separating disturbance-induced successional changes from climate-induced changes (Archer et al. 1995; Brubaker 1986; Davis 1986; Gear and Huntley 1991; Ogden 1985; Prentice 1992; Veblen and Lorenz 1987, 1988; Veblen and Markgraf 1988; Veblen and Stewart 1982). However, it is precisely at this time scale of years to a few decades for which the need for understanding the effects of climatic variation on vegetation change is most urgent (Mellillo et al. 1996; Overpeck et al. 1990; Solomon 1986).

As a part of a broader study of landscape dynamics in relation to human activities and climatic variations (Veblen and Lorenz 1988; Veblen et al. 1992a; Kitzberger et al. 1997), in this study we specifically investigated the effect of climatic variation on the demographic parameters (establishment and mortality) of tree populations along the forest–steppe border in northern Patagonia. Most studies of the effects of climatic variation on forest dynamics have focused on tree regeneration (Brubaker 1986), very few have incorporated the effects of climate-altered disturbance regimes (e.g., Bradshaw and Zackrisson 1990; Clark 1988, 1990; Grimm 1983; Johnson and Larsen 1991; Swetnam 1993), and the effects of climate on tree mortality have rarely been considered (Betancourt et al. 1993; Elliot and Swank 1994; Szeicz and MacDonald 1995). As the death of mature trees is vitally important in releasing resources potentially available for the new establishment of trees of the same or different species (Franklin et al. 1987), determining the influences of climatic variation on tree mortality is essential for understanding the effects of climatic variation on forest dynamics. In the course of completing this study, we were also interested in evaluating the influences of climatic variations at different frequencies (i.e. annual- versus decadal-scales) on forest establishment and mortality. Are establishment and death of trees influenced by the same type of climatic fluctuations, or are they differentially sensitive to climatic variations of different frequency and duration? To properly assess the impact of future climatic changes on forests, it is important to determine any difference in demographic responses to climatic variations of different duration.

The woodland ecotone between forest and the Patagonian steppe, between 37° and 44°S latitude in Argentina (Figure 1), provides an excellent environment for the study of the effects of climatic variations on forests for several reasons: 1) tree populations may respond more obviously to climatic variability at treeline sites, due both to the occurrence of threshold climatic conditions and the availability of open sites free from competition from adult trees; 2) the tree species along this gradient provide excellent material for dendroclimatic reconstructions and dendrochronological dating of past events of tree establishment and mortality; and 3) previous work has elucidated the main features of stand development patterns and responses to natural and anthropogenic disturbances. In this chapter, we examine regional patterns of tree establishment and mortality of the coniferous *Austrocedrus chilensis* in woodlands near the ecotone with the steppe. Instrumental climatic records and tree-ring proxy records are used to relate variations in age structures and occurrence of episodic tree death to both annual- and decadal-scale climatic variations. Differential effects of low- versus high-frequency variations in climate on forest demography are observed. The potentially confounding role of disturbances in masking relationships between climate and forest dynamics in northern Patagonia is also considered.

2. Climate and vegetation in Northern Patagonia

The main determinants of the climate of northern Patagonia are the circum-Antarctic cyclonic belt to the south, the southeastern Pacific high-pressure cell to the northwest, and the north-south-trending mountain barrier of the Andes (Aceituno 1988; Miller 1976; Prohaska 1976). The Andean Cordillera is an effective barrier to the westerlies so that mean annual precipitation declines from 4000–6000 mm on the Chilean side of the Andes to less than 200 mm approximately 100 km east of the crest of the Andes in Argentina (Figure 2; Almeyda and Saez 1958; Barros *et al.* 1983). Mean annual temperatures vary from 10°C on the west side of the Andes, to 6°C in the subalpine deciduous forest near treeline, and 8°C at the steppe–forest transition east of the Andes (Almeyda and Saez 1958; Donoso 1981; Gallopín 1978).

Storm frequency in the northern Patagonian Andes is highest in winter when on average 40% of the precipitation occurs (Miller 1976). Seasonal and annual variations in precipitation are strongly influenced by changes in the intensity and latitudinal positions of the southeastern Pacific high-pressure cell, which result in shifts in the storm tracks. Maximum winter precipitation is related to a northward shift of the westerly storm tracks along the coast of Chile (Pittock 1980; Villalba 1990).

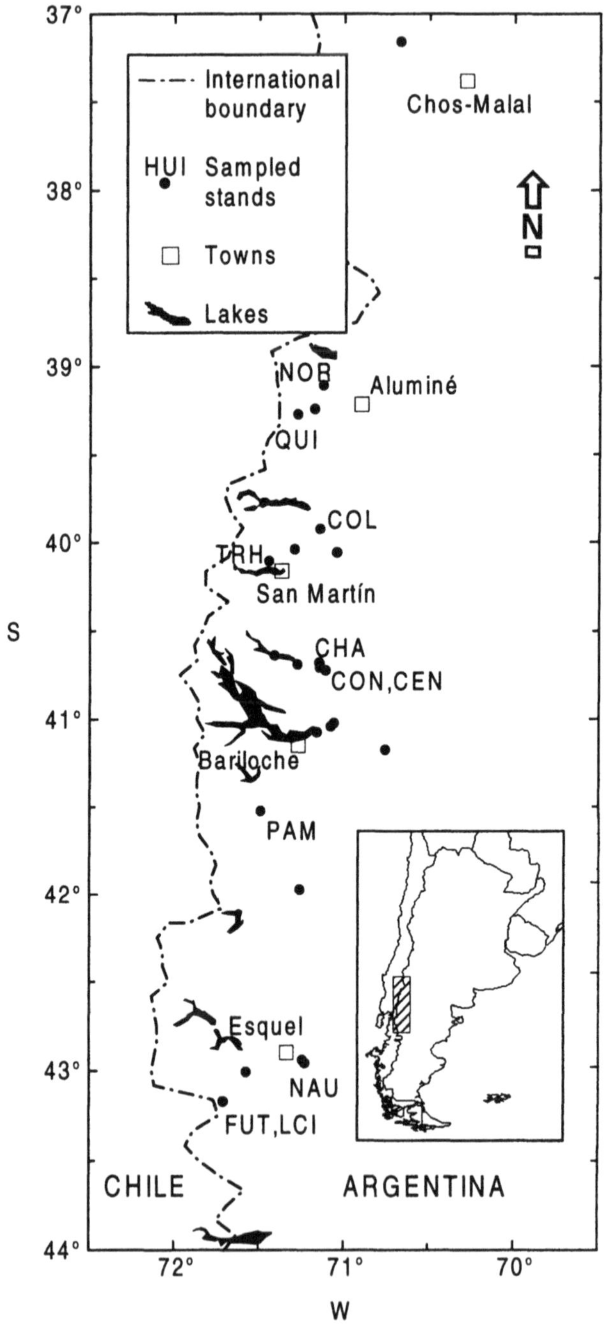

Fig. 1. Map showing locations of stands sampled for dating of tree establishment and mortality in northern Patagonia, Argentina.

Physiographically, from west to east, the northern part of the Argentinean Patagonia is subdivided into the main Andean cordillera, the pre-cordillera of foothills and glacial lakes, and the Patagonian plains. Between 39 and 43°S latitude in the Andes, extensive volcanic ash deposits overlie Pleistocene glacial topography and are the soil-forming parent materials throughout the region. Following the strong precipitation gradient from west to east, there is a dramatic vegetation gradient from temperate rainforests in Chile to the Patagonian steppe in Argentina (Veblen et al. 1995). In the western area of the Argentinean Andes, the montane slopes are dominated by tall forests of the evergreen *Nothofagus dombeyi*, with dense understories of the bamboo *Chusquea culeou*. Higher elevations (> c. 1000 m) are dominated by deciduous *Nothofagus pumilio* forests, and tundra above treeline. Further east, where precipitation declines to about 1500 mm, *Austrocedrus chilensis* and *N. dombeyi* form extensive codominant stands. In response to more xeric conditions to the east, *A. chilensis* forms pure stands to finally open up into *A. chilensis* woodlands codominated by sclerophyllous shrubs and small trees. Further east, *A. chilensis* disappears and steppe elements such as spiny shrubs and bunchgrasses (*Stippa* spp. and *Festuca* spp.) dominate.

Austrocedrus chilensis, the dominant tree species at the forest–steppe ecotone, is a conifer of highly variable form and size. It occurs as multi-stemmed poorly formed individuals at dry, rocky sites and as a conical-shaped tree more than 35 m tall in dense stands on mesic sites (Tortorelli 1956). In Argentina, *Austrocedrus* occurs discontinuously from 36°30' to 39°S and more continuously from 39°30'S to 43°35'S (Seibert 1982). The distribution of *Austrocedrus* reflects its ability to withstand more xeric conditions than those tolerated by *Nothofagus* species and other rainforest species of Southern Chile (Veblen et al. 1995). In the northern Patagonian Andes, between 39° and 43°S, *Austrocedrus* occurs in pure, dense stands where the annual precipitation is less than 1700 mm; eastward these stands open up into sparse woodlands adjacent to the Patagonian steppe (Seibert 1982). Scattered trees on rocky outcrops, occur in the Patagonian forest–steppe ecotone under mean annual precipitation as low as 500 mm.

In general, stand development patterns and tree population responses to natural and anthropogenic disturbance along the gradient east of the Andes from mesic *Nothofagus dombeyi* forests to xeric woodlands *Austrocedrus* at the steppe ecotone are well known. Along this gradient there have been studies of stand responses to earthquake-related disturbance (Veblen et al. 1992a; Kitzberger et al. 1995), small-scale treefalls (Veblen 1989), fire (Veblen and Lorenz 1987, 1988; Veblen et al. 1992a; Kitzberger et al. 1997), and browsing by large mammals (Veblen et al. 1992b).

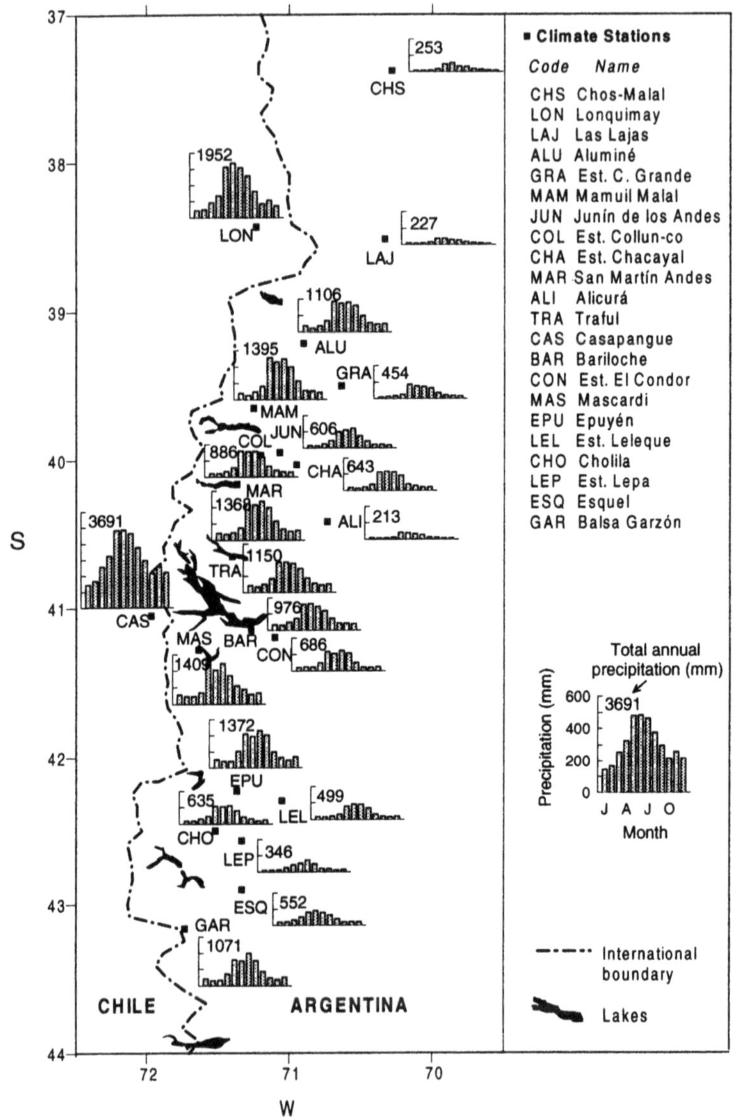

Fig. 2. Annual precipitation distribution for 22 meteorological stations in northern Patagonia. Sources: Servicio Meteorológico Nacional, Argentina; Corporación Forestal, Chile; Hidronor Nor-Patagonia; Parques Nacionales, Argentina; Centro Nacional Patagónico (CONICET).

3. Field sampling

3.1 Age structure

Austrocedrus ages were sampled over a north-south gradient from 39° to 43° S (Figure 1; Table 1). Our goal was to develop a large, regionally extensive sample of trees that had established recently enough to permit determination of their total ages. If most trees were so old that their ages could not be reliably determined, then it would not be possible to relate tree age and climatic variation at a fine scale of temporal resolution. By establishing relationships between the abundances of trees and climatic variation at an annual resolution over the past c. 30 years, it should be possible to relate patterns in less precisely aged older populations to climatic variation. Consequently, areas in which tree sizes and spatial patterns implied a recent invasion of the steppe were preferentially selected for sampling. Areas were selected for sampling which covered a wide range of latitude and topographic positions within the forest–steppe ecotone. These areas also include a wide range of stand ages, but there was a clear bias towards the sampling of younger stands to permit the analysis of climatic influences on tree establishment at an annual level of resolution. Within each subjectively determined homogeneous area, one to several 400 to 6000 m^2 plots were randomly located (Table 1).

					Age structure		Mortality	
Site	Plot code	Latitude (°S)	Longitude (°W)	Elevation (m)	No. of plots	No. of trees	No. of plots	No. of trees
Ñorquinco	ÑOR	39°05'	71°07'	1130			3	138
A. Malal-co	QUI	39°17'	71°16'	1130	2	86		
Collun-co	COL	40°03'	71°17'	975-1060	3	132		
P. Trhompul	TRH	40°07'	71°26'	1060	1	76		
P. Córdoba	PCO	40°39'	71°09'	830-1010	6	272	1	20
Chacabuco	CHA	40°40'	71°00'	905	1	132		
Confluencia	CON	40°42'	71°09'	810-1075	3	128	2	101
El Centinela	CEN	40°44'	71°06'	805-900	6	470	1	20
P. del Toro	PAM	41°32'	71°29'	1140	1	91		
Nahuel-Pan	NAU	42°58'	71°13'	850	3	221		
Los Cipreses	LCI	43°10'	71°41'	550			1	30
Futaleufú	FUT	43°11'	71°42'	470			1	18

Table 1. Characteristics of sites sampled for age structures and tree mortality.

For all trees in each plot we attempted to determine dates of germination to an interannual level of precision. In each plot, all small tree seedlings (i.e., <100 cm tall) were uprooted. Trees too large to uproot were cored as close as possible to the root collar either by excavating "coring holes" in the ground surface or by angling the increment borer downwards. For each core, the sampling height and its position in relation to other cores from the same tree (i.e., parallel, perpendicular or opposite) were recorded. For larger trees that could not be cored at the root collar due to mechanical limitations, cores were taken as low on the stem as possible and a mean age to coring height was added to estimate total tree age (Villalba and Veblen 1997c). In most cases repetitive core samples were taken until the pith was intercepted. In those cases where repetitive samples failed to intercept the pith, two or more cores were taken and a combination of graphical and growth rate methods was used to estimate total tree ages (Villalba and Veblen 1997c). Minimum ages of trees with rotten centres were determined but not entered into the quantitative analysis of climatic influences on establishment. Cores were mounted, sanded and dated following standard dendrochronological methods (Stokes and Smiley 1968). Given adequate preparation of the samples and optical magnification, intra-annual bands (or false rings) in *Austrocedrus* are easily distinguished from annual tree rings. In contrast to the abrupt boundaries between annual rings, the intra-annual bands are recognized by the gradual transition in cell size on both margins of the band (Villalba and Veblen 1996). Based on 25 chronologies from *Austrocedrus*, which include a total of 129,511 cross-dated tree- ring measurements, the percentage of missing rings was only 0.147% (i.e., only one missing ring out of 680 rings; Villalba 1995).

For those sites where tree population structures were determined, tree-ring width chronologies were used to compare periods of tree establishment with fluctuations in radial growth. To supplement meteorological records from sites several to many kilometers distant from the sampling sites, local chronologies were used for relating tree recruitment with climatic variations evident in tree growth at the site itself. For these comparisons, a set of 27 tree-ring chronologies derived from *Austrocedrus* along the forest–steppe ecotone in northern Patagonia were used. Geographical locations, site characteristics, and procedures used to standardise the chronologies are given in Villalba and Veblen (1997a).

3.2 Tree mortality

Cross sections were extracted from dead trees in 9 stands including some of the 26 stands in which all seedlings and trees were aged and mapped for the regional stand structure study (see previous section, Fig. 1 and Table 1). Trees were carefully examined to try to determine probable causes of death. Trees showing evidence of past fires (fire scars or charcoal) were not included. Dead *Austrocedrus* trees uprooted by wind were also excluded. Most sample sites were located on rocky, steep slopes, which are not affected by a fungal pathogen that appears to kill trees at some poorly drained sites (Havrylenko et al. 1989).

In each of the 9 plots, all dead-standing trees > c. 5 cm diameter at breast height (dbh) and sound logs were sampled. Additional samples from dead trees were collected near the age-structure plots to increase sample sizes. Complete or partial cross sections were taken with a manual saw and all samples included ≥ 50 tree rings to facilitate cross dating. Typically, two samples were taken from each tree at different heights to reduce problems associated with missing or partial rings.

Samples of dead trees were segregated into four categories: 1) cross section with bark, 2) cross section without bark but showing no evidence of ring erosion, 3) cross section without bark showing incipient ring erosion, and 4) cross section with pronounced ring erosion. The estimated accuracies of dates of tree deaths associated with these categories are: close to annual accuracy (except where one or two rings were missing prior to death) for the first two categories, 3 to 5 years for the third, and greater than 5 years for the fourth. The calendar date of the outermost ring on dead trees was determined by cross dating against master chronologies developed from the live trees in each stand. Both visual cross dating (Stokes and Smiley 1968) and quantitative cross dating (program COFECHA, Holmes 1983) were used to cross date samples of dead trees against master chronologies. Because tree rings contain many clues to cross dating (e.g., light rings, intraseasonal bands, frost rings) in addition to ring width (Fritts 1976), we also used visual observations when making cross-dating decisions. For those sites where tree mortality was determined, ring-width chronologies were used both to cross date the outermost rings on the dead samples and to describe local patterns of tree growth (Villalba 1995).

4. Tree establishment and climatic variations

The effects of climatic fluctuations on tree establishment were analyzed at two scales: (1) the time scale covered by the instrumental climate record from c. 1910 to the present, and (2) the past 200 years for which climatic fluctuations were inferred from tree-ring records.

4.1 Tree age structures

Tree age frequency distributions for the 26 stands considered individually reflect the tendency of this relatively shade-intolerant species to form broadly bell-shaped age frequency distributions. In northern Patagonia, these age structures may develop in either of two ways (Veblen and Lorenz 1987, 1988): 1) Crown fires may destroy relatively dense stands of *Austrocedrus* and trigger a multi-decadal period of new tree establishment; or, 2) cessation of formerly frequent surface fires may allow sparse woodlands of *Austrocedrus* to increase in density and to invade previously treeless sites in the steppe. The former pattern occurs at slightly more mesic sites (e.g. more westerly locations, in valley bottoms and on south-facing slopes), whereas the latter pattern is typical of more xeric sites (Villalba and Veblen 1997b). Most of the sites sampled in this study were close to the ecotone with the steppe and represented recent invasions of the steppe by *Austrocedrus*.

The age structures imply longer periods of tree recruitment at more westerly, humid sites and also in rocky, extremely harsh environments. At the former type of site, however, tree establishment appears to have been more continuous. In contrast, at drier sites located further east or on north-facing slopes age structures are less continuous. Seedling survival at the marginal, drier sites is probably controlled by the adverse climatic conditions that prevail most of the time. The most common pattern of tree recruitment observed at sites with greater soil moisture availability (due to aspect, soil depth and/or more westerly location) is one of short but continuous periods of tree establishment (Villalba and Veblen 1997b).

When the ages of the c. 1300 *Austrocedrus* trees from all 26 sites are combined, several peaks in age structure are evident (Figure 3). The abundance of trees dating from c. 1962 to 1979 is the most obvious feature of the regional pattern. Compared to a relatively small number of trees dating from the late 1950s, there is a sharp increase in the abundance of trees from 1961 to 1963, then a gradual decrease until 1967, and again a peak between 1974 and 1977. Trees dating from the 1980s are extremely rare. Relatively few trees date from the 1900 to 1920 period in comparison with the greater number of trees from the late 19th century and after the 1920s (Figure 3). When only older stands are considered (i.e., stands that initiated prior to 1900), a sharp decline in the number of trees dating from the early 1900s is more evident (Figure 4). For these older stands, most surviving trees established over a ca. 30-year period beginning about 1865. However, because the errors associated with estimates of total age increase for older (larger) trees, the dates assigned to most trees in these older stands may vary by ± 15 years or more.

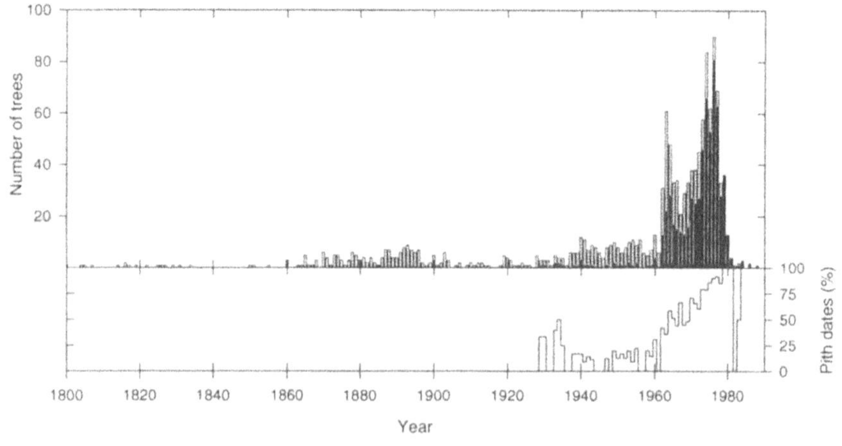

Fig. 3. The age–frequency distribution from 1312 *Austrocedrus* trees along the forest–steppe ecotone in northern Patagonia, Argentina, for the interval 1800–1989. In the lower part of the figure, the percentage of samples exactly dated (pith dates) is indicated. Black and grey bars refer to pith and estimated dates, respectively.

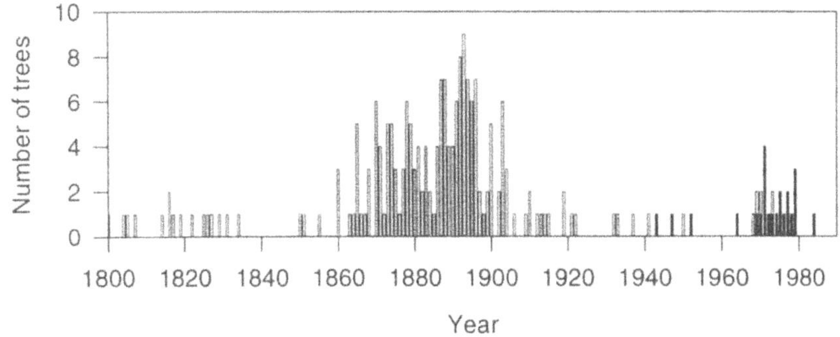

Fig. 4. Age frequency distributions of *Austrocedrus* in stands where establishment began prior to 1900. Black and grey bars refer to pith and estimated dates, respectively.

4.2 Instrumental records

Regional climatic records of precipitation and temperature were created by averaging monthly standard deviations from temperature and precipitation records from nearby climate stations (Figure 2). Average monthly standard deviations for each site were computed and then averaged across the sites. As a consequence of using normalised standard deviations, each station, independent of its particular values of temperature and precipitation, has the same weight in the averaged regional records. The precipitation record included the Collun-co, Chacayal, San Martín de los Andes, Bariloche, El Condor, Leleque and Esquel weather stations (Figure 2). The regional temperature record consists of the Collun-co, Bariloche and Esquel stations (Villalba 1995). For the period 1906 to 1989, de Martonne's (1926) aridity index was computed from monthly temperature and precipitation data. Regional records of temperature and precipitation were grouped into seasonal data sets of three to six months for comparison with records of tree establishment and mortality. October through March showed the strongest association with both tree establishment and mortality, and, consequently only this season is used in the following analyses. The radial growth of *Austrocedrus* is consistently and strongly related to climatic variations during October through March (Villalba 1995; Villalba and Veblen 1997a).

On a regional scale, relatively few trees date from the first 20 years of the 20th century when warm-dry conditions prevailed (Figure 5). Warm-dry conditions were especially marked from 1907 to 1918. Alternating wet and dry conditions were recorded from 1919 to 1924. From 1925 to 1933, late spring–early summer climatic conditions were substantially wetter and cooler, particularly in the years 1925, 1926 and 1932. However, there is no obvious peak in tree age structure that can be associated with the cool-wet summers from 1925 to 1933. After the average to dry summers from 1934 to 1937, cool-wet conditions resumed in 1938 and lasted until 1941; they were abruptly interrupted by the extremely hot-dry summers of 1942 and 1943. However, there is no obvious decline in the numbers of trees dating from the 1942–43 drought. Cool-wet conditions resumed from

1944 through 1946, and afterwards warm-dry vs. cool-wet conditions alternated until 1953. Severe hot-dry conditions were recorded from 1956 to 1962 concurrent with a slight decrease in the number of surviving trees.

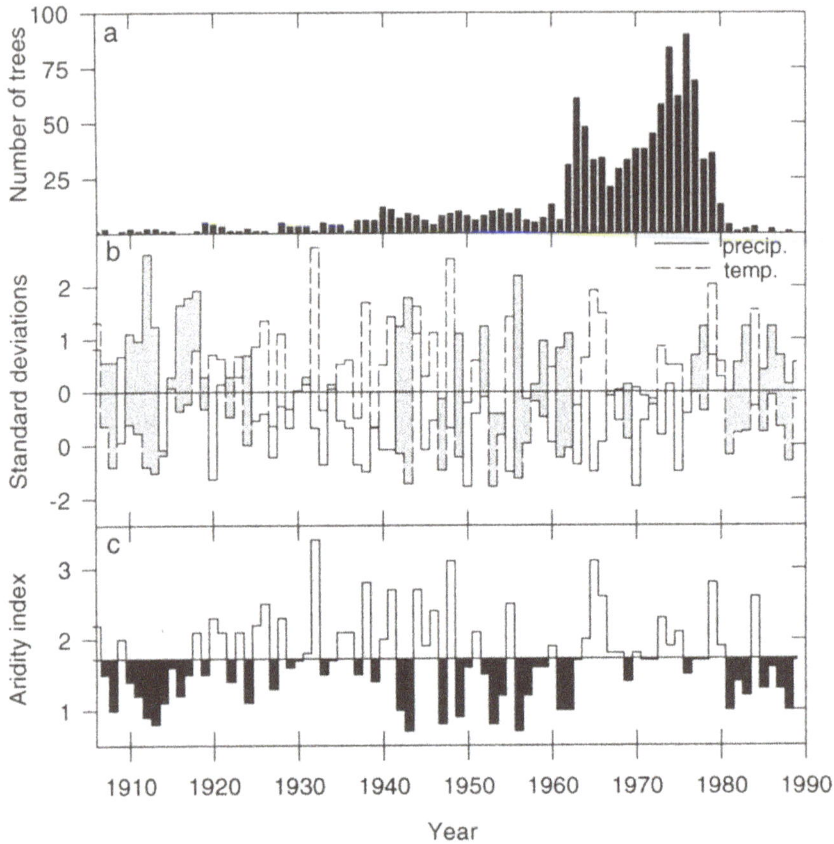

Fig. 5. Composite age–frequency distribution of *Austrocedrus* from all 26 stands for the interval 1907–1989 (a), regional deviations in temperature and precipitation (b), and De Martonne's (1926) aridity index (c) for October through March for the same interval. Grey and black histograms in parts (b) and (c) indicates years of moisture deficit.

Following the warm-dry interval from 1956 to 1962, the climatic pattern changed drastically (Figure 5). Cool-wet conditions started in 1963 and prevailed until 1979. During this interval, massive *Austrocedrus* establishment is reflected in the regional age structure, and peaks of surviving trees coincide with significantly cooler and wetter summers in 1964–66 and 1973–75 (Figure 6a). A substantial change in summer climatic conditions occurred again around 1980. With the exception of the summer of 1984, a warm-dry pattern prevailed throughout the 1980s. The abrupt decrease since 1979 and the absence of trees dating from the 1980s coincides with this change towards a warmer and drier climate (Figure 6).

The relationships between tree age structure and climatic variation are clearly more consistent since 1960. On a decadal scale, spring–summer climatic patterns show a greater contrast since 1953–56 (Figure 5). We compared the mean monthly aridity indexes for the warm-dry intervals of 1953–62 and 1980–87 and those for the cool-wet interval from 1963 to 1979 (Figure 6). Substantial differences in water availability are indicated for the spring and summer months, and these differences are statistically significant ($P<0.05$; *t* test) for the months of November, December, and February. Potential relationships between abundance of surviving trees and climatic variation for the pre-1955 period could not be determined due to both the greater interannual variability in climate during the 1940s and early 1950s and to the greater errors in estimating the ages of trees more than 40 years old.

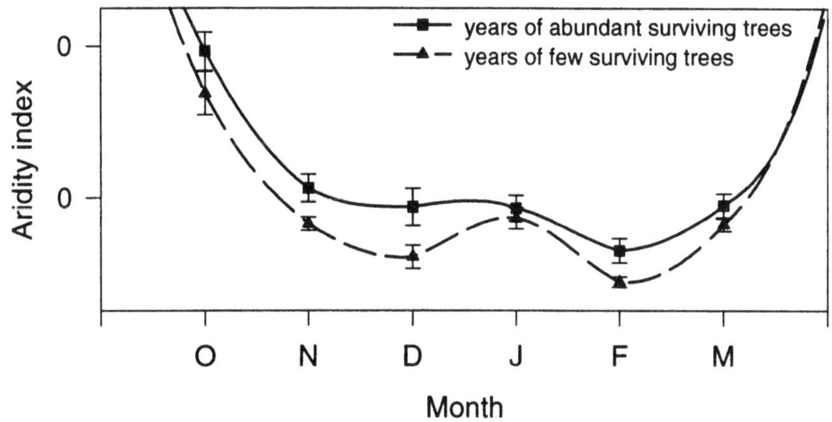

Fig. 6. Differences in mean monthly De Martonne's (1926) aridity indexes between years of abundant surviving trees (1963–1979) and years of few surviving trees (1956–61 and 1980–88) in northern Patagonia. There are statistically significant differences in the mean patterns between years of abundant and few surviving trees for November, December and February.

4.3 Tree-ring records

At a regional scale, relationships between radial growth and age structure are also evident. To produce a regional record of variation in tree radial growth, we used the amplitudes of the first principal component resulting from a principal component analysis (Cooley and Lohnes 1971) of 15 chronologies from *Austrocedrus* covering the interval 1790–1989 (Figure 7). Mean amplitudes of the first principal component were compared for periods of contrasting numbers of surviving trees (Table 2). A period of sustained radial tree growth above the long-term mean occurred from 1868 to 1892. Two sub-periods, 1868–1874 and 1887–1892, have tree-ring indexes similar in magnitude to those recorded during the most recent period of massive establishment in the 1970s (Figure 7). The difference in the mean amplitude of the first principal component for the interval 1868–1892 *versus* 1964–1977 is not great enough to reject the possibility that the two intervals are sub-samples from the same population (Table 2). The increase in abundance of trees starting in the 1860s and lasting until about 1900 may reflect favourable climatic conditions. However, the relationship between radial growth and tree age structure is necessarily speculative because of the relatively small numbers of surviving trees and the inaccuracy of determining germination dates for older trees.

Period	Mean	Std. Dev.	Tree survival	Comparison with 1964-1977 mean*	
				t	probability
1964-1977 (n = 14 yrs.)	1.31	1.73	abundant	---	---
1868-1892 (n = 25 yrs.)	1.23	2.20	present	-0.116	p = 0.9083
1901-1917 (n = 17 yrs.)	-1.94	1.77	reduced	-5.15	p = < 0.0001
1925-1941 (n = 17 yrs.)	2.19	2.27	present	-1.19	p = 0.2430
1957-1963 (n = 7 yrs.)	-4.03	1.75	reduced	-6.65	p = < 0.0001

* Due to the lag in the response of tree growth to climatic variations, tree growth periods are lagged one year in relation to major climatic patterns from the instrumental record.

Table 2. Comparison of mean tree growth (t test) during the 1964 to1977 period of abundant tree establishment with other intervals during the past 150 years. The amplitudes of the first principal component from 15 tree-ring chronologies covering the interval 1800-1989 were used for comparison

The interval 1899–1917 is the most extended period of below-average tree growth over the last 300 years (Villalba 1995; Villalba and Veblen 1997a). Concurrent with this period of below-average growth, the number of surviving trees declined (Figure 7). The differences in the mean amplitudes of the first principal component for the intervals 1901–1917 *versus* 1964–1977 is greater than would be expected by chance ($p < 0.0001$; Table 2). Tree-growth indexes gradually increase during the 1930s, peak in 1941 and suddenly decrease to extremely low values in 1943 and 1944 (Figure 7). The growth indexes from 1933 to 1941 represent the most conspicuous peak in tree growth recorded in the twentieth century. Although a slight increase in the number of surviving trees occurs after 1936, the magnitude of this change is not great relative to the increase in radial growth (Figure 7). Similarly, when stands are considered individually, few have age structures suggestive of above average tree establishment during the late 1930s to early 1950s. However, the extremely hot-dry summers of 1942 and 1943 correspond with a regionally extensive episode of mortality of mature trees (see next section on mortality) during which an increase in seedling mortality is also likely to have occurred. Thus, any favourable response of tree populations to the moister conditions of the 1930s may have been obscured by increased mortality of young trees in the early 1940s.

Tree-ring growth decreases to its lowest values during the dry, warm period from 1957 to 1962. This period is also a time of relatively scarce surviving trees, at least in comparison with the following period of increased radial growth and abundant surviving trees that begins about 1963 (Figure 7). The increase in surviving trees in the 1963 to 1979 period parallels an increase in tree radial growth. Radial tree growth was relatively low in 1978–79 but has increased since 1980 and remained above the mean between 1984 and 1987 (Figure 7). The relationship between tree growth and climate appears to have changed during the 1980s, so that despite the warm and dry conditions indicated by the instrumental record tree growth did not decline commensurately (Villalba 1995). However, this period of warm and dry conditions is clearly a time of scarce tree establishment (Figure 5).

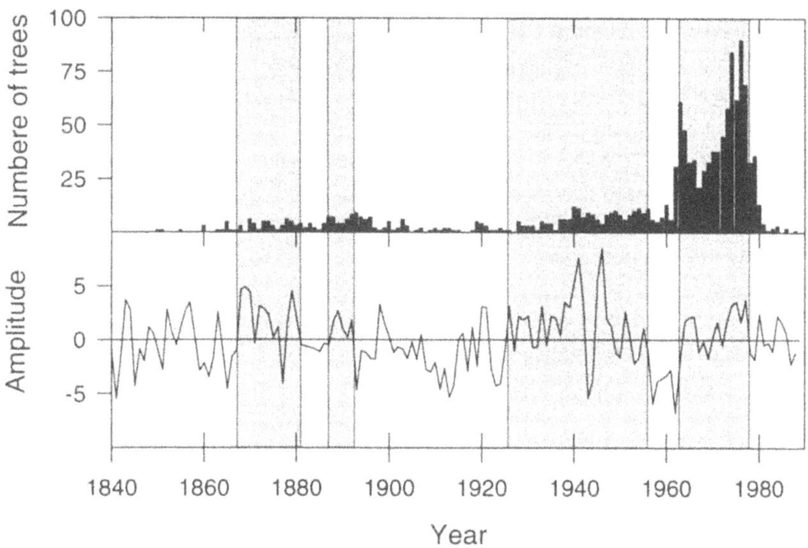

Fig. 7. The age–frequency distribution from 1312 *Austrocedrus* tree along the forest–steppe ecotone in northern Patagonia, Argentina (a) and the radial tree growth summarized as the amplitude of the first principal components from 15 tree-ring chronologies covering the interval 1800–1989 (b).

Although the age structure patterns of *Austrocedrus* strongly coincide with climatic variation for the period since the early 1960s, for earlier time periods the influence of climatic variation is less obvious. For example, the tree-ring record indicates that the 1860s to 1890s and 1930s to early 1950s were characterized by spring and summer moisture availabilities at least as great as those of the 1963–1979 period of abundant tree survival. However, the regional tree age structure indicates only modest increases in the abundances of trees surviving from these earlier periods of favourable climate. Due to the cumulative effects of mortality on older cohorts, it is not surprising that past periods of favourable climate become increasingly difficult to detect as the cohort ages. For example, as the 1962–1979 cohort ages and mortality reduces the abundance of surviving trees, the recognition of this period as one of abundant tree establishment will become more difficult. Over the longer term, variations in disturbance regimes, some related to climatic variations and others not (e.g. fire *versus* grazing), have also undoubtedly influenced the regional pattern of *Austrocedrus* age structures (Veblen et al. 1992a).

5. Tree mortality and climatic variations

When considered individually, sample sites show clear episodes of tree mortality most of which are coincident for nearby stands. For example, in three nearby sites at Lake Ñorquinco (ÑORQ1, 2 and 5) frequency distributions of mortality dates indicate large percentages of tree deaths occurred in the 1910s (Figure 8). Cross sections from trees which died in this interval showed only incipient erosion of the outer rings, so that the true dates of death are probably no more than 2 to 10 years later. The tree-ring chronology from this site shows years of radial tree growth substantially below the long-term mean in 1913–14, 1916 and 1919, approximately coincident with the tree-mortality episode (Figure 8).

When dates of tree death for all 9 sites are combined, periods of abundant tree mortality are evident in 1907–1918, 1942–43, and the 1950s (Figure 9a). For the earliest period, the percentage of cross sections with bark is low (Figure 9b), and, consequently many of the tree deaths may have occurred a few years later (i.e. many of the 1907–1910 dates may indicate death in ≈1912–1914). For the two more recent periods of mortality, high percentages of samples with bark assure relatively precise dating of these events.

Periods of abundant tree mortality are consistently associated with drought. The instrumental record shows that the 1910s, 1942–43, 1953–54 and 1956–1962 were periods of above average temperatures and below average precipitation during spring and summer. Extremely warm-dry conditions were recorded in the springs and summers of 1912–1913, 1942–1943, and 1956 (Figure 9c). The more precise coincidence of tree mortality and drought for the 1940s and 1950s probably reflects less accurate dating of tree death during the 1910s episode. Less severe or shorter droughts in 1924, 1953, 1961–62, and 1981–83 (Figure 9c) approximately coincide with less marked episodes of tree mortality in 1922–1924, 1953–54, 1962–63, and 1983, respectively (Figure 9a). The occurrence of years of regional drought during these periods of increased tree mortality is consistent with the reduced radial growth of surviving trees near each site.

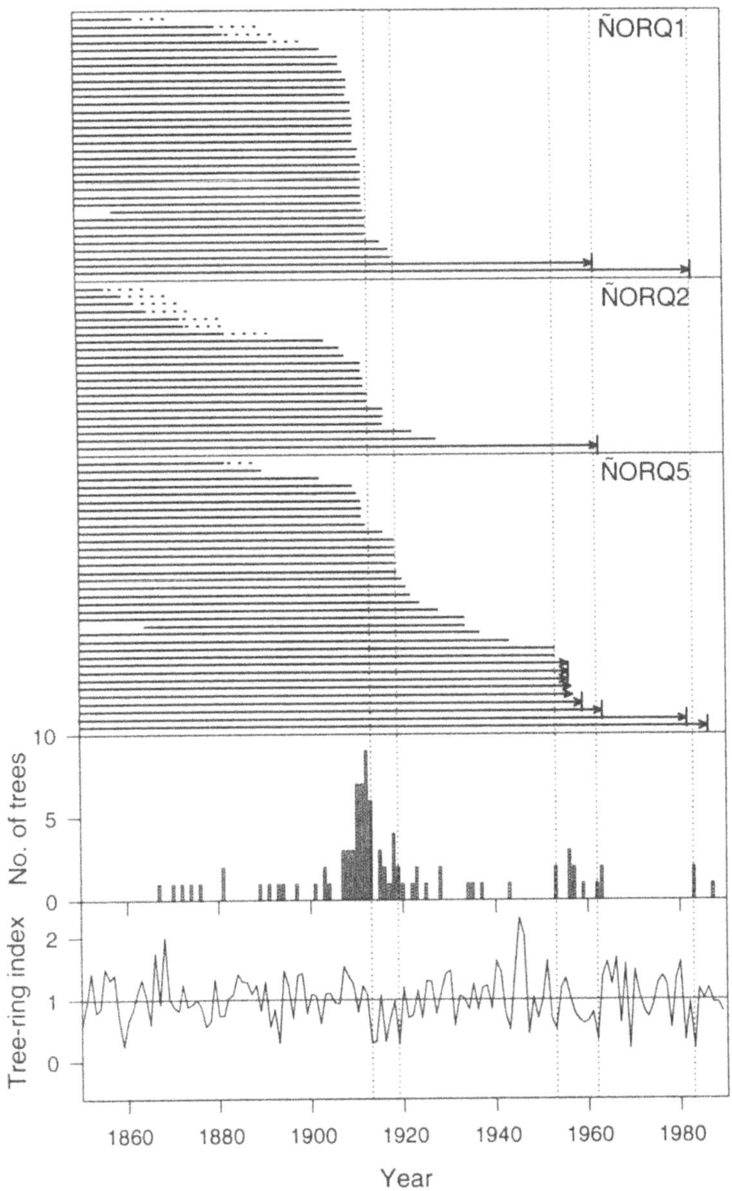

Fig. 8. Results of dendrochronological dating of dead *Austrocedrus* at ÑORQ1, NORQ2, and NORQ5. The tree-ring indexes are from the Ñorquinco chronology (n = 50). Samples of dead trees are segregated into four categories: cross sections with bark (——>|), cross sections without bark but showing no signs of ring erosion (——>), cross sections without bark showing incipient ring erosion (———), and cross sections with pronounced ring erosion (— – –).

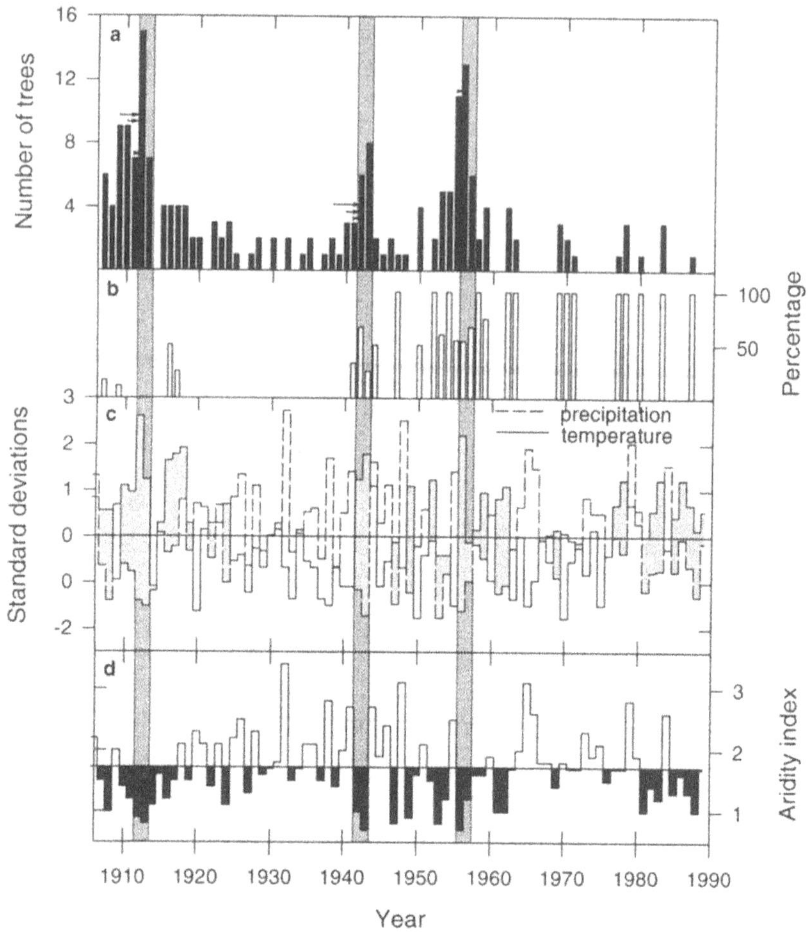

Fig. 9. Number of *Austrocedrus* deaths per year (a), percentage of cross sections with bark for each year (b), regional deviations in temperature and precipitation for spring–summer (October to March) (c), and DeMartonne's aridity index for spring–summer (d). Grey vertical bars indicate years of severe droughts associated with the three most important peaks in tree mortality since 1906. Some mortality dates appear to lead severe droughts by one to several years due to erosion of rings and failure of some trees to produce rings during years of extreme droughts; such dates are joined to the appropriate drought years by horizontal arrows.

De Martonne's aridity index (1926), which is an indicator of water deficiency, shows severe droughts concurrent with regional peaks in tree mortality during 1912–13, 1942–43 and 1956–57 (Figure 9d). Less severe water deficits occurred in 1916, 1924, 1947, 1949, 1953, 1961–62, and during the 1980s related to secondary peaks in tree death. Years with increased mortality are characterized by a substantial reduction in the amount of water available for tree growth during the growing season. Mean monthly aridity indexes for years with increased mortality shows more severe water stress from October to March than the for years with regular mortality (Fig 10).

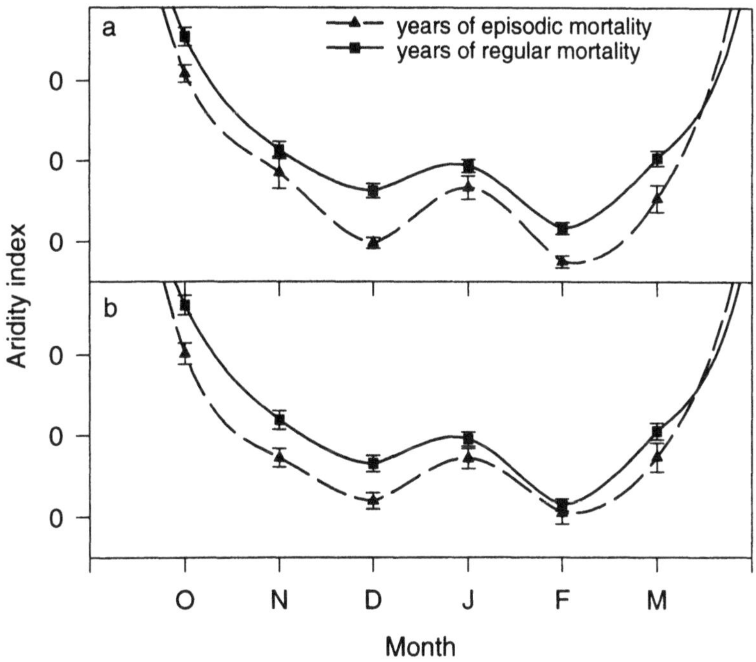

Fig. 10. Differences in mean monthly De Martonne's (1926) aridity indexes between years of episodic versus regular mortality of *Austrocedrus* in northern Patagonia. During the growing season (October to March), water stress is more intense for years of abundant tree mortality (i.e. lower values of De Martonne's aridity index). Differences in mean monthly indexes are shown for major regional mortality events (1912–13, 1942– 43, and 1956–57) (a), and major and secondary peaks of tree mortality (1912–13, 1916–17, 1923–24, 1942–43, 1953, 1956–67 and 1961–62) (b). Statistically significant differences in the mean patterns between years of episodic *versus* regular mortality are observed for October and December in both (a) and (b) patterns.

6. High *versus* low frequency climatic variations and forest demography

Based on the regionally extensive sample of *c.* 1300 *Austrocedrus* ages along the forest–steppe ecotone in northern Patagonia, there is an association of tree survival with climatic variation at an approximately decadal scale. The strongest association between tree age structure and climatic variation is for the period since about 1960. The instrumental climatic record indicates that the 1957 to 1962 period of extremely warm and dry summers was followed by wet and moist conditions from 1963 to 1979 which in turn are followed by warm and dry summers throughout the 1980s. Coincident with the shift towards a more favourable climate there was a sharp increase in numbers of surviving trees that established between 1962 and 1979. Similarly, a nearly total absence of trees dating from the 1980s is associated with this period of warm and dry summers. Given that in each stand all seedlings were aged and that most sample sites were open stands where the regeneration of *Austrocedrus* would not be limited by a closed tree canopy (Veblen and Lorenz 1987, 1988), the scarcity of surviving trees that established after 1979 is remarkable.

Tree ages during the time period when most individuals could be precisely dated (i.e. since the early 1960s) suggest that single years of favourable climate do not necessarily result in enhanced establishment or survival. For example, despite the occurrence of wet and moist conditions in 1984 there is no peak of tree survival during the 1980s decade of drought. Multi-year periods of favourable climate appear to be required for abundant seedling survival. In contrast to findings in other environments (Baker 1990; Little et al. 1994), at the forest–steppe ecotone in northern Patagonia, a single year of favourable climate appears insufficient for successful tree establishment.

Spring and summer drought is clearly the most critical climatic condition that limits both the radial growth of *Austrocedrus* (Villalba 1990, 1995; Villalba and Veblen 1997a) and seedling establishment. Episodic mortality of adult trees also coincides with the severe droughts of 1912–13, 1942–43 and 1956–57 (Villalba 1995), and it is likely that these droughts also increased the mortality rates of seedlings and saplings.

In contrast to *Austrocedrus* establishment, peaks in tree mortality are associated with short (one or two years), extreme climatic events. Although the mortality peaks in 1912–13 occurred during the two driest years within almost a decade-long dry period, the mortality peak in 1942–43 was encompassed by two important wet intervals, and the mortality event in 1956–57 was immediately preceded by a wet year. Annual climatic variations, which control year-to-year water availability, appear to have greater effects on tree survival than do long-lasting droughts characterized by gradual transitions from previous wet intervals.

To further characterise the differential responses of *Austrocedrus* establishment and mortality to different scales of climatic variability, the regional meteorological records from northern Patagonia were spectrally decomposed using singular spectrum analysis (SSA; Vautard and Ghil 1989; Vautard et al. 1992). Unlike the spectral analysis based on the Blackman–Tukey method, SSA is a fully non-parametric technique based on principal component analysis and is thus ideally suited for the analysis of short, chaotic series. SSA uses a number of lagged copies of a centred time series, sampled at equal increments to calculate eigenvalues and eigenvectors of their covariance matrix (Vautard and Ghil 1989, Vautard et al. 1992). Following the principal component analysis, eigenvalues are ranked according to the percentage of variance accounted for in the original time series. Because of the shortness of climatic records in northern Patagonia and the interest in identifying only the principal modes of temporal variation, the SSA was performed on the climatic records using a lag window of five (i.e., 17% of the series lengths). The choice of the window width is a compromise between the amount of information to be retained and statistical significance. Information increases with larger lag windows, but the significance decreases.

The principal mode of temperature variations reconstructed from the SSA, which retained more than 27% of the total variance, is associated with a low-frequency, oscillatory mode of c. 40 years (Figure 11a). The other four components, which represent the high-frequency variations in the temperature series, show waves peaking between 2.2 and 6.4 years. High-frequency components were added to obtain a single series that contains the short-term variations in temperature (Figure 11b). The summer precipitation record was decomposed similarly. Low-frequency variations in precipitation are also represented by a unique oscillatory mode at 40 years, which explains almost 18% of the total variance (Figure 11a). The remaining waves, peaking between 2 and 6.4 years, were averaged to obtain a single record that represents the high-frequency variations of summer precipitation (Figure 11b).

Contrasting patterns in the low-frequency variations for summer temperature and precipitation are observed since 1900 (Figure 11a). Particularly since the late 1950s, when the percentage of exactly dated trees increased, warm-dry *versus* cool-wet patterns are strongly associated with periods of sparse *versus* abundant tree establishment, respectively (Figure 3). This relationship is also observed early in the 1900s but is less obvious. Warm and dry conditions prevailed at the beginning of the 20th century in association with a decrease in the rate of tree recruitment at that time (Figure 3). Afterwards, summer conditions turned to a cool-wet pattern lasting from the early 1920s to 1950, being particularly cool around 1938 (Figure 11a). Although a secondary peak in tree establishment was suggested for the interval 1936 to 1956 (Figure 3), its existence is doubtful. Errors in age estimate (few samples reached the pith during this interval) and/or human-induced disturbances (surface fires and grazing) may have either obscured or prevented the expected tree establishment event from 1930 to 1950. In comparison with the massive tree recruitment from 1963 to 1977, tree establishment was low from 1957 to 1961 (Figure 3), and almost nil during the 1980s. Warm-dry summer conditions were the dominant climatic pattern during the two intervals with reduced tree recruitment. In contrast, cool-wet summers prevailed from 1963 to 1976 (Figure 11a). Consequently, low-frequency variations in summer climate, with decadal oscillatory modes, are a major climatic influence on tree establishment along the forest–steppe ecotone in northern Patagonia.

In contrast to recruitment, mortality of adult trees is clearly associated with high-frequency variations in climate. Although some mortality events, such as those recorded in the 1910s and late in the 1950s, were concurrent with long-lasting intervals of warm-dry climatic conditions, other episodes (such as 1942–43) occurred during a long-term wet interval (Figure 11a). In addition, no widespread mortality event has yet been recorded for the long-lasting climatic event during the 1980s characterized by warm and dry summer conditions. Major and secondary peaks of tree death correspond better with short-term variations in climate. Peaks in mortality are concurrent with extreme warm-dry climatic events such as those occurring in 1912–13, 1917, 1924, 1942–43, 1949, 1952–53, 1956–57, 1961–62, 1978, and 1983 (Figure 11b). Interestingly, tree mortality episodes during the last 90 years are concentrated into two periods of high climatic variability from 1910 to 1924 and from 1940 to 1962. Relatively stable warm-dry conditions prevailed during the 1980s, and although tree recruitment was severely limited there was no episode of massive tree mortality. A gradual transition from cool-wet to warm-dry conditions appears to be associated with less tree mortality.

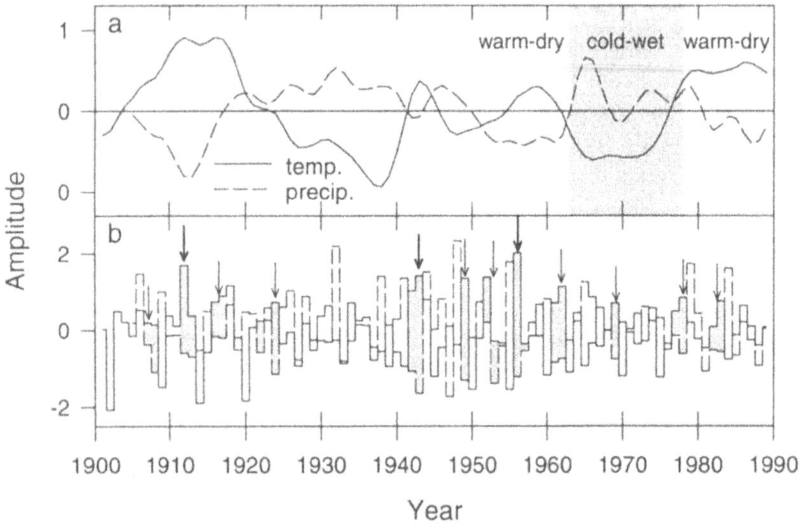

Fig. 11. Waveforms of summer (October to March) temperature and precipitation deviations for northern Patagonia during the period 1901–1987, estimated using singular spectral analysis. 11(a), Low-frequency variations in temperature and precipitation with oscillatory modes close to 40 years. 11(b), High-frequency variations in temperature and precipitation with oscillatory modes between 2 and 6 years. Major and secondary peaks of tree mortality recorded along the forest–steppe border are indicated by thick and thin arrows, respectively. Grey bars indicate warm-dry summers associated with mortality of adult trees.

7. Concluding remarks

Climate-induced vegetation change is often confused with disturbance-induced vegetation change, and to minimise this source of confusion, it is important to thoroughly consider the effects of both climate-related and non-climatic disturbances. Separating climate-induced from human-caused vegetation changes constitutes the major challenge to properly assess the influence of climatic variations on vegetation dynamics.

The difficulties of identifying the influences of climatic variations on vegetation dynamics also include temporal and spatial scaling problems. At a stand-scale, forests are often undergoing changes in response to non-climatic disturbances that may obscure any climatic influence on forest dynamics (Veblen and Stewart 1982; Ogden 1985). At regional scales the relative contribution of climate-related changes in vegetation are more manifest than at local-scales (Swetnam and Betancourt 1990; Ricklefs 1987). Because tree establishment may depend more on decadal-scale climatic variations whereas tree mortality is associated with interannual climatic variations, the link between mortality and climate may be more evident. It is also important to note that tree establishment and forest recovery after disturbances have similar time–frequency responses (decadal-scale). Consequently, to separate climatic-induced from disturbance-induced episodes of tree establishment is sometime problematic.

Synchronous episodes of tree establishment and mortality along a 450 km north–south transect in the woodland–steppe ecotone strongly imply a broad-scale climatic influence on *Austrocedrus* demography. Human influences through changes in fire regimes or grazing could account for synchrony at a multi-decadal time scale but are unlikely to explain synchrony at shorter time scales over such a large area. Climatically-driven changes in tree establishment and mortality appears to be a primary mechanism of vegetation change on regional and larger spatial scales in northern Patagonia.

Climatic variations of different frequencies have different causes and different physical and biological effects (McDowell et al. 1990). Tree demographic processes respond differentially to high *versus* low frequency variations in climate. For example, in our study seasonal-scale droughts increased mortality rates, but decadal or longer cool-wet periods are required for widespread increases in tree establishment. The distinction between the effects of high- *versus* low-frequency variations of climate on forest dynamics is crucial to properly predict forest response to climatic changes. Forests will respond differentially to climatic changes characterized by a gradual change in mean conditions or by an increase in interannual variability. Our observations suggest that future changes in species distributions associated with global warming, (e.g. tree invasion), may require favourable climatic conditions lasting a decade or longer. In contrast, forest mortality, and consequently forest contraction, may result from short-term extreme climatic events.

Acknowledgments. We thank Alberto Ripalta, Guillermo Debandi, Guillermo DiGiuseppe, and Alberto Iandolino for research assistance. The Nahuel Pan and Pampa del Toro sample sites were suggested by Omar Picco and Thomas Kitzberger, respectively. We are greatly indebted to Carlos Martín and Mónica Mermoz of Parques Nacionales, Argentina for facilitating the research. For permission to conduct this study on their properties, we thank the owners and managers of Estancia Collun-co, Cerro Los Pinos, Chacabuco, Cerro Leones, San Ramón, and Siete Condores. For commenting on this research we thank Thomas Kitzberger, Vera Markgraf, Henry Diaz and Julio Betancourt. Research was supported by NASA, the National Science Foundation of the U.S.A., the National Geographic Society, the University of Colorado, and the National Research

Council of Argentina (CONICET). Lamont-Doherty Earth Observatory Contribution Number 0000.

8. References

Aceituno P (1988) On the functioning of the Southern Oscillation in the South American sector, Part I: Surface climate. Monthly Weather Rev 116:505–524

Almeyda AE, Saez SF (1958) Recopilación de datos climáticos de Chile y mapas sinópticos respectivos. Ministerio de Agricultura, Santiago

Archer S, Schimel DS, Holland EA (1995) Mechanisms of shrubland expansion: land use, climate or CO^2? Climatic Change 29:91–99

Baker WL (1990) Climatic and hydrological effects on the regeneration of Populus angustifolia James, along the Animas River, Colorado. J Biogeogr 17:59–73

Barros V, Cordón V, Moyano C, Méndez R, Forquera J, Pizzio O (1983) Cartas de precipitación de la zona oeste de las provincias de Río Negro y Neuquén. Unpublished report, Facultad de Ciencias Agrarias, Universidad Nacional del Comahue, Cinco Saltos, Río Negro

Bartlein PJ, Prentice C, Webb T, III (1986) Climate response surface from pollen data for some eastern North America taxa. J Biogeogr 13:35–57

Betancourt JL, Pierson EA, Aasen Rylander K, Fairchild-Parks JA, Dean JS (1993) Influence of history and climate on New Mexico Piñon–Juniper woodlands. In: Aldon EF, Shaw DW (eds). Managing Piñon–Juniper ecosystems for sustainability and social needs. Gen. Tech. Rep. RM-236, Fort Collins, Colorado, USDA Forest Service, Rocky Mountain Forest & Range Exp. Station, pp 42–62

Bradshaw RHW, Zackrisson O (1990) A two thousand year history of a northern Swedish boreal forest stand. J Veg Sci 1:519–528

Brubaker LB (1986) Responses of tree populations to climate change. Vegetatio 67:119–130

Clark JS (1988) Effect of climate change on fire regimes in northwestern Minnesota. Nature (London) 334:233–235

Clark JS (1990) Fire and climate change during the last 750 yrs in northwest Minnesota. Ecol Mon 60:135–159

COHMAP Members (1988) Climatic changes of the last 18,000 years: Observations and model simulations. Science 241:1043–1052

Cooley WW, Lohnes PR (1971) Multivariate data analysis. Wiley, New York

Davis MB (1986) Climatic instability, time lags, and community disequilibrium. In: Diamond J, Case TJ (eds) Community Ecology. Harper and Row, New York, pp 269–284

Davis MB (1989a) Lags in vegetation response to greenhouse warming. Climatic Change 15:75–82

Davis MB (1989b) Insights from paleoecology on global change. Bull Ecol Soc Am 70:222–228

Donoso C (1981) Tipos forestales de los bosques nativos chilenos. Proyecto CONAF/FAO/PNUD. Documento de trabajo No. 38. Santiago.

Elliot KJ, Swank WT (1994) Impacts of drought on tree mortality and growth in a mixed hardword forest. J Veg Sci 5:229–236

Franklin JF, Shugart HH, Harmon ME (1987) Tree death as an ecologcal process. BioScience 38:550–556

Fritts HC (1976) Tree-rings and climate. Academic Press, London, England

Gallopin GC (1978) Estudio ecológico integrado de la Cuenca del Río Manso Superior (Río Negro, Argentina). I. Descripción general de la cuenca. Anales de Parques Nacionales, Argentina, 14:161–230

Gear AJ, Huntley B (1991) Rapid changes in the range limits of Scots Pine 4000 Years Ago. Science 251:544–547

Graham RL, Turner MG, Dale VH (1990) How increasing CO^2 and climate change affect forests. BioScience 40:575–587

Grimm EC (1983) Chronology and dynamics of vegetation change in the prairie and woodland region of southern Minnesota, USA. New Phytol 93:311–350

Havrylenko M, Rosso PHA, Fontenla SB (1989) *Austrocedrus chilensis*: contribución al estudio de su mortalidad en Argentina. Bosque 10:29–36

Holmes RL (1983) Computer-assisted quality control in tree-ring dating and measurement. Tree-Ring Bull 43:69–75

Johnson EA, Larsen CPS (1991) Climatically induced change in fire frequency in the southern Canadian Rockies. Ecology 72:194–201

Kitzberger T, Veblen TT, Villalba R (1995) Tectonic influences on tree growth in northern Patagonia, Argentina: the roles of substrate stability and climatic variation. Can J For Res 25:1684–1696

Kitzberger T, Veblen TT, Villalba R (1997) Climatic influences on fire regimes along a rainforest-to-xeric woodland gradient in northern Patagonia, Argentina. J Biogeogr 24:35–47

Little RL, Peterson DL, Conquest LL (1994) Regeneration of subalpine fir (Abies lasiocarpa) following fire: effects of climate and others factors. Can J For Res 24:934–944

Martonne E de (1926) Une nouvelle fonction climatologique: l'indice d'aridité. Météorologie 2:449–458

McDowell PF, Webb T, III, Bartlein PJ (1990) Long-term environmental change. In: Turner BL, Clark WC, Kates RW, Richards JF, Mathews JT, Meyer WB (eds) The Earth as transformed by human action. Global and regional changes in the biosphere over the past 300 years. Cambridge University Press, Cambridge, pp 143–162

Melillo JM, Prentice IC, Farquhar GD, Schulze E-D, Sala OE (1996) Terrestrial biotic responses to environmental change and feedbacks to climate. In: Houghton JT, Meira Filho LG, Callander BA, Harris N, Kattenberg A, Maskell K (eds), Climate Change 1995. The Science of Climate Change. Cambridge University Press, Cambridge, pp 447–481

Miller A (1976) The climate of Chile. In: Schwerdtfeger W (ed) World Survey of Climatology. Vol. 12, Chapter 3, Climates of Central and South America. Elsevier, Amsterdam.

Neilson RP (1986) High-resolution climatic analysis and southwest biogeography. Science 232:27–34

Ogden J (1985) An introduction to plant demography with special reference to New Zealand trees. N Z J Bot 23:751–772

Overpeck JT, Rind D, Goldberg R (1990) Climate-induced changes in forest disturbance and vegetation. Nature (London) 343:51–53

Pittock AB (1980) Patterns of climate variation in Argentina and Chile– I Precipitation, 1931–1960. Monthly Weather Rev 108:1347–1361.

Prentice IC (1986) Vegetation response to past climate variation. Vegetatio 67:131– 141

Prentice IC (1992) Climate and long-term vegetation dynamics. In: Glenn-Lewin DC, Peet RA, Veblen TT (eds) Plant Succession: Theory and Prediction. Chapman and Hall, London, pp 293– 339

Prohaska F (1976) The climate of Argentina, Paraguay and Uruguay. In: Schwerdtfeger W (ed) World Survey of Climatology. Vol. 12, Chapter 2: Climates of Central and South America. Elsevier, Amsterdam.

Ricklefs RE (1987) Community diversity: relative roles of local and regional processess. Science 235:167–171

Seibert P (1982) Carta de vegetación de la región de El Bolsón, Río Negro y su aplicación a la planificación de uso de la tierra. Doc Phytosoc 2:1–120

Shugart HH (1984) A theory of forest dynamics. Springer, New York.

Sirois L, Payette S (1991) Reduced postfire tree regeneration along a boreal forest–tundra transect in northern Quebec. Ecology 72:619–629

Solomon AM (1986) Transient response of forests to CO_2-induced climate change: simulation modeling experiments in eastern North America. Oecologia 68:567–579

Stokes MA, Smiley TL (1968) An Introduction to Tree-Ring Dating. University of Chicago Press, Chicago.

Swetnam TW (1993) Fire history and climate change in giant sequoia groves. Science 262:885–889

Swetnam TW, Betancourt JL (1990) Fire–Southern Oscillation relations in the southwestern United States. Science 249:1017–1020

Szeicz JM, MacDonald GM (1995) Recent white spruce dynamics at the subarctic alpine treeline of north-western Canada. J Ecol 83:873–885

Tortorelli LA (1956) Maderas y bosques argentinos. Editorial ACME, Buenos Aires.

Vautard R, Ghil, M (1989) Singular spectrum analysis in nonlinear dynamics, with applications to paleoclimatic time series. Physica D 35:395–424

Vautard R, Yiou P, Ghil M (1992) Singular-spectrum analysis: a toolkit for short noisy chaotics signals. Physica D 38:95–126

Veblen TT (1989) Tree regeneration responses to gaps along a transandean gradient. Ecology 70:543–545

Veblen TT, Lorenz DC (1987) Post-fire stand development of *Austrocedrus– Nothofagus* forests in Patagonia. Vegetatio 78:113–126

Veblen TT, Lorenz DC (1988) Recent vegetation changes along the forest/steppe ecotone in northern Patagonia. Ann Assoc Amer Geog 78:93–111

Veblen TT, Markgraf V (1988) Steppe expansion in Patagonia? Quatern Res 30:331–338

Veblen TT, Stewart GH (1982) On the conifer regeneration gap in New Zealand: the dynamics of *Libocedrus bidwillii* stands on South Island. J Ecol 70:413–436

Veblen TT, Kitzberger T, Lara A (1992a) Disturbance and forest dynamics along a transect from Andean rain forest to Patagonian shurbland. J Veg Sci 3:507–520

Veblen TT, Mermoz M, Martín C, Kitzberger T (1992b) Ecological impacts of introduced animals in Nahuel Huapi National Park, Argentina. Conservation Biol 6:71–83

Veblen TT, Burns BR, Kitzberger T, Lara A, Villalba R (1995) The ecology of the conifers of southern South America. In: Enright N, Hill R (eds.) Ecology of the Southern Conifers. Melbourne Univ. Press, Melbourne, pp 120–155

Villalba R (1990) Latitude of the surface high-pressure belt over western South America during the last 500 years as inferred from tree-ring analysis. Quaternary of South America and Antarctic Peninsula 7:273–303

Villalba R (1995) Climatic influences on forest dynamics along the forest–steppe ecotone in northern Patagonia. Ph.D. dissertation, Department of Geography, University of Colorado at Boulder

Villalba R, Veblen TT (1996) A tree-ring record of dry spring-wet summer events in the forest–steppe ecotone, northern Patagonia, Argentina. In: Dean JS, Meko DM, Swetnam TW (eds) Tree Rings, Environment and Humanity. Radiocarbon (1996) 107–116

Villalba R, Veblen TT (1997a) Spatial and temporal variation in Austrocedrus growth along the forest–steppe ecotone in northern Patagonia. Can J For Res 27:580–597

Villalba R, Veblen TT (1997b) Regional patterns of tree population age structure in northern Patagonia: climatic and disturbance influences. J Ecol 85:113–124

Villalba R, Veblen TT (1997c) Improving estimates of total ages based on increment core samples. Ecoscience 4:534–542

Webb T, III (1987) The appearance and disappearance of major vegetation assemblages long-term vegetational dynamics in eastern North America. Vegetatio 69:177–187

High-altitude forest sensitivity to global warming: results from long-term and short-term analyses in the Eastern Italian Alps*

Marco Carrer, Tommaso Anfodillo, Carlo Urbinati, Vinicio Carraro

Abstract. Dendroecological (long-term) analysis and ecophysiological (short-term) monitoring were used interactively to study the responses of tree-ring growth to climate in timberline mixed forests (consisting of *Larix decidua* Mill., *Picea abies* (L.) Karst. and *Pinus cembra* (L.)) in the Italian Eastern Alps (2000–2100 m a.s.l.).

Climate–growth linear response functions (LRF) revealed that warm temperatures in June and July have a positive effect on radial growth whereas precipitation during the vegetation period has no effect. Monitoring of the intra-annual radial growth dynamics using band dendrometers confirmed that the radial growth rate of the three species in June and July was greater when air temperatures were higher. Tree-ring formation lasted about 50–60 days (from mid-June to the beginning of August).

Tree responses to climatic factors were better defined by using Neural Network Response Functions (NLRF) and by assessing water relations. NLRF highlighted the presence of a clear temperature threshold effect, especially in June and July. Above 13–16 °C, the three species seemed unable to take full advantage of warm and sunny days.

We believe that these thresholds are mainly due to the very high stomatal sensitivity to vapour pressure deficit in the trees. High water availability during summer resulted in species becoming adapted to moist conditions, so a mild water deficit may have a major impact on trees, depending on the drought resistance strategies that have been developed. *Picea abies* and *Pinus cembra* are water-saving species whereas *Larix decidua* could be classified as water-using because of its high capacity for water uptake. These different responses should be taken into account when considering the effects of global change on timberline trees.

Keywords. alpine timberline, band dendrometer, dendroecology, climate warming effects, neural networks, response functions, water relations.

1. Introduction

The role of upper treeline ecotones as indicators of environmental change is widely recognized. One of the generally accepted consequences of the increase of greenhouse gases (CO_2, CH_4, N_20 etc.) in the atmosphere is a rise in the Earth's temperature. As temperature is one of the most important limiting factors at the alpine timberline (Baig and Tranquillini 1976; Tranquillini 1979; Havranek 1993; Holtmeier 1993), the effects of a warmer climate might be particularly pronounced in this ecotone.

The assessment of temperature effects is often a demanding task for various and often inter-correlated reasons such as: a) uncertainty over future scenarios of

* This research was carried out with the financial support of the Ministry of University and Scientific and Technological Research (MURST) funds at 40%. The authors wish to thank the Regole of Cortina d'Ampezzo for having allowed the study in their forests. Special thanks to Fausto Fontanella, Roberto Menardi and Giuseppe Sala of the Centre of Alpine Environment for their valuable technical support. We also thank the Alberti and Alverà family, owners of the 5 Torri and Palmieri Refuges, for the kind hospitality offered throughout the work.

air temperature changes, particularly in mountainous regions (Wanner and Beniston 1995); b) spatio-temporal heterogeneity of the changes; c) differential importance of temperature values when considered as annual, seasonal or daily means and extremes (Körner 1995); d) species-specific responses; e) asynchronous reactions of trees to forcing factors (Solomon and Cramer 1993); f) effective variations in plant temperature (degree of aerodynamic coupling between the plant layer and free atmosphere); g) the number of processes which are temperature-mediated, such as frost tolerance, mineral nutrient supply, photosynthetic rate, rate of cell division, and rate of mitochondrial respiration (Körner and Larcher 1988); h) variation of snow cover patterns and soil temperature; i) noise introduced by human disturbance.

Climate warming has been thought to be the cause of an upwards altitudinal shift in the distributions of alpine plants (Grabherr et al. 1994), for the displacement of the arctic treeline and for an increase in stem growth in the krummholz zone (Lescop-Sinclair and Payette 1995). However, no effects of the recent higher summer temperatures on the altitudinal range of alpine *Pinus sylvestris* and *Pinus cembra* have been recorded (Hättenschwiler and Körner 1995).

In view of such complexity, this work was implemented to better understand the relationships between temperature and the radial growth of three species of high-altitude trees (*Larix decidua* Mill., *Picea abies* (L.) Karst. and *Pinus cembra* L.) in the Italian Eastern Alps.

Two approaches were undertaken, which integrated methods and disciplines operating at different spatio-temporal scales. We carried out both long-term and short-term analyses, the former based on dendroecological studies and the latter on ecophysiological techiques. We were able, by retrospective analysis at the forest community level, to identify the type of environmental factors producing the most significant variations in growth response, their temporal distribution (i.e. in which season these effects are enhanced) and, in some cases, the operating thresholds for the factors. Then, after three years of *in situ* monitoring and measurement of several physiological parameters (hourly radial growth dynamics, xylem water potential, transpiration, etc.), we were able to increase the resolution of the time-scale and better explain the results obtained from the long-term analyses.

2. Materials and Methods

2.1 Long- term analysis

2.1.1 Sites

The study areas are located near Cortina d'Ampezzo (Dolomite Region, Eastern Italian Alps: 46°27' N and 12° 08' E) on the Croda da Lago and Becco di Mezzodì Mountains, with an E–NE aspect and 30% slope (Fig. 1). The forest stands consist of a timberline *Larici–cembretum* (2000–2100 m a.s.l.) growing on dolomite and limestone bedrock and rendzina brown soils (Rendzic Leptosols).

Their composition is 72% *Larix decidua* Mill., 23% *Pinus cembra* L. and 5% *Picea abies* (L.) Karst.

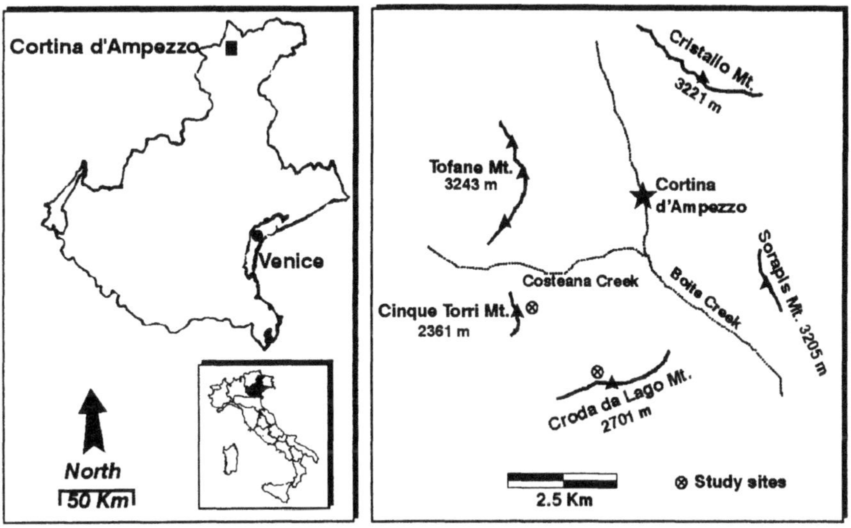

Fig. 1. Location of the research area (see text for the geographic coordinates).

2.1.2 Climatic data

Total monthly precipitation and mean maximum and minimum monthly temperature data were collected for the period 1925–1994 at the meteorological station in Cortina d'Ampezzo (1275 m a.s.l.), the nearest station to the sampling sites. This station has the longest and most continuous and homogeneous climate records in this sector of the Dolomites.

2.1.3 Sampling procedure and chronology development

Sampling followed the methods described by Schweingruber et al. (1989). Two cores were extracted using Pressler increment borers from each of 112 living trees (Table 1) at breast height (about 130 cm), one each from the uphill and cross-slope sides (at least 90° apart from each other). Cores were prepared following standard procedures (Stokes and Smiley 1968; Swetnam et al. 1985) and ring widths were measured to the nearest 0.01 mm using the Aniol CCTRMD measuring system. Individual ring width series were checked, corrected and dated using the computer program CATRAS (Aniol 1983, 1987).

The long-term growth trend present in the ring-width series, which is mainly attributable to the trees' ageing trend, disturbance regimes, etc., was removed by standardization (Fritts 1976; Cook et al. 1989). As this procedure is one of the least objective in dendroecological analyses, two different standardization methods were performed so as to reduce the level of subjectivity: ARMA modelling and Gaussian low-pass filtering.

Sixteen different ARMA (p,q) models (from ARMA (0,0), corresponding to the raw data, to ARMA (3,3)) and sixteen Gaussian filters (with a window ranging from 0 (i.e. raw data) to 30 years with a 2-yr step between every filter), were chosen from among the most commonly used and were then computed.

Averaging the standardized tree-ring series with the arithmetic mean, 31 different chronologies (15 from the ARMA modeling, 15 from the Gaussian low-pass filtering and 1 corresponding to the raw data) were obtained for each species.

Species	N	Period	MA	mA	MRW	SD	MS	R1
Larix decidua	56	1515–1994	277	200	73	29.5	.32	.57
Pinus cembra	28	1494–1994	246	96	85	26.0	.15	.87
Picea abies	28	1594–1994	214	109	92	27.9	.17	.79

N , number of trees; *MA*, mean age of the samples; *mA*, minimum age; *MRW*, mean ring width in 1/100 mm; *SD*, standard deviation; *MS*, mean sensitivity; *R1*, first order autocorrelation.

Table 1. Descriptive statistics for raw data chronologies

We carried out a sensitivity analysis to test the results of the bootstrapped response functions computed for all 31 chronologies (Fig. 2) (Carrer 1997) and to choose the most significant standardization methods.

2.1.4 Analyses of long-term climate–growth relationships

We first computed a linear response function analysis (LRF) (Fritts 1976; Blasing et al. 1984; Guiot 1990) to quantify the long-term influence of climate factors on tree growth at the timberline. In this approach the ring-width indices are compared to precipitation and temperature records. The predictors are monthly maximum or minimum temperatures and total monthly precipitation, according to the biological year generally adopted for southern Europe and the Mediterranean area, i.e. from October of the previous growing season (t-1) to September of the year in which the ring was formed (current growing season, t).

This type of response function has a basic weakness in that it can only assess linear relationships between ring-widths and climate, whereas it is likely that biological processes do not often follow this pattern. To overcome this bias, we used non-linear response functions (NLRF) based on Artificial Neural Networks (ANN) (Guiot et al. 1995; Keller et al. in press; Carrer 1997).

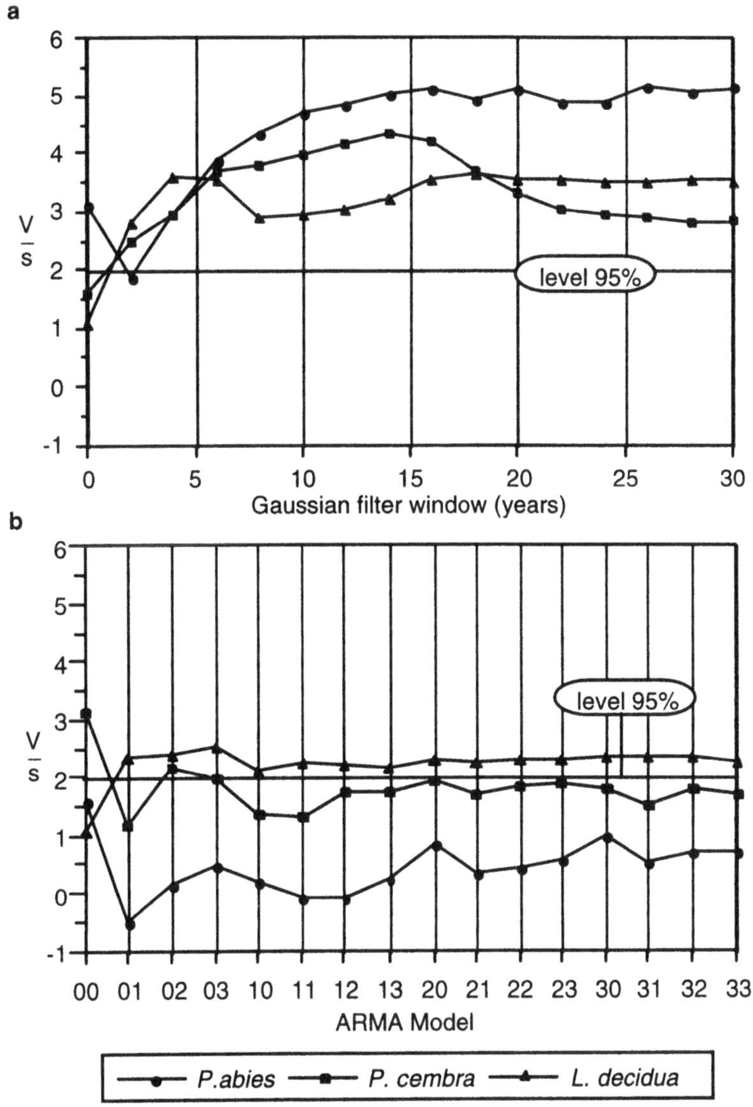

Fig. 2. Global significance of LRFs computed with different types of chronologies standardized with Gaussian filters (a) and ARMA models (b). *V*, verification correlation mean; *s*, standard deviation (50 replications). Horizontal line defines 95% significance level.

2.1.5 Artificial Neural Networks

An Artificial Neural Network (ANN) is a non-algorithmic but adaptive information processing system based on parallel calculation methods (Caudill and Butler 1992). Increasing numbers of applications are concerned with ANN. In computer science they are used to expand the properties of non-symbolic information processing and the automatic learning system in general. In several engineering fields they are used to exploit signal processing, automatic control and robotics (Kröse and Van Der Smagt 1993; Kappen 1996). In statistics, applications include the creation of flexible, non-linear regressed and classification models (Sarle 1994), and in neuro-physiology and cognitive science they are used to describe and explore the functions of the brain (Sarle 1997).

An ANN consists of a number of very simple processing units (neurodes), organized in layers and interconnected by communication channels (connections). Each unit receives several input signals but sends only a single output signal, consisting of a synthesis of the previous ones, to the following layers. Every input and output signal changes during its run through the network owing to the weight of the connections that control each input and to the neurode's activation function, which controls each unit output. The sigmoid activation function is one of the most widely used, and gives the network its non-linear behaviour.

The most important feature of an ANN is its training and learning potential, which is the mechanism by which the net adapts itself to its environment. The result of this adaptation process is that the ANN obtains a representation of this environment. This plasticity of the system is encoded within the power to modify the connection weights and the activation functions in order to minimize the estimation error, i.e. the differences between the output and the target function (Smith 1993; Heskes and Kappen 1991). Back-propagation of the error is one of the most widely used tools to assess ANN learning. Using this method the hidden nodes are able to determine how to change their weights on receiving (back-propagating) error information from each of the output nodes.

The ANN architecture involved in statistical analysis consists of an input layer where the independent variables enter the nets, and an output layer where the results are compared with the target variable and one or more hidden layers. The numbers of input and output units are determined by the number of independent and dependent variables, but the number of hidden layers and their respective units must be decided by the user. A network with a single hidden layer can approximate any target function, whereas the number of hidden nodes determines the complexity of the output function. An ANN can approximate a target function of any complexity if it has sufficient hidden nodes (Smith 1993).

The great power inside the ANN must be used very carefully as it conceals over-fitting, the most serious potential problem with the method. Given a limited sample size and sufficient noise in the data, a net with too many hidden units can over-fit. This means that it could end up modelling the stochastic structure of the noise in the sample as well as the inherent structure of the target function (Smith 1993; Sarle 1995). There are several methods to avoid over-fitting, with the best being the simplest, namely limiting the number of hidden units and stopping training when over-fitting begins (Smith 1993; Sarle 1997).

We first performed the LRF with three different sets of monthly regressors (Table 2): precipitation and mean maximum temperature (PTMAX), precipitation and mean minimum temperature (PTMIN) and all the previous variables together (PTMAXTMIN). Because of the high significance level reached with the PTMAX combination, in this first phase of the analysis we have limited the interpretation of the results to this set of variables.

Regressors	sV *Larix decidua*	R^2 *Larix decidua*	sV *Pinus cembra*	R^2 *Pinus cembra*	sV *Picea abies*	R^2 *Picea abies*	NR
PTMAX	3.457	71.4	5.415	78.5	4.333	77.1	24
PTMIN	2.279	61.5	2.83	68.1	2.057	59.1	24
PTMAXTMIN	3.382	78.1	4.367	81.9	5.214	81.9	36

Table 2. Results of LRFs computed with different sets of monthly variables

PTMAX, set of regressors including monthly precipitation and mean maximum temperature; *PTMIN*, set of regressors including monthly precipitation and mean minimum temperature; *PTMAXTMIN*, set of regressors including monthly precipitation and mean maximum and minimum temperatures; *sV*, significance of the correlation coefficient for verification of the LRF model (values • 1.96, 2.58, 3.29 are significant respectively at 95.0, 99.0 and 99.9% level); R^2, % of variance explained by climate; *NR*, number of regressors

Both linear and non-linear response functions were tested for significance and stability using a bootstrap procedure[2] (Efron 1979; Guiot 1990).

2.2 Short-term analysis

2.2.1 Site

Experiments were conducted at a timberline ecotone at 2080 m a.s.l. on the Cinque Torri Mountain (less than 4 km from the long-term analysis stands). The site has a southerly aspect and 30% slope. The timberline here consists of relatively young, mixed stands of *Larix decidua*, *Pinus cembra* and *Picea abies*, which are invading the edges of recently abandoned pasture lands (Del Favero et al. 1985).

Six similar trees were selected (two of each species) with diameter, age and height ranging respectively between 24.2 and 33.4 cm, 35 and 58 years, and 7.2 and 11.1 m.

2.2.2 Intra-annual growth dynamics

Changes in circumference were measured by band dendrometers. The sensing element was a variable resistor which, by applying a 2 V excitation voltage, gave an output signal ranging from 0 to 2000 mV. The precision resistor and datalogger

---

2 All the dendrochronological and dendroclimatic analyses were made using the software package PPPBASE developed by Joël Guiot and Claude Goeury at the IMEP-CNRS, Marseille.

(see below) enabled a 0.01 mm variation in circumference to be detected. The resistor support device was fixed to the stem by two 7 cm long screws. Friction between the bark and the stainless band was reduced by a thin teflon foil placed between the bark and the band.

2.2.3 Water relations

Xylem water potential was measured weekly with a pressure chamber on one-year-old shoots of *Larix decidua* and *Picea abies* and on one-year-old needle bundles of *Pinus cembra*. Four samples were collected at 2 m height (two on the south- and two on the north-facing side of the crown) on each tree from just before dawn until sunset at two-hour intervals. Data were then averaged for each species, as no statistically significant differences were recorded between the two trees or crown aspects. The experiment lasted from 29 May to 6 October 1996.

Xylem sap flux density (Fd, $dm^3 \ dm^{-2} \ h^{-1}$) was measured in each tree using 2 cm long continuously heated sap flowmeters (Granier 1985). Sensors were inserted into the xylem (NW aspect) at 1.5–2 m height on the stem

All the different sensors and the standard meteorological factors were monitored every minute, and averaged and stored every 15 minutes with a data logger (Campbell Ltd CR10) connected to two multiplexers (Campbell Ltd, AM32). Power was provided by two solar panels (Helios technology, 50 W) and batteries (140 Ah). Technical and logistic support was ensured by the Centre of Alpine Environment of the University of Padua, located 20 km away in S. Vito di Cadore.

3. Results and discussion

3.1 Long-term analysis and intra-annual growth dynamics

Fig. 2 clearly shows that the LRF significance is always lower with ARMA modelling than with Gaussian filtering. This is particularly evident for the evergreen species (*Picea abies* and *Pinus cembra*), where no ARMA model (except (0.2) and (0.3) for *P. cembra*) led to a significant LRF.

This result may be attributable to an excessive effect of the ARMA modelling in removing the long-term variance of the chronologies and may also imply a cut-off of some short-term climatic signal. This is especially evident for the evergreen species, where a climatic input could have inertial effects for several years. The ARMA-modelled LRFs of *Larix decidua* (the species with the highest mean sensitivity and the lowest first order autocorrelation, because of its deciduous behaviour) are, in contrast, all significant, whereas the filtered ones have the lowest lag. The LRF profiles are shown in Fig. 3.

We used the standardization option for each species that yielded the best results in the previous sensitivity analysis, namely the 18-yr, 16-yr and 12-yr Gaussian filters respectively for *Larix decidua*, *Picea abies* and *Pinus cembra*.

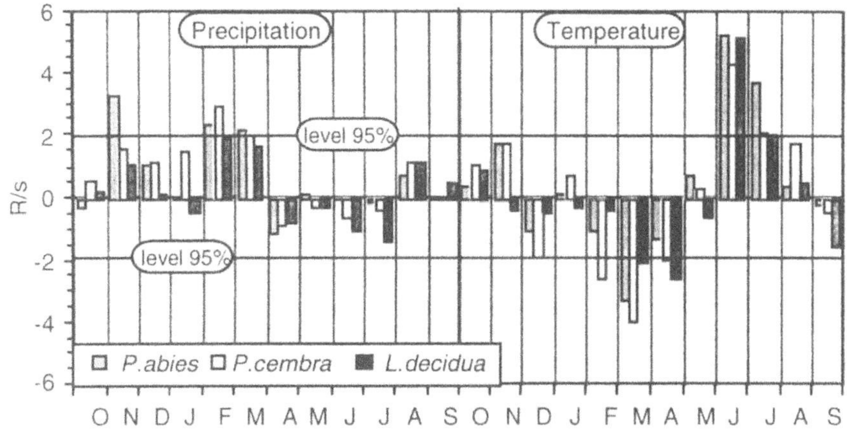

Fig. 3. Linear response functions computed for the three species (PTMAX). The ring width–climate relationships are significant at the 95% level when ratio R/s = 1.96. *R*, regression coefficient mean; *s*, standard deviation (for 300 simulations).

The high significance of the LRFs and the high variance explained by climate (Table 2) confirmed the key role played by climate in determining interannual growth dynamics. In general, precipitation had less influence on radial growth than temperatures, as would be expected at these high altitudes. Late winter precipitation (February and March) had a positive influence, although with slight differences between species. This might be due to the insulating effect of snow cover in preventing low soil temperatures (which would lead to a longer period of root inactivity), rather than in increasing water availability, which is rarely a limiting factor in this vegetation zone (Tranquillini, 1979). This is confirmed by the precipitation regime.

Temperatures seemed to have a strong impact in two different periods, namely at the end of the winter dormancy of the trees, from February until April, and during the two months of major tree-ring formation (June and July). Negative effects of warmer temperatures in the pre-vegetative period in high-altitude environments have also been reported by other authors (Eckstein and Aniol 1981; D'Arrigo et al. 1992; Colenutt and Luckman 1996). The general explanation is that warm spring temperatures may promote photosynthesis and respiration while roots are still too cool to be efficient in water uptake, inducing stress dehydration and loss of carbon reserves. Frost drought has been often reported as affecting timberline evergreen species such as *Picea abies* and *Pinus cembra* (Tranquillini 1979), while in *Larix decidua* a significant respiration rate has been measured from stems and branches in warm late winters (Havranek 1985).

The positive responses in June and July confirm the importance of maximum temperatures for timberline species and indicate the shortness of the radial growth period. June temperature had the highest weight, highlighting the strong opportunistic behaviour and the special adaptation of all three species to the onset of seasonal growth. Körner (1995) and Colenutt and Luckman (1996) concluded that the first growth phase may be the most significant in determining the total ring width.

The LRF results were consistent with the records of intra-annual radial growth dynamics. Further, the intra-annual growth data allowed us to assess the relationships among climatic factors and tree responses better. Cumulated stem radial growth (expressed as circumference variations) of *Larix decidua* and *Picea abies* versus daily air temperature for 1995 to 1997 growing seasons were compared (Fig. 4). The tree-ring formation was synchronous in both species and trees. Its onset appeared to be strongly affected by temperature. High June temperatures (as in 1996) led to an acceleration of the process while a cold June (1995) caused a growth delay of 20 days as compared to the following year. As the recorded period for xylem formation is about 60 days, a 20-day lag is important.

In addition to its effect on the timing of the onset of stem growth, temperature was an essential factor determining the total annual growth. In July 1995 mean air temperature was significantly higher than in the two following years. Trees reacted by sharply increasing their radial growth rate, which resulted in the widest tree ring in the three years. On the other hand, the smallest tree ring corresponded to the coldest July temperatures (1997).

Seasonal growth appears to be stimulated by high temperatures only if they occur in the first or middle part of the growing season, while a favourable phase late in the season does not produce any positive effect because of the more conservative behaviour of the species at the end of the vegetation period (Körner 1995). During 1997 we recorded a similar pattern, with an increasing trend in mean daily temperature during summer (the highest yearly values occurred in the first week of September) not being associated with an increase in radial growth (Fig. 4).

Some conclusions can be inferred from these two analyses (LRF and intra-annual growth dynamics). Temperature plays a basic role in determining the long- and short-term radial growth patterns. Trees seem very quick to take advantage of favourable conditions occurring during the early and middle phases of the growing season and, in these cases, the higher the temperatures the wider the ring, regardless of the length of the period of radial growth.

From these first results an increase in radial growth could be envisaged within a global warming scenario for the timberline species that were studied. However, physiological responses are usually not linear and show some saturation or adaptation effects, especially when the environmental conditions are changing. Neural Network Response Functions (NLRF) were therefore of great help in assessing non-linear climate–growth relationships.

A more reliable model of NLRF can be obtained by introducing a reduced number of independent variables, so we fixed the set of climatic regressors to seven. So as to better compare the results from the different response function approaches, a further LRF was computed with this new independent data set (Fig. 5).

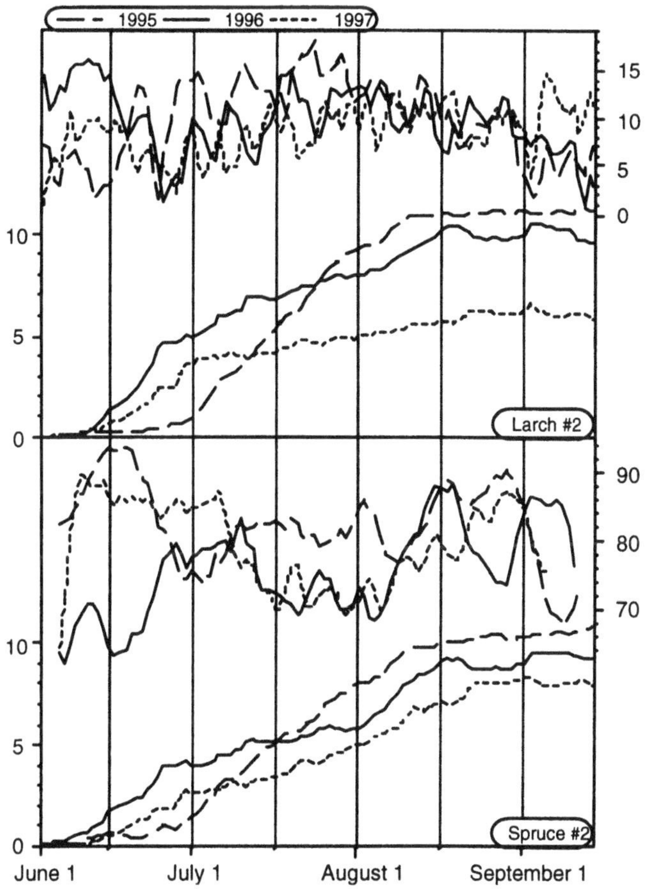

Fig. 4. Intra-annual circumference growth patterns for *Larix decidua* #2 and *Picea abies* #2, mean daily air temperature and air humidity in the three years of observations. Air humidity values were smoothed with a 10-days running mean.

The more realistic representation of biological processes reached by NLRF was confirmed by its higher level of explained variance and significance in comparison to the LRF (Table 3). The results are particularly interesting, especially from an ecological point of view, when the analysis is limited to the months of radial growth (June and July). It is possible to distinguish clear differences in the climatic responses of the three species which were not detectable with the LRF (Figs. 5–6).

Fig. 6 highlights the presence of a threshold effect in all but one case (*Pinus cembra* in June). For example, the radial growth of *Picea abies* would benefit from increased temperatures in June up to a threshold of 13°C. Any increase above this value is not predicted to cause further increases in ring width. *Larix decidua* is more responsive to high temperatures, with its equivalent threshold being around 16°C.

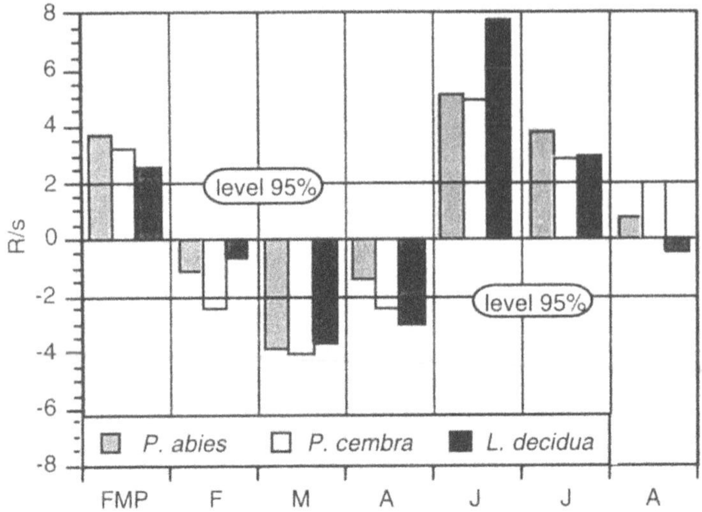

Fig. 5. Linear response functions computed with the same number and type of variables used with NLRF. *FMP*, sum of February and March precipitation.

These results enable further interpretation of the growth dynamics of the trees and suggest that additional factors besides temperature could affect radial growth at the timberline. In order to provide a better basis for the ecological interpretation of the NLRF results and to understand which could be the other factors involved in the timberline growth–climate relationships, we carried out a short-term analysis using an ecophysiological approach.

Type	sV *Larix decidua*	R^2 *Larix decidua*	sV *Pinus cembra*	R^2 *Pinus cembra*	sV *Picea abies*	R^2 *Picea abies*
LRF	7.15	57.3	7.51	61.0	7.89	64.9
NLRF	27.4	63.4	25.0	64.2	27.2	78.3

sV, significance of the correlation coefficient for verification of the model (values ≥ 1.96, 2.58, 3.29 are significant respectively at 95.0, 99.0 and 99.9% level); R^2, % of variance explained by climate; *NR*, number of regressors

Table 3. Comparison between LRFs and NLRFs results

3.2 Short-term analysis

3.2.1 Water relations

In general, in this alpine area, summer is the wettest season (mean precipitation of the last century about 470 mm). In 1996, at the end of July, there was an unusual dry period of 10 days with less than 0.4 mm/day rain. We will call this a "mild water deficit period" (MWDP), as only four similar periods were recorded between 1960 and 1990.

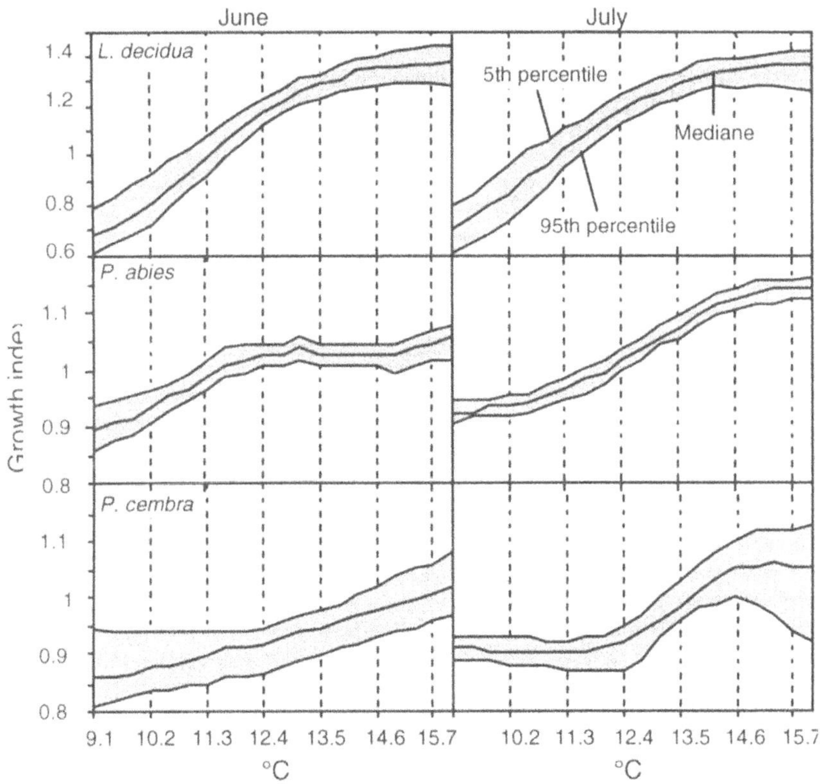

Fig. 6. Non-linear response function results for June and July mean maximum temperature. Grey bands show 90% confidence limit (for 50 replications); *I*, ring width indices.

Variations in minimum water potential (ψm) throughout the season highlighted the different behaviour of the evergreen and deciduous species. In *Picea abies* and *Pinus cembra,* ψm related well to precipitation variations (Fig. 7). The lowest values (-1.18 MPa; and -1.49 MPa, respectively) occurred at the end of the MWDP, the highest (-0.52 MPa and -0.60 MPa) on 22 June (day 174). *P. cembra* had lower ψm values, probably because of the sampling method (needles instead of twigs). In both species, the seasonal minimum ψm appeared well above the threshold of incipient plasmolysis, which is about -2.8 MPa in *P. abies* (Anfodillo & Casarin, 1993) and about -2.3 MPa in *P. cembra* (data not yet published). In addition, the ψm of *P. abies* was higher at the timberline than in lower altitude stands (Lu et al. 1995).

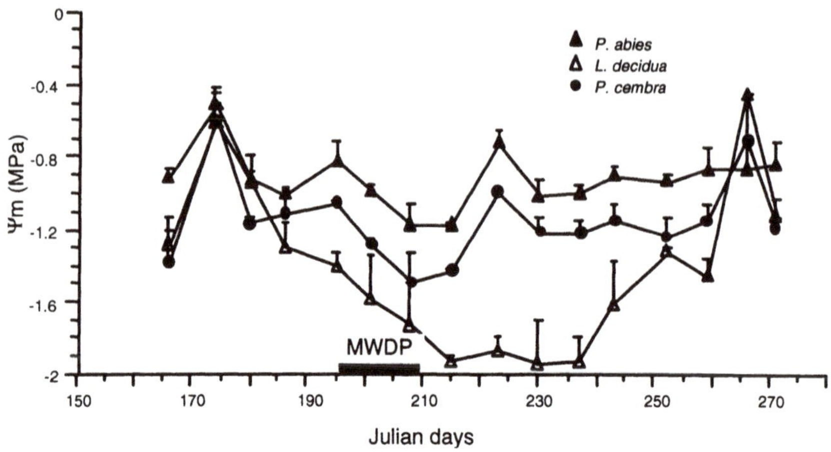

Fig. 7. Seasonal variations of minimum xylem water potential in investigated trees. Error bars = ±1SD

In *Larix decidua* ψm decreased regularly from June (-0.7 MPa) to the end of July (day 215), stabilising at about -1.9 MPa until the end of August (day 237). Afterwards, ψm again increased, reaching the same values as at the beginning of the season. This pattern (which was not linked to precipitation events) seems to be due mainly to an osmotic adjustment, as osmotic potential at full turgor varied throughout the season from about -0.5 to -1.5 MPa (data not yet published, derived from analysis of PV curves).

The variations of ψm may suggest that *Picea abies* and *Pinus cembra* suffered a higher impact of the MWSP than *Larix decidua* did and that they developed a marked water-saving strategy (also more accentuated than in lowland ecotypes), leading to relatively low variations in ψm. *L. decidua*, on the other hand, by dropping the water potential through osmoregulation, was able to maintain turgor (i.e. high stomatal conductance) at lower ψ (Morgan 1984), as well as being able to increase its water uptake capacity from the soil. This is consistent with the deciduous strategy, as the highest possible photosynthetic capacity has to be maintained during the short growing season despite any variation in water availability. In addition, *L. decidua* develops a deep root system which gives it access to water sources in the deepest and wettest soil layers, as demonstrated using hydrogen stable isotope analysis (Valentini et al. 1994).

Seasonal variations in the Fd daily sum in each species and the average diurnal vapour pressure deficit (VPD) (from 0600 to 2000) are shown in Fig. 8. *Larix decidua* had the highest values of Fd (mean maximum range 3–3.5 $dm^3\ dm^{-2}\ h^{-1}$) while in *Picea abies* and *Pinus cembra*, Fd maxima were respectively 0.8–1.0 and 0.6–0.8 $dm^3\ dm^{-2}\ h^{-1}$, suggesting a more efficient control of water loss. The lowest Fd values were found in *P. cembra*, which is the most drought-resistant species (with a water-saving strategy) at the alpine treeline (Tranquillini 1979).

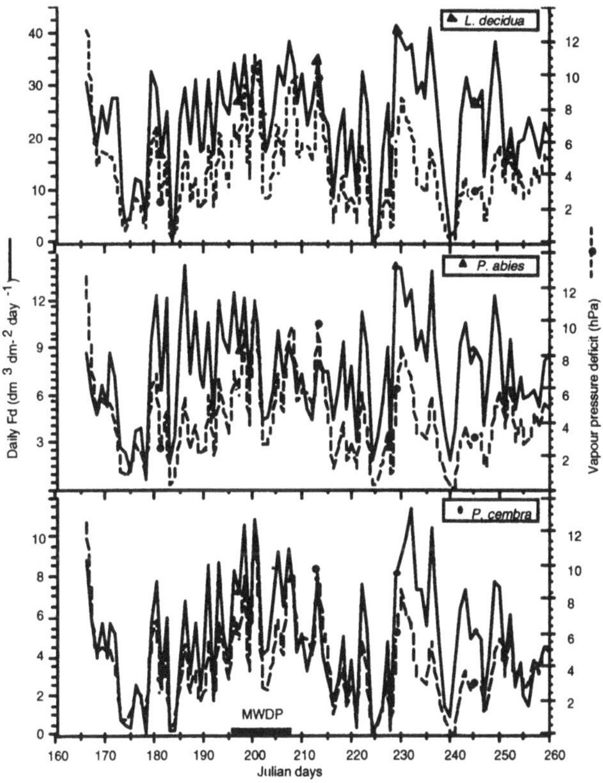

Fig. 8. Seasonal variations of Fd for the three measured species and VPD. VPD curve has been drawn in each plot for better visual appreciation.

The effect of MWDP on Fd differed between the three species (Fig. 9). Comparing the cumulated Fd for seven days with high water availability against the Fd for seven days during MWDP, no clear variation was observed in *Larix decidua*. In contrast, Fd in *Picea abies* and *Pinus cembra* decreased by 25 and 32%, respectively, suggesting that these species are unable to maintain a sufficient leaf water content after a few dry days. This also suggests that despite the elevated summer precipitation, soils at high altitudes may also become physiologically dry because of their small water storage capacity.

All species showed Fd variations coupled with VPD but only below a threshold of 7–8 hPa. In days with high VPD (15–20 hPa), a small increase in Fd was recorded as compared to low VPD days (5–8 hPa). Figure 10 shows the relationship between daily maximum Fd and VPD, when the Fd maximum occurred. Regardless of species and tree, Fd increases with VPD from 0 to 4–5 hPa, with a stabilizing trend. Over 8 hPa there is no further increase, suggesting that strong stomatal control may occur at very low VPD values. This is consistent with other studies that have shown that high-altitude trees have a higher stomatal sensitivity to drought than ones from low altitudes (Barton and Teeri 1993; Lu et al. 1995).

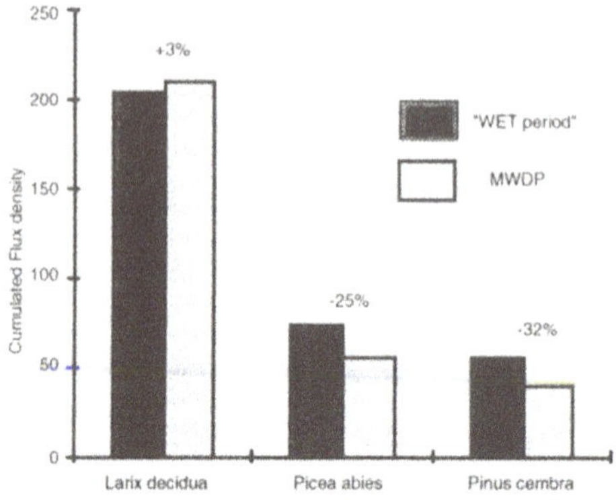

Fig. 9. Effect of MWDP on cumulated transpiration in each species over a period of seven dry days compared to moist days.

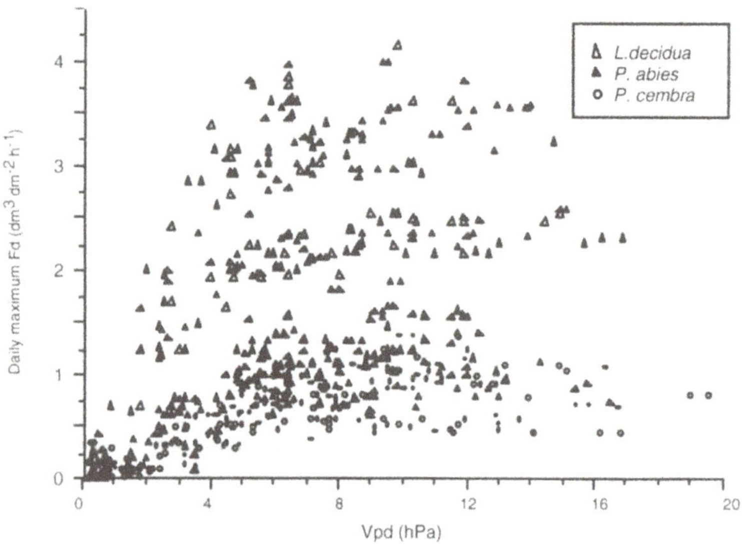

Fig. 10. Relation between maximum daily Fd and VPD. VPD values correspond to maximum Fd measurements.

The effect of MWDP on the growth dynamics of *Larix decidua* and *Picea abies* is shown in Fig. 4. On dry days, *L. decidua* seemed less affected than *P. abies* (even though the behaviour of the two *L. decidua* trees differed) which is apparent from the more obvious reduction in the variation in circumference. We do not know at present whether this effect is due to stem shrinkage caused by dehydration or to a true decrease of radial growth. Further analysis will be undertaken in this direction.

High temperatures seem to have a highly positive influence in stimulating growth processes, especially when associated with adequate water availability or low VPD (high relative humidity). The 1995 season reveals this well. It was the warmest of the three years but also had the highest relative humidity (Fig. 4). Bright days with high VPD and moderate water deficit could, in contrast, negatively affect growth processes.

4. Conclusions

As expected, the LRFs confirmed that temperature rather than precipitation is the main control on the radial growth of trees at the upper timberline. Summer (June–July) temperatures have the greatest effects on growth dynamics. These results are supported by data from intra-annual band dendrometers, which recorded that tree-ring formation lasted only 50–60 days (from mid-June to the beginning of August). This attests to the opportunistic behaviour and good adaptation of the tree species present in such an environment.

The better definition of tree responses to climatic factors obtained through the NLRF indicated the presence of a clear temperature threshold effect, especially in June and July. Above 13–16 °C, the three species seem unable to take full advantage of warm and sunny days.

The different responses recorded in *Picea abies* and *Pinus cembra* in comparison to *Larix decidua* suggest that in the case of an increase in air temperature (and VPD), the latter could be favoured in competition against the former. However, predictions of a change in stand composition must be carefully evaluated in view of the possible future scenarios of precipitation and cloudiness associated with climate warming, as well as the possible effects of higher CO_2 concentration on tree growth (Nicolussi et al. 1995). If precipitation rate, regimes and cloudiness change towards more xeric conditions, as has been hypothesised (e.g. Wanner and Beniston 1995), a long-term change in species composition in timberline would be possible.

Finally, the responses of timberline trees to climate are highly variable because of the different operating thresholds and the high adaptation of species to this environment. Therefore, concepts such as "a rise in temperature means an increase in (radial) growth" should be considered with care when developing models of future scenarios in a global warming environment.

5. References

Anfodillo T, Casarin A (1993) Variazioni stagionali nelle relazioni idriche di rametti di abete rosso lungo un gradiente altitudinale. In Anfodillo T, Urbinati C (eds) Ecologia delle foreste di alta quota. Proc XXX Corso di Cultura in Ecologia, University of Padova, pp 143–171

Aniol RW (1983) Tree-ring analysis using CATRAS. Dendrochronologia 1:45–53

Aniol RW (1987) A new device for computer measurament of tree-ring widths. Dendrochronologia 4:135–141

Baig MN, Tranquillini W (1976) Studies on upper timberline: morphology and anatomy of Norway spruce (*Picea abies*) and stone pine (*Pinus cembra*) needles from various habitat conditions. Can J Bot 54:1622–1632

Barton AM, Teeri JA (1993) The ecology of elevational position in plants: drought resistance in five montane pine species in Southern Arizona. Am J Bot 80:15–25

Blasing TJ, Solomon AM and Duvick DN (1984) Response functions revisited. Tree-Ring Bulletin 44:1–15

Caudill M, Butler C (1992) Understanding Neural Networks: Computer Explorations. The MIT Press Cambridge, Massachusetts, Vol. 1

Carrer M (1997) Analisi dendroecologica e della struttura spaziale in una cenosi del limite superiore nelle Alpi orientali. PhD Thesis, Univerity of Padova

Colenutt ME, Luckman BH (1996) Dendroclimatic characteristics of alpine larch (*Larix lyallii* Parl.) at treeline sites in Western Canada. In Dean JS Meko DM, Swetnam TW (eds) Tree Rings, Environment and Humanity, Radiocarbon pp.143–154

Cook E, Briffa K, Shiyatov S, Mazepa V (1990) Tree-ring standardisation and growth-trend estimation. In Cook ER and Kairiukstis LA (eds) Methods of dendrochronology: application in the enviromental science. Kluwer Academic Publishers, Dordrecht, pp. 104–123

D'Arrigo RD, Jacoby GC, Free RM (1992) Tree-ring width and maximum latewood density at the North American tree line: parameters of climate change. Can J For Res 22:1290–1296.

Del Favero R, De Mas G, Lasen C, Paiero P (1985) Il pino cembro nel Veneto. Regione del Veneto, Dip Foreste

Eckstein D, Aniol RW (1981) Dendroclimatological reconstruction of the summer temperatures for an alpine region. Mitt Forst Bundesversuchsanst.Wien 142:391–398

Efron B (1979) Bootstrap methods: another look at jacknife. The annals of statistics 7:1–26

Fritts HC (1976) Tree rings and climate, Academic Press, New York

Grabherr G, Gottfried M, Pauli H (1994) Climate effects on mountain plants. Nature (London) 369:448

Granier A (1985) Une nouvelle méthode pour la mesure de flux de sève brute dans le tronc des arbres. Ann Sci For 42:193–200

Guiot J (1990) Methods of calibration. In Cook ER and Kairiukstis LA (eds) Methods of dendrochronology: application in the enviromental science. Kluwer Academic Publishers, Dordrecht, pp 165–177

Guiot J, Keller T, Tessier L (1996) Relational databases in dendroclimatology and new non-linear methods to analyse the tree response to climate and pollution. Proc Inter Workshop on Asian and Pacific Dendrochronology, FFPRI Scientific Meeting Report 1:17–23

Hättenschwiler S, Körner C (1995) Responses to recent climate of *Pinus sylvestris* and *Pinus cembra* within their montane transition zone in the Swiss Alps. J Veg Sci 6:375–368

Havranek WM (1985) Gas exchange and dry matter allocation in larch at the alpine timberline on Mount Patscherkofel. In Turner H, Tranquillini W (eds) Establishment and tending of Subalpine Forest: Research and Management. Proc 3rd IUFRO Workshop P 1.07-00, 1984. Eidg Anst forstl Versuchsw Ber 270:135–141

Havranek WM (1993) The significance of frost and frost-drought for the alpine timbeline. In Anfodillo T, Urbinati C (eds) Ecologia delle foreste di alta quota. Proc XXX Corso di Cultura in Ecologia, University of Padova, pp 115–127

Heskes TM, Kappen B (1991) Learning processes in neural networks. Physical Review A 44 (4):2718–2726

Holtmeier FK (1993) The upper timberline: ecological and geographical aspects. In Anfodillo T, Urbinati C (eds) Ecologia delle foreste di alta quota. Proc XXX Corso di Cultura in Ecologia, University of Padova pp 1–26

Kappen HJ (1996) An overview of neural network applications. Proc. 6th Intern. Congress for Computer Technology in Agriculture, Wageningen, pp 75–79

Keller T, Guiot J, Tessier L Climatic effect of atmospheric CO_2 doubling on tree-radial-growth in South Eastern France. J Biogeogr (in press)
Körner C, Larcher W (1988) Plant life in cold climates. In Long SP, Woodward FI (eds) Symposia of the Society of Experimental Biology 42:25–57
Körner C (1995) Impact of atmospheric changes on mountain vegetation: the ecophysiological perspective. In Guisan et al. (eds) Potential ecological impacts of climate change in the Alps and Fennoscandian mountains. Publication of Conservatoire et Jardin botanique de la Ville de Genève n. 8, pp 113–120
Kröse BJA, Van Der Samgt PP (1993) An introduction to Neural Networks. University of Amsterdam, Amsterdam
Lescop-Sinclair K, Payette S (1995) Recent advance of arctic treeline along the eastern coast of Hudson Bay. J Ecol 83:929–936
Lu P, Biron P, Bréda N, Granier A (1995) Water relations of adult Norway spruce (*Picea abies* (L.) Karst.) under soil drought in the Vosges mountains: water potentials, stomatal conductance and transpiration. Ann Sci For 52:117–129
Morgan JM (1984) Osmoregulation and water stress in higher plants. Ann Rev Plant Physiol 35:299–319
Nicolussi K, Bortenschlager S, Körner C (1995) Increase in tree-ring width in subalpine *Pinus cembra* from the central Alps that may be CO_2 related. Trees 9:181–189
Sarle WS (1994) Neural networks and statistical models. Proc Annual SAS Users Group International Conference, Cary, NC, SAS Institute, 1538–1550
Sarle WS (1995) Stopped training and other remedies for overfitting. Proc 27th Symposium on the Interface, SAS Institute, 1–4
Sarle WS ed. (1997) Neural Network FAQ, parts 1–3, periodic posting to the Usenet newsgroup comp.ai.neural-nets, URL: ftp://ftp:sas.com/pub/neural/FAQ.html.
Schweingruber FH, Kairiukstis L, Shiyatov S (1990) Primary data. Sample Selection. In Cook ER and Kairiukstis LA (eds) Methods of dendrochronology: application in the enviromental science. Kluwer Academic Publishers, Dordrecht, pp 23–35
Smith M (1993) Neural Network for Statistical Modeling. Van Nostrand Reinhold NY.
Solomon AM, Cramer W, (1993) Biospheric implications of global environmental change in Solomon AM & Shugart HH (eds.) Vegetation dynamics & global change. Chapman & Hall, New York, London, pp 25–52
Stokes MA & Smiley TL (1968) An introduction to tree-ring dating. The University of Chicago Press, Chicago
Swetnam TW, Thompson MA, Sutherland EK (1985) Using dendrochronoogy to measure radial growth of defoliated trees. Agriculture Handbook, USDA, Forest Service 639:1–37
Tranquillini W (1979) Physiological ecology of the alpine timberline. Ecological Studies 31
Valentini R, Anfodillo T, Ehlringer J (1994) Water sources utilization and carbon isotope composition ($\delta^{13}C$) of co-occurring species along an altitudinal gradient in the Italian Alps. Can J For Res 24:1575–1578
Wanner H, Beniston M (1995) Approaches to the establishment of future climate scenarios for the Alpine region. In Guisan et al. (eds) Potential ecological impacts of climate change in the Alps and Fennoscandian mountains. Publication of Conservatoire et Jardin botanique de la Ville de Genève n. 8, pp 87–95

Climate, limiting factors and environmental change in high-altitude forests of Western North America

David L. Peterson

Abstract. The relationships between forest dynamics and climate are predictable for high-altitude forest ecosystems in western North America and other mountainous regions. The duration of snowpack interacts with spring and summer temperature to determine when a snowfree soil surface and sufficiently high soil temperatures for physiological activity occur. Regeneration of tree seedlings varies spatially and temporally as mediated by the duration of the snowpack, which affects the length of the growing season on high-precipitation sites and the soil moisture supply on low-precipitation sites. Regeneration is favoured by climatic conditions that produce a mesic soil moisture regime rather than extremes and by summer temperatures that are sufficiently high to facilitate carbon gain in seedlings. Relatively short-term climatic trends can have major impacts on regeneration patterns, particularly after disturbances. Tree growth in high-snowfall environments (under a marine climate and near the treeline) is generally limited more by precipitation than by temperature, with growth being negatively correlated with snowpack depth. There are many sources of spatial and temporal variation in growth response to climate, most of which are not included in modeling efforts at large spatial scales. Growth response varies between species and within species, depending on subregional climate (high vs. low precipitation in the same mountain range), altitude (treeline vs. lower elevation), aspect (north vs. south) and genotype. The effects of climatic variation on high-altitude forests are distinct from effects in low-altitude ecosystems, and models based on low-altitude forests are not necessarily applicable at higher altitudes. The potential for vegetational inertia—long lag times in response to environmental variation—needs to be considered when evaluating the response of high-altitude forests to climatic change.

1. Introduction

The impacts of changes in the atmospheric environment on high-altitude forest ecosystems of western North America have attracted less scientific attention than lower elevation forests (Beniston 1994; Beniston and Fox 1996). This is surprising, given that mountains dominate the regional landscape, but it may be the result of the fact that most modeling efforts (Henderson-Sellers and McGuffie 1995; Leemans et al. 1996), decision making and policy focus on large spatial scales. General circulation models (GCM) "see" mountain ranges as convexities in a larger landscape rather than as discrete entities with specific properties; mountain ranges such as the Olympic Mountains (ca. 4,000 km^2 area, 2,300 m elevation) of Washington state are barely visible on a GCM grid. Current mesoscale modeling efforts may partially resolve this scale issue with respect to mountainous landscapes.

Mountains provide an ideal landscape for examining the response of forest ecosystems at various spatial scales, because they have a wide variety of life zones in close proximity to one another. Differences in species and vegetative associations that are found over hundreds of kilometers longitudinally or latitudinally at low elevation can be observed over 2–3 kilometers in altitudinal distance (Peterson et al. 1997). This compression of life zones and habitats provides the opportunity to observe past and future environmental changes over a relatively small geographic area with diverse meso- and micro-climates.

Sufficient information exists to discern the climatic factors that limit growth and regeneration of high-altitude forests (or subalpine forests, located at the highest elevation where tree species dominate the vegetation) throughout most of the mountainous regions of western North America. In this paper, data on growth, demography and paleoecology are synthesized to infer the potential response of subalpine forests in western North America to future climatic change.

2. "Migrating" Trees: Treeline and Tree Invasion

Altitudinal treeline – the uppermost extent of tree species in mountains – has been a topic of great interest in the climatic change literature of the past decade (Stevens and Fox 1991; Slatyer and Noble 1992). Indeed, documenting a change in the location of ecotones is one of the few tangible ways in which the effects of climate on vegetation distribution and abundance can be measured. Despite dramatic claims that vegetation zones and tree species potentially may shift several hundred meters upward in elevation under typical climatic change scenarios of a 2–3° C increase (Peters 1990; Franklin et al. 1992), paleoecological and demographic data suggest more subtle changes.

A wide range of paleoecological studies indicates that the treeline has rarely varied by more than 100 m elevation during the Holocene (Rochefort et al. 1994), including the warm climatic period of approximately 10,000–6,000 years B.P. On some particularly stable landscapes, vegetation composition has changed relatively little for thousands of years (Gavin 1997). This is not because vegetation assemblages are insensitive to climate. In fact, they are highly sensitive to subtle and relatively small-scale temporal variations in climate and other environmental conditions, as can be seen in almost any pollen diagram derived from a sediment core (Brubaker and McLachlan 1995). However, identifying the climate–vegetation relationship requires that more than just one factor – typically mean air temperature in most climatic change scenarios – be evaluated in the larger context of competition and disturbance.

A number of studies have documented increased tree establishment (or tree invasion in some articles) in the subalpine zone of mountains in western North America during the 20th century (Rochefort et al. 1994; Taylor 1995; Woodward et al. 1995; Hessl and Baker 1997). This appears to be a real phenomenon, although most of the increased establishment is into meadows that are part of the forest–meadow mosaic rather than at upper treeline (Weisberg and Baker 1995). Tree establishment tends to occur during discrete periods of time and is not synchronous among different regions of North America. Tree establishment is spatially and temporally variable between species and within species at different locations.

This variation can be explained by analyzing the role of limiting climatic factors at each location. The mountains in western North America contain three general climatic zones: maritime, continental and mediterranean. Maritime climate is predominant in the northwestern United States and southwestern British Columbia (Canada). It is characterized by high coastal rainfall, a long winter period of rain and snow and extensive cloud cover. Winter temperature is cool, summer temperature is mild and summer precipitation is low. Continental climate is predominant in the western interior of the United States and Canada. Much of the winter precipitation is snow, and summer precipitation occurs as thunderstorms in the central and southern part of the range. Winter temperature is cold, and summer temperature varies from cool in the north to warm in the south. The predominant climate in the southwestern and Pacific coastal USA is mediterranean. Annual precipitation is low, and most of it falls during winter. Summer is hot and dry with droughts in many years, and winter temperature is mild.

Each of these climatic zones has distinctive limiting factors for tree establishment (Table 1). In the maritime climate of the Pacific Northwest, snowpack is the limiting factor in areas of high snowfall (primarily on the west side of the mountains). If snowpack is reduced, then there is greater opportunity for tree seedlings to germinate and have sufficient time to grow. In the continental climate of the Rocky Mountains and eastern Cascade Mountains, extremely cold winter temperatures and relatively low soil moisture due to low snowpacks and dry summer weather are most likely to limit tree establishment. If air temperatures are warmer and moisture supply is reliable throughout the year, there will be an increased probability of tree seedling germination and growth. In the mediterranean climate of the Sierra Nevada, low soil moisture will always be a limiting factor. Increased moisture supply via the snowpack or summer precipitation can ameliorate this limitation sufficiently to allow regeneration.

	Summer		Winter		
Climatic zones	Precipitation	Temperature	Precipitation	Temperature	Limiting factor
Maritime	Low	Mild	High	Cool	High spring and summer snowpack
Continental	Low	Cool-Mild	Low	Cold	Low summer soil moisture, low summer temperature
Mediterranean	Very low	Hot	Low	Mild	High spring snowpack, low summer soil moisture

Table 1. Climatic characteristics and limiting factors for tree establishment in mountain climatic zones of western North America.

The role of limiting factors can be seen in each of these climatic zones. In the Olympic Mountains of Washington (maritime but with continental influence on the east side), tree regeneration during the 20th century was studied along a precipitation gradient ranging from extremely wet (west side, high snowpack) to dry (east side) (Woodward et al. 1995). The results indicate that: (1) *Tsuga mertensiana*, a subalpine dominant in high snowpack regions, established in wet sites during a period of low snowpack (1920–1950), (2) *Abies lasiocarpa*, a subalpine dominant in drier regions, established in dry sites during a period of high snowpacks (1950–1990), and (3) both species established in mesic sites during a period of overlap between the high and low snowpack periods cited above (Fig. 1).

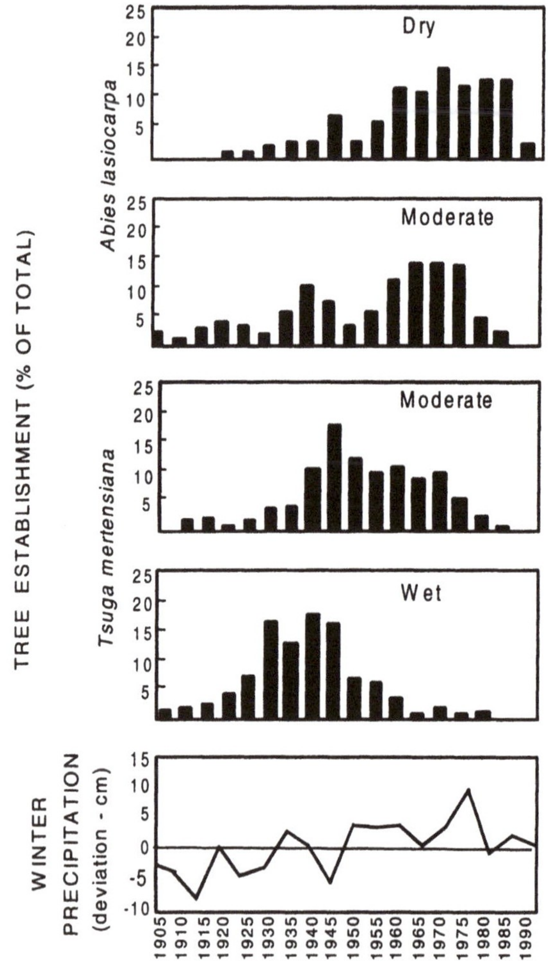

Fig. 1. *Abies lasiocarpa* and *Tsuga mertensiana* establishment (5-year totals) and precipitation (5-year running mean) in the Olympic Mountains during the 20th century. Sample sizes are: *A. lasiocarpa*, 546 at dry sites, 186 at moderate sites; *T. mertensiana*, 259 at moderate sites, 375 at wet sites. Note different temporal establishment patterns between species and between locations with different winter precipitation. Adapted from Woodward et al. (1995), reprinted with permission.

This evidence indicates that establishment occurred when the limiting factors – high snowpack on the wet west side, and low soil moisture on the dry east side – were alleviated. Climatic conditions that ameliorate extremes and promote mesic conditions maximize opportunities for tree regeneration (Agee and Smith 1984; Rochefort et al. 1994; Woodward et al. 1995; Rochefort and Peterson 1996). This is true at various scales, including the meso-climatic scale illustrated above, as well as the microclimatic scale affected by differences in elevation, aspect and landform (Fig. 2).

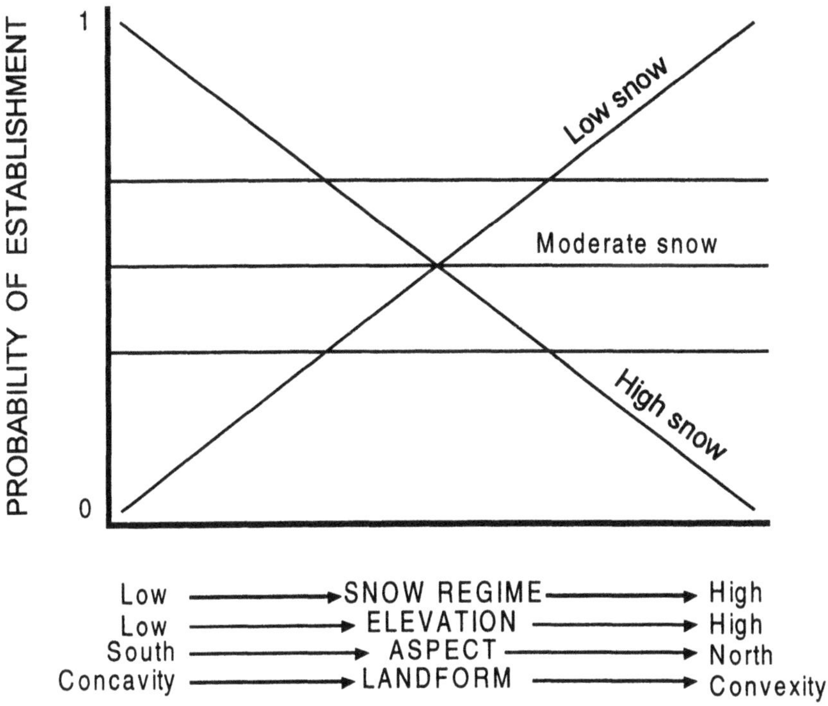

Fig. 2. Conceptual model of the effects of different patterns of winter snowfall (low, moderate and high, for a few years to decades) on probability of regeneration in maritime subalpine forests. Regeneration is affected at different spatial scales by meso-climate (snow regime) and micro-climate (elevation, aspect, landform). The horizontal dashed lines indicate variation around the mean response to moderate snowfall.

In the Colorado Rocky Mountains (continental climate), a major period of regeneration by *Abies lasiocarpa* and *Picea engelmannii* from 1940 to 1970 was correlated with warmer temperatures and higher snowpacks, conditions which are apparently rare during the past few centuries (Hessl and Baker 1997). Establishment occurred when the limiting factors – cold temperature and low soil moisture – were alleviated. In the southern Cascade Range of northern California (mediterranean climate), *Tsuga mertensiana* establishment was greatest during periods of high precipitation (spring and summer) and above-normal temperature

(1890–1920, 1955–1972, 1988–1990) (Taylor 1995). Establishment occurred when the limiting factor – duration of snowfree growing season and available soil moisture – was alleviated. These additional examples from different climatic zones also indicate that moderation of climatic extremes is a key to subalpine tree regeneration.

Disturbance, primarily fire, is an important component of the regeneration dynamics of subalpine tree species (Veblen et al. 1991, 1994; Bessie and Johnson 1995). In some subalpine ecosystems, fire occurs frequently enough that meadows are maintained free of woody vegetation, particularly where meadows occur in concavities which contain late snowpacks. In addition, there can be substantial lags (100 years or more) in post-fire regeneration (Shankman and Daly 1988) until limiting factors such as duration of snowpack can be sufficiently ameliorated to permit tree germination and subsequent growth (Fig. 3, Little et al. 1994). On the other hand, fire often creates the appropriate soil conditions and reduced plant competition necessary for germination.

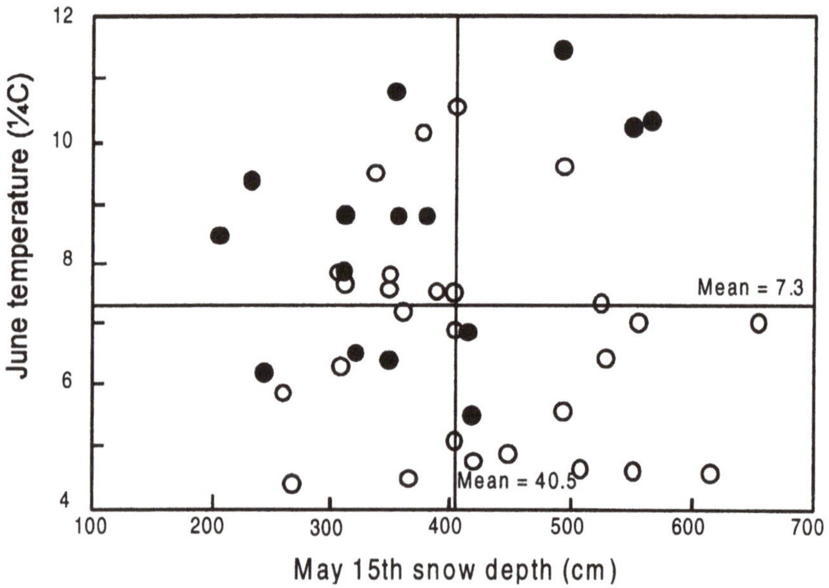

Fig. 3. Scatterplot of postfire *Abies lasiocarpa* establishment data for a site in the Cascade Mountains. Data are for 1948–1991, where years with successful (•) and unsuccessful (o) establishment are indicated. Regeneration is primarily in years with lower spring snowpack. From Little et al. (1994), reprinted with permission.

3. Tree Growth Response to Climate: Limiting Factors and Sources of Variation

Tree growth is commonly used as a measure of fitness relative to climatic regimes in forest ecosystems. Several authors have noted increased growth in subalpine species during the 20th century (Graumlich et al. 1989; Innes 1991; Peterson et al. 1991; Briffa 1992; Graybill and Idso 1993; Peterson 1994), the cause of which has been ascribed to increased temperature, increased atmospheric CO_2 or other factors. Unfortunately it is difficult to infer definitive cause-and-effect relationships from these studies.

Analyses of growth–climate relationships typically focus on the mean response of a species in a particular location to climate, rather than the range of responses among sample trees. This variation in response to climate, and the factors responsible for the variation, are potentially valuable in assessing the full range of the impacts of climatic change at different scales. However, dendroecological (as opposed to dendroclimatological) studies cannot obtain this information if sampling is conducted on a particular elevation, aspect or biogeographic subregion of a particular species, or if study sites are selected without regard to these factors.

Variation in micro-site affects the response of tree growth to climate (Kienast et al. 1987), with specific sources of variation being elevation (Hansen-Bristow 1986), aspect, landform and soil characteristics. The exact nature of this variation differs among species (Colenutt and Luckman 1991). In the northern Cascade Mountains of Washington, growth response of *Picea engelmannii*, *Abies lasiocarpa*, and *Larix lyallii* to climate varied considerably at different elevations, on different aspects, and on different landforms (ridge, backslope, valley) (Peterson and Peterson 1994). In the Colorado Rocky Mountains, growth response of *P. engelmannii*, *A. lasiocarpa* and *Pinus contorta* to climate also varied considerably with respect to both elevation and topographic position (Villalba et al. 1994); variation was significant over relatively short distances. These data on similar species in diverse regions confirm that there are strong patterns of growth response at relatively small spatial scales that need to be considered.

In a detailed study of growth response to climate in the Olympic Mountains, Ettl and Peterson (1995a,b) sampled along elevation gradients from treeline to the lowest extent of *Abies lasiocarpa* distribution in wet west-side forests and dry east-side forests. Winter precipitation (snowfall) was negatively correlated with tree growth at most locations. However, the magnitude of correlations vary from $r= -0.5$ at wet mid-elevation sites to -0.2 at dry low-elevation sites (Ettl and Peterson 1995b). Site chronologies are similar to each other in many respects with 62% of the variance explained by the first principal component (PC). A plot of the first two PCs shows that the sites are separated along PC2 (20% of the variance), where low- and middle-elevation dry sites have negative weightings, and high-elevation and all wet sites (high snow environments) have positive weightings (Fig. 4). This suggests that duration of snowpack dominates growth response to climate by affecting the length of growing season (Fig. 5).

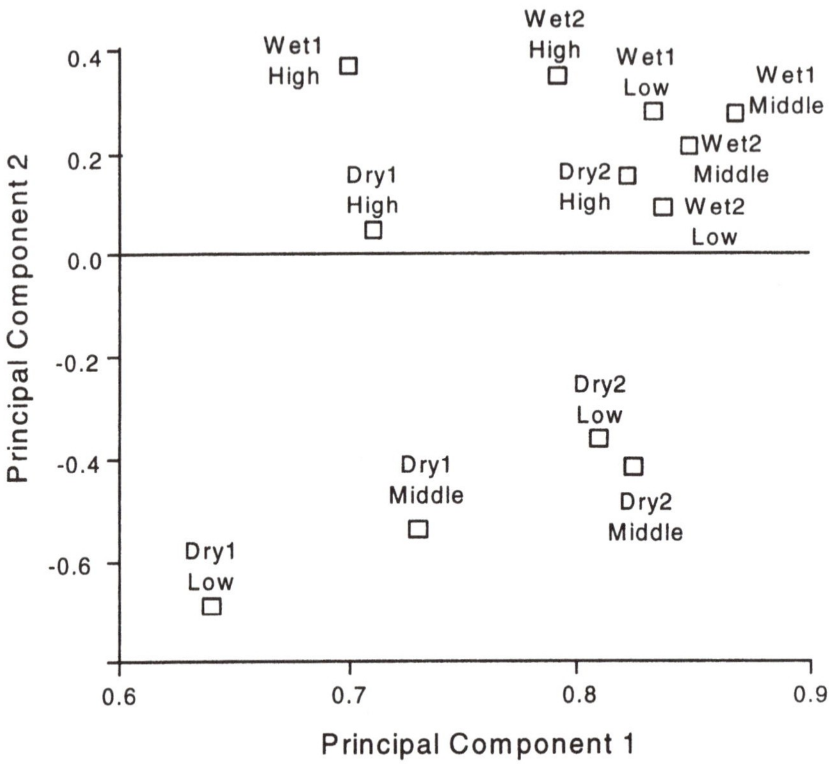

Fig. 4. First and second principal components of mean site chronologies for *Abies lasiocarpa* in the Olympic Mountains, showing groupings of wet and high elevation sites versus low- and middle-elevation dry sites. Adapted from Ettl and Peterson (1995), reprinted with permission.

A snow-free soil surface that facilitates soil warming is critical for the inception of positive net photosynthesis (Fig. 6, DeLucia and Smith 1987). Current summer temperature is also significantly (positively) correlated with *A. lasiocarpa* growth in the Olympics with correlations generally higher at higher elevations and higher at the high-snowpack, west-side sites in early summer (Ettl and Peterson 1995b). It has also been shown that growth–climate correlations of this species vary significantly between high-snow and low-snow years (Fig. 7, Peterson and Peterson 1994).

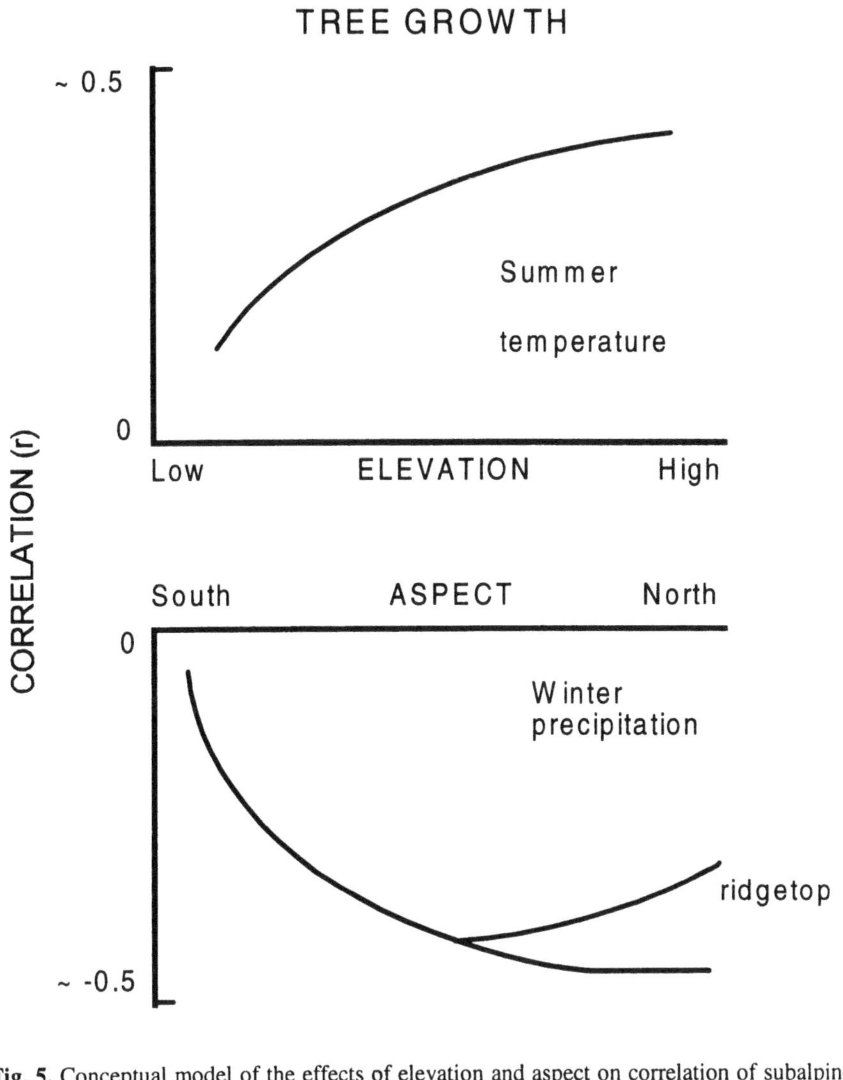

Fig. 5. Conceptual model of the effects of elevation and aspect on correlation of subalpine tree growth with winter precipitation and spring temperature. Correlations generally are greater at higher elevations and on more northerly aspects. Ridgetop locations that have snow removed by wind (dashed line) can have lower correlations between growth and precipitation than lower elevation sites.

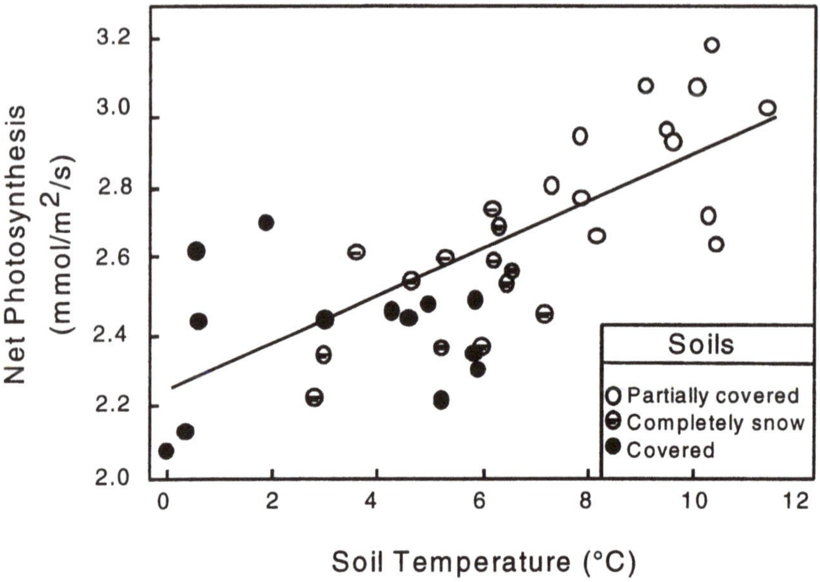

Fig. 6. Photosynthetic response of *Picea engelmannii* in the Medicine Bow Mountains, Wyoming, as a function of soil temperature, with snow cover conditions indicated. From DeLucia and Smith (1987), reprinted with permission.

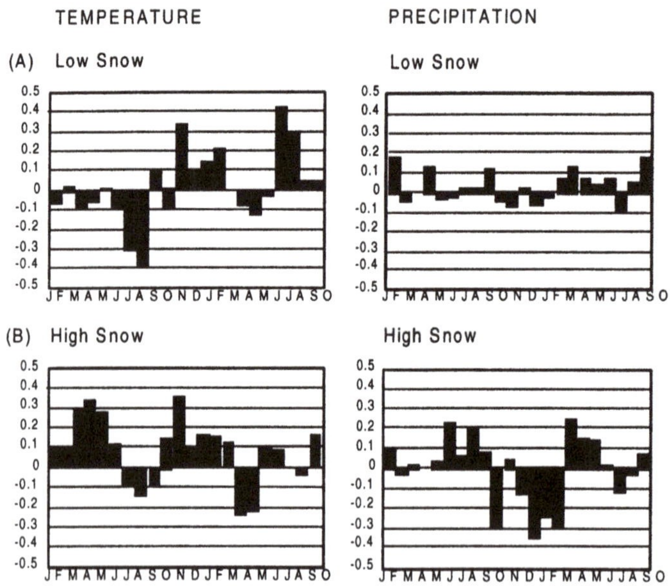

Fig. 7. Correlations between annual *Abies lasiocarpa* growth and monthly temperature and precipitation for the northern Cascade Mountains of Washington, 1896–1990, comparing years with winter precipitation above and below the median value. From Peterson and Peterson (1994), reprinted with permission.

In a subsequent study of *Tsuga mertensiana* throughout its range in the Cascade Mountains of Washington and Oregon, Peterson and Peterson (manuscript submitted) found patterns of growth response to climate similar to that of *Abies lasiocarpa*. Correlations of growth with winter precipitation (negative), spring temperature (positive) and Palmer Drought Severity Index (negative) were higher at high-elevation sites than at low-elevation sites (Fig. 8), a trend that was consistent across a latitudinal range of 800 km. Despite some variation, there are clearly general patterns of growth response to climate that have coherence across species and large geographic areas.

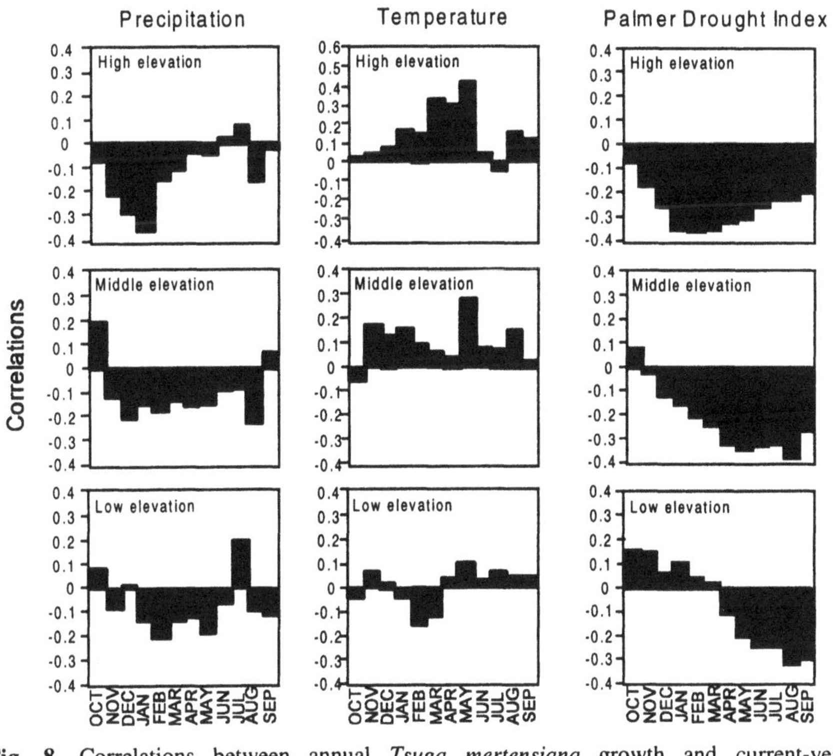

Fig. 8. Correlations between annual *Tsuga mertensiana* growth and current-year precipitation, temperature and Palmer Drought Severity Index on Mt. Hood, Oregon. Elevational differences in growth response to climate at this site are typical of most sites sampled across an 800-km latitudinal range in the Cascade Mountains.

Ettl and Peterson (1995a) also assessed the effects of interannual climatic extremes on *Abies lasiocarpa* growth, and found response to be highly individualistic. Growth was not significantly higher or lower on the majority of sites for extreme-climate years examined. In addition, few climatic variables are correlated with growth of the majority of individuals on most sites, suggesting that some individuals are relatively unresponsive to climate. For example, the percentage of trees (mean for two sites) for which growth was significantly negatively correlated with winter precipitation ranges from 66 to 94% at wet sites and from 17 to 45% at dry sites (Table 2); the percentage of trees significantly positively correlated with August temperature ranges from 2 to 40% at wet sites and from 2 to 49% at dry sites (high elevation sites had higher correlations of growth with temperature). Individual differences in growth–climate response probably vary due to micro-site variation in soil depth, soil moisture, wind and insolation (Kienast et al. 1987; Villalba et al. 1994; Peterson and Peterson 1994; Ettl and Peterson 1995a). This range of variation in growth response to climate is critical in evaluating how tree species will respond to climatic change in a particular region.

	n	Elevation	Winter precipitation	August temperature
West-side – wet	50	High	66	40
	49	Middle	94	14
	49	Low	77	2
East-side – dry	57	High	45	49
	61	Middle	29	2
	60	Low	17	15

Table 2. Percentage of *Abies lasiocarpa* in the Olympic Mountains correlated with winter precipitation (negative) and August temperature (positive) (Ettl and Peterson 1995a). Data are summarized for westside (wet) sites and eastside (dry) sites at different elevations.

Genotype is an additional source of variation in growth response to climate. While this topic has been assessed for small trees in many common-garden provenance studies, it has rarely been addressed for mature trees under natural conditions. Allozyme analysis of *Abies lasiocarpa* populations in the Olympic Mountains found that observed heterozygosity and percentage of polymorphic loci of west-side, wet site trees is significantly lower than that of eastside, dry site trees (Ettl and Peterson, manuscript in prep.). There are also significant differences in genetic structure among all populations and in allele frequencies between populations at different elevations for most loci. Radial growth of *A. lasiocarpa* is partially related to genetic differences among individuals at five loci. Growth is positively associated with heterozygosity in three allozymes (Fig. 9) and homozygosity in two allozymes. This contrasts with previous studies which found consistent positive correlations between growth and heterozygosity (Ledig et al. 1983, Jelinski 1993).

Regardless of the exact nature of genetic impacts on growth response to climate, it is clear that there is sufficient genetic variation present in *A. lasiocarpa* to affect growth at different scales – within an individual stand, along an elevational gradient and between sub-regional areas (west-side vs. eastside). Because other subalpine tree species can be expected to have similar levels of genetic variation, growth response to climatic change will likely be spatially variable within all species. Furthermore, there is the possibility of micro-evolutionary changes in populations due to differential responses of genotypes to an altered climate and elevated atmospheric CO_2 (Thomas and Jasienski 1996). This could have a significant impact on the trajectory of specific population characteristics.

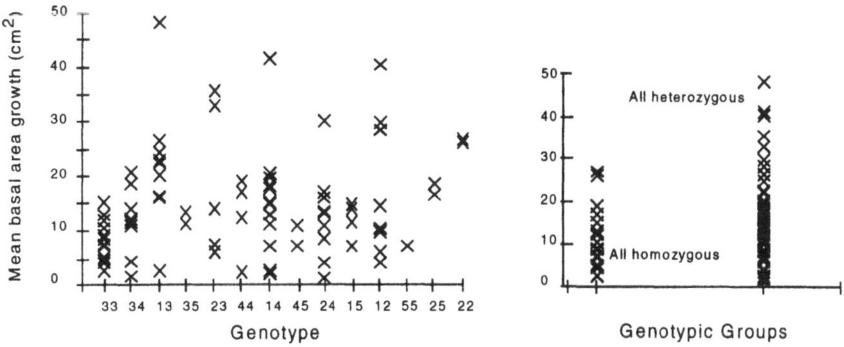

Fig. 9. Basal area growth of *Abies lasiocarpa* in the Olympic Mountains during 1950–1990, summarized by genotype (each symbol represents one tree) (Ettl 1995, Ettl and Peterson, manuscript in prep.). The graph on the left depicts genotypes for phosphoglucoisomerase 2 (PGI-2) arranged in order from most common (left) to least common; a Kruskal-Wallis test indicates that differences in growth among genotypes are significant (p=.07). The graph on the right summarizes growth for heterozygous and homozygous (for PGI-2) trees; a Mann-Whitney U-test indicates that heterozygous individuals grow significantly faster (p=.07).

4. Effects of Climatic Change in Subalpine Forests: Biogeographic Implications of Variation at Different Scales

Plant species respond individualistically to climatic change (Delcourt and Delcourt 1991; Fischlin and Gyalistras 1997), and the paleoecological literature for the subalpine regions of western North America provides a long-term perspective on biogeographic response to climatic variation. In general, vegetational patterns are the product of climatic and nonclimatic limiting factors operating at different spatial and temporal scales (Delcourt and Delcourt 1988; Schoonmaker 1998), and the impact of an altered limiting factor depends on the scale at which the impact is assessed (Peterson and Parker 1998). Climatic changes associated with the retreat of Laurentide and Cordilleran ice and the amplification of the seasonal cycle of radiation contribute to broad spatial patterns of vegetation on millennial time scales. Differences in geomorphic features, soils and migration history affect the distribution and abundance of species at smaller spatial and time scales (Whitlock 1993). Most of the tree species present at high-altitude locations today have occupied those sites since 10,000 B.P. or earlier. Mountain ranges south of the continental ice or with nunataks above the ice during glaciation have maintained forests with the full complement of contemporary species during much of the Quaternary (Buckingham et al. 1995; Peterson et al. 1997).

Although response of subalpine species to variation in Holocene climate varies among mountain ranges, there are some similarities in distribution and abundance over large spatial and temporal scales. It is particularly instructive to look at palynological data from the time segment 10 000–6 000 BP. This period was significantly warmer than any other multimillennial portion of the Holocene and was considerably warmer than the full glacial. Although the paleoecological record provides little information on variation in precipitation, it may provide clues about which species will be dominant in a warmer climate in the centuries ahead. For example, in the Canadian Rocky Mountains of Alberta, Abies lasiocarpa, Picea engelmannii and Alnus species (probably A. sinuata and A. tenuifolia at high elevations) were more common in paleocological samples than they are today, and Pinus species (particularly P. albicaulis) were common throughout this period (Luckman and Kearney 1986; MacDonald 1989). Abies lasiocarpa and P. engelmannii are relatively tolerant of low soil moisture; Alnus regenerates well after disturbances and often survives them (e.g., avalanches) as well. Farther south in the Yellowstone region of the Rocky Mountains in northwestern Wyoming, A. lasiocarpa and P. engelmannii became dominant after glacial recession, followed by increased Pinus contorta dominance (and presumably increased fire frequency, because P. contorta regenerates well after fire) during peak warming (Whitlock 1993); these three species continue to dominate the subalpine zone of this region today (Despain 1990).

Palynological data from the eastern Olympic Mountains of Washington indicate that A. lasiocarpa dominated the coniferous forest component during the Holocene warming (Brubaker and McLachlan 1995). Alnus sinuata (regenerates well after fire and tolerates hydrologic disturbance) and Chamaecyparis nootkatensis (a member of the Cupressaceae confined to the Pacific Northwest; tolerant of landslides and avalanche) were also common during this period. The distribution and abundance of taxa in the pollen record during this warm period are distinctive from any other portion of the Holocene (Fig. 10).

Results of an empirically-based modeling effort in the Olympics (Zolbrod and Peterson, submitted manuscript) agree quite well with paleoecological data from the Olympics and to some extent from other mountainous regions. In both warmer/wetter (+2°C, +20%) and warmer/drier (+2°C, -20%) climatic scenarios, the dominance of *A. lasiocarpa* in east-side subalpine forests increases over time; *Pinus contorta* also increases, especially in dry scenarios with higher fire frequency. In westside forests, the dominance of another *Abies* species, *A. amabilis*, also increases. Although not definitive proof, the agreement of empirically-based, warm-scenario modeling (Zolbrod and Peterson, submitted manuscript) with paleoecological data from the Holocene warming (Brubaker and McLachlan 1995) suggests that some species–temperature relationships can indeed be predicted with confidence. However, the role of precipitation in modifying modeled species–temperature relationships needs greater emphasis. Some studies predict that warming will cause considerable expansion of *Pseudotsuga menziesii* into the current subalpine zone (Franklin et al. 1992), although this is unlikely unless warming is accompanied by much lower snowfall, because this species is particularly sensitive to breakage from snow. In fact, our model output indicates that in a warmer climate, *P. menziesii* increases in dominance only with decreased winter precipitation (Zolbrod and Peterson, manuscript submitted).

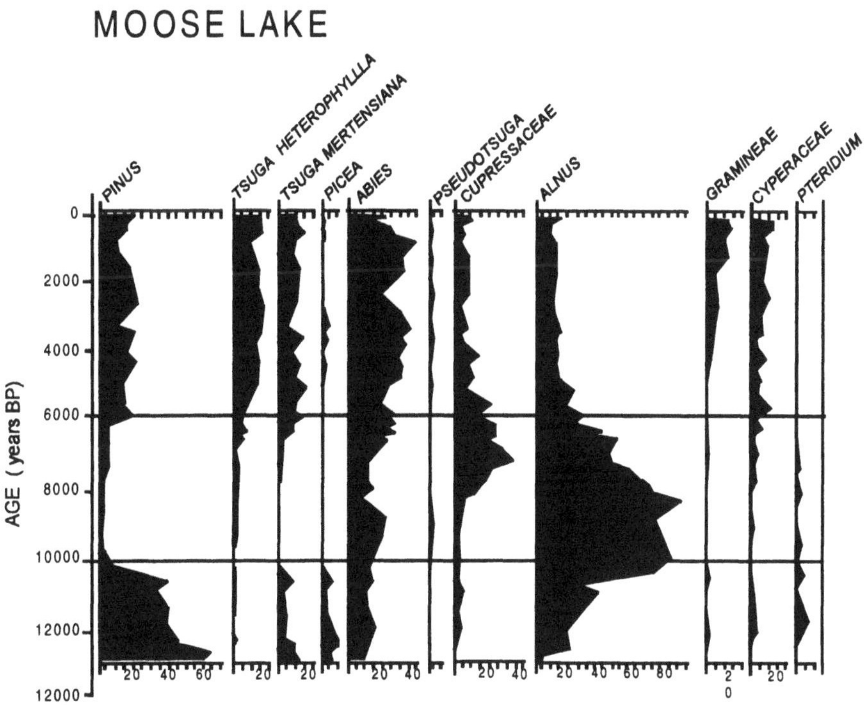

Fig. 10. Fossil pollen diagram for Moose Lake, at 1500 m elevation in the eastern Olympic Mountains, showing post-glacial variations in major pollen taxa. From Brubaker and McLachlan (1995), reprinted with permission.

Demographic, dendroecological and physiological data indicate that changes in the distribution and abundance of subalpine tree species in western North America in a warmer climate will depend as much or more on precipitation patterns as on temperature. In high-altitude environments, temperature is clearly a limiting factor for gas exchange (DeLucia and Smith 1987) and other physiological functions (Tranquillini 1979; Hadley and Smith 1986), and warmer summer temperatures can optimise photosynthetic production in mature trees. Warmer temperatures are beneficial to seedlings as well, but only on sites where shallow-rooted seedlings have an adequate moisture supply (Cui and Smith 1991;Little et al. 1994).

Regardless of temperature conditions, trees are not able to conduct gas exchange effectively unless the soil surface is sufficiently snowfree to allow soil warming. Therefore, ecological assessments of the potential impacts of climatic change on subalpine forests must consider how altered magnitude and seasonality of precipitation will affect regeneration and tree growth. This approach will provide more accurate predictions than models that consider only temperature changes.

Scientists may need to be patient in their quest to detect significant changes in altitudinal treeline and climatically-linked trajectories of tree regeneration and growth. Mountainous regions are filled with "relict species" and endemic populations which survive millennia of climatic fluctuations. Mature trees survive several centuries of climatic variation. Idso (1989) points out that "vegetational inertia" (Pearsall 1959; Smith 1985) can create a lag in tree establishment above current treelines. The *potential* for a higher altitudinal treeline and expansion of trees into meadows may simply require that limiting factors be ameliorated sufficiently for successful germination and growth. Given a long enough period of time, this may occur periodically regardless of directional climatic trends. Discriminating the effects of "normal" responses of natural resources to climatic variation from human-caused responses is perhaps the greatest challenge facing mountain ecologists.

Acknowledgments. I thank Donald McKenzie, David W. Peterson and former members of the Field Station for Protected Area Research lab for their ideas and review of this paper. Darci Horner assisted with preparation of figures. Research was supported by the Global Change Research Program of the U.S. Geological Survey–Biological Resources Division.

5. References

Agee JK, Smith L (1984) Subalpine tree establishment after fire in the Olympic Mountains, Washington. Ecology 65:810–819

Beniston M (ed) (1994) Mountain environments in changing climates. Routledge, London

Beniston M, Fox DG (1996) Impacts of climate change on mountain regions. In: Watson RT, Zinyowera MC, Moss RH (eds) Climate change 1995, Cambridge University Press, Cambridge, UK, pp 191–213

Bessie WC, Johnson EA (1995) The relative importance of fuels and weather on fire behavior in subalpine forests. Ecology 76:747–762

Briffa KR (1992) Increasing productivity of "natural growth" conifers in Europe over the last century. In: Bartholin TS, Berglund BE, Eckstein D, Schweingruber FH (eds) Tree rings and the environment. Lundqua Report 34, Lund University, Lund, Sweden, pp 64–71

Brubaker LB, McLachlan JS (1995) Landscape diversity and vegetation response to long-term climate change in the eastern Olympic Peninsula, Pacific Northwest, U.S.A. In Walker B, Steffen, W (eds) Global change and terrestrial ecosystems. Cambridge University Press, Cambridge, UK, pp 184–203

Buckingham NE, Schreiner EG, Kaye TN, Burger JE, Tisch EL (1995) Flora of the Olympic Peninsula. Northwest Interpretive Association, Seattle

Colenutt ME, Luckman BE (1991) Dendrochronological investigation of *Larix lyallii* at Larch Valley, Alberta. Can J For Res 21:1222–1233

Cui M, Smith WK (1991) Photosynthesis, water relations and mortality in *Abies lasiocarpa* seedlings during natural establishment. Tree Physiol 8:37–46

Delcourt HR, Delcourt PA (1988) Quaternary landscape ecology: relevant scales in space and time. Landsc Ecol 2:23–44

Delcourt HR, Delcourt PA (1991) Quaternary ecology: a paleoecological perspective. Chapman and Hall, London

DeLucia EH, Smith WK (1987) Air and soil temperature limitations on photosynthesis in Engelmann spruce during summer. Can J For Res 17:527–533

Despain DG (1990) Yellowstone vegetation: consequences of environment and history in a natural setting. Roberts Rinehart Publishers, Boulder, Colo., USA

Ettl GJ (1995) Growth of subalpine fir (*Abies lasiocarpa*) in the Olympic Mountains, Washington: response to climate and genetic variation. Dissertation, University of Washington, Seattle

Ettl GJ, Peterson DL (1995a) Extreme climate and variation in tree growth: individualistic response in subalpine fir (*Abies lasiocarpa*). Global Change Biol 1:231–241

Ettl GJ, Peterson DL (1995b) Growth response of subalpine fir (*Abies lasiocarpa*) to climate in the Olympic Mountains, Washington, USA. Global Change Biol 1:213–230

Fischlin A, Gyalistras D (1997) Assessing the impacts of climatic change on forests in the Alps. Global Ecol Biogeogr Letters 6:19–37

Franklin, JF, Swanson FJ, Harmon ME (1992) Effects of global climatic change on forests in northwestern North America. In: Peters RL, Lovejoy TE (eds) Global warming and biological diversity. Yale University Press, New Haven, Conn., USA, pp 244–257

Gavin D (1997) Soil pollen records of a subalpine meadow in the northeastern Olympic Mountains, Washington. Master's thesis, University of Washington, Seattle

Graumlich L, Brubaker LB, Grier CC (1989) Long-term growth trends in forest net primary productivity: Cascade Mountains, Washington. Ecology 70:405–410

Graybill DA, Idso SB (1993) Detecting the aerial fertilization effect of atmospheric CO_2 enrichment on tree-ring chronologies. Global Biogeochem Cycles 7:81–95

Hadley JL, Smith WK (1986) Wind effects on needles of timberline conifers: seasonal influence on mortality. Ecology 67:12–19

Hansen-Bristow K (1986) Influence of increasing elevation on growth characteristics at timberline. Can J Bot 64:2517–2523

Henderson-Sellers A, McGuffie K (1995) Global climate models and "dynamic" vegetation changes. Global Change Biol 1:63–75

Hessl AE, Baker WL (1997) Spruce and fir regeneration and climate in the forest–tundra ecotone of Rocky Mountain National Park, Colorado, U.S.A. Arct Alp Res 29:173–183

Idso SB (1989) A problem for climatology? Quatern Res 31:433–434

Innes JL (1991) High-altitude and high-latitude tree growth in relation to past, present and future global climate change. The Holocene 1:174–180

Jelinski DE (1993) Associations between environmental heterogeneity, heterozygosity, and growth rates of *Populus tremuloides* in a Cordilleran landscape. Arct Alp Res 25:183–188

Kienast F, Schweingruber FH, Bräker OU, Schär E (1987) Tree-ring studies on conifers along ecological gradients and the potential of single-year analyses. Can J For Res 17:683–696

Ledig FT, Guries RP, Bonefield BA (1983) The relation of growth to heterozygosity in pitch pine. Evolution 37:1227–1238

Leemans R, Cramer W, Van Minnen JG (1996) Prediction of global biome distribution using bioclimatic equilibrium models. In: Breymeyer AI, Hall DO, Melillo JM, Ågren GI (eds) Global change: effects on coniferous forests and grasslands. John Wiley and Sons, Chichester, UK, pp 413–450

Little RL, Peterson DL, Conquest LL (1994) Regeneration of subalpine fir (*Abies lasiocarpa*) following fire: effects of climate and other factors. Can J For Res 24:934–944

Luckman BH, Kearney MS (1986) Reconstruction of Holocence changes in alpine vegetation and climate in the Maligne Range, Jasper National Park, Alberta. Quatern Res 26:244–261

MacDonald GM (1989) Postglacial palaeoecology of the subalpine forest–grassland ecotone of southwestern Alberta: new insights on vegetation and climate change in the Canadian Rocky Mountains and adjacent foothills. Paleogeog, Palaeoclim, Palaeoecol 73:155–173

Pearsall WH (1959) The ecology of invasion: ecological stability and instability. New Biol 29:95–101

Peters RL (1990) Effects of global warming on forests. For Ecol Manage 35:13–33

Peterson DL (1994) Recent changes in the growth and establishment of subalpine conifers in western North America. In: Beniston M (ed) Mountain environments in changing climates. Routledge, London, pp 234–243

Peterson DL, Parker VT (1998) Dimensions of scale in ecology, resource management, and society. In: Peterson DL, Parker VT (eds) Ecological scale: theory and applications. Columbia University Press, New York, in press

Peterson DW, Peterson DL (1994) Effects of climate on radial growth of subalpine conifers in the North Cascade Mountains. Can J For Res 24:1921–1932

Peterson DL, Arbaugh MJ, Robinson LJ, Derderian BR (1991) Growth trends of whitebark pine and lodgepole pine in a subalpine Sierra Nevada forest, California, U.S.A. Arct Alp Res 22:233–243

Peterson DL, Schreiner EG, Buckingham NM (1997) Gradients, vegetation and climate: spatial and temporal dynamics in the Olympic Mountains, U.S.A. Global Ecol Biogeogr Letters 6:7–17

Rochefort RM, Little RL, Woodward A, Peterson DL (1994) Changes in sub-alpine tree distribution in western North America: a review of climatic and other causal factors. The Holocene 4:89–100

Rochefort RM, Peterson DL (1996) Temporal and spatial distribution of trees in subalpine meadows of Mount Rainier National Park, Washington, U.S.A. Arct Alp Res 28:52–59

Schoonmaker P (1998) Paleoecological perspectives on ecological scale. In: Peterson DL, Parker VT (eds), Ecological scale: theory and applications. Columbia University Press, New York, in press

Shankman D, Daly C (1988) Forest regeneration above tree limit depressed by fire in the Colorado Front Range. Bull Torrey Bot Club 115:272–279

Slatyer RO, Noble IR (1992) Dynamics of montane treelines. In: Hansen AJ, di Castri F (eds), Landscape boundaries: consequences for biotic diversity and ecological flows. Springer-Verlag, New York, pp 346–359

Smith AG (1985) Problems of inertia and threshold related to postglacial habitat changes. Phil Trans Royal Soc London B 161:331–342

Stevens GC Fox JF (1991) The causes of treeline. Ann Rev Ecol Syst 22:177–191

Taylor AH (1995) Forest expansion and climate change in the mountain hemlock (*Tsuga mertensiana*) zone, Lassen Volcanic National Park, California, U.S.A. Arct Alp Res 27:207–216

Thomas SC, Jasienski M (1996) Genetic variability and the nature of microevolutionary responses to elevated CO_2. In: Körner C, Bazzaz FA (eds) Carbon dioxide, populations and communities. Academic Press, New York, pp 51–81

Tranquillini W (1979) Physiological ecology of the alpine timberline. Springer-Verlag, New York.

Veblen TT, Hadley KS, Nel EM, Kitzberger T, Reid M,Villalba R (1994) Disturbance regime and disturbance interactions in a Rocky Mountain subalpine forest. J Ecol 82:125–135

Veblen TT, Hadley KS, Reid M (1991) Disturbance and stand development of a Colorado subalpine forest. J Biogeogr 18:707–716

Villalba R, Veblen TT, Ogden J (1994) Climatic influences on the growth of subalpine trees in the Colorado Front Range. Ecology 75:1450–1462

Weisberg PJ, Baker WL (1995) Spatial variation in tree regeneration in the forest–tundra ecotone, Rocky Mountain National Park, Colorado. Can J For Res 25:1326–1339

Whitlock C (1993) Postglacial vegetation and climate of Grand Teton and southern Yellowstone National Parks. Ecol Monogr 63:173–198

Woodward A, Schreiner EG, Silsbee DG (1995) Climate, geography, and tree establishment in subalpine meadows of the Olympic Mountains, Washington. Arct Alp Res 27:217–225

Managing Swiss forests: when climate intervenes

Jean Combe

Abstract. For more than a century, forest management in Switzerland has been based on a philosophy attuned to nature. Silviculture which imitates natural cycles and processes is believed to promote the growth of healthy, stable forests. Because of the diversity of species and ages, which are often mixed on a small scale, forest stands are able to resist many extremes of climate and, in the event of damage (mainly due to windthrow and snow breakage), the natural forest ecosystem can recover rapidly from total destruction thanks to abundant natural regeneration. Over the long term, Swiss foresters have become accustomed to harvesting 20% of the annual production as "storm timber". The most serious natural disasters are mainly due to extremes of one single climatic factor; e.g. Hurricane Vivian in 1990, which toppled over almost five million m^3 of timber – the equivalent of Switzerland's average yearly timber harvest.

For a forest manager, however, it is interesting to analyse the less spectacular damage which occurs annually in natural forest stands and accounts for between 12 and 20% of the annual harvest. It appears that a combination of two or three climatic factors has a much greater effect on forest stability than an extreme of one single factor. Wind speeds of up to 100 km/hour may often be harmless to forests, whereas lower speeds may uproot mature trees over a wide area if the soil is waterlogged following a long period of rain. This kind of damage can occur throughout the year in coniferous stands, whereas broadleaved species growing in waterlogged soils may resist winds if this climatic combination comes when the vegetation period is over and they have lost their leaves. A number of such patterns of damage to forest stands is presented and their dependence on the occurrence of a combination of specific climatic factors is proposed.

1. Introduction

The following observations reflect the concerns of a practicing forester responsible for the management of a municipal forest in the Swiss Jura. They are based on personal observations gleaned from 20 years of work in a forestry domain which once represented the wealth of our ancestors and will, hopefully, bring security and happiness to future generations. In presenting these observations within a scientific context, I would like to:

- contribute towards the recognition and better understanding of the dangers presented by the climatic extremes which may be lying in wait for our forests, and to
- anticipate these major climatic risks successfully, so that we can prevent damage from occurring.

2. Forest disasters are mainly forestry disasters

Information on the extent of past forest disasters can be obtained from the annual reports of forest managers, accounts of the activities of the Federal Forestry Directorate and from the digests of Federal Statistics. By "disaster" I mean the necessary harvesting of timber due to natural disturbances, in the form of biotic or abiotic damage. The forester terms this "salvage", "forced" or "compulsory" felling ("chablis" in French, "Zwangsnutzungen" in German).

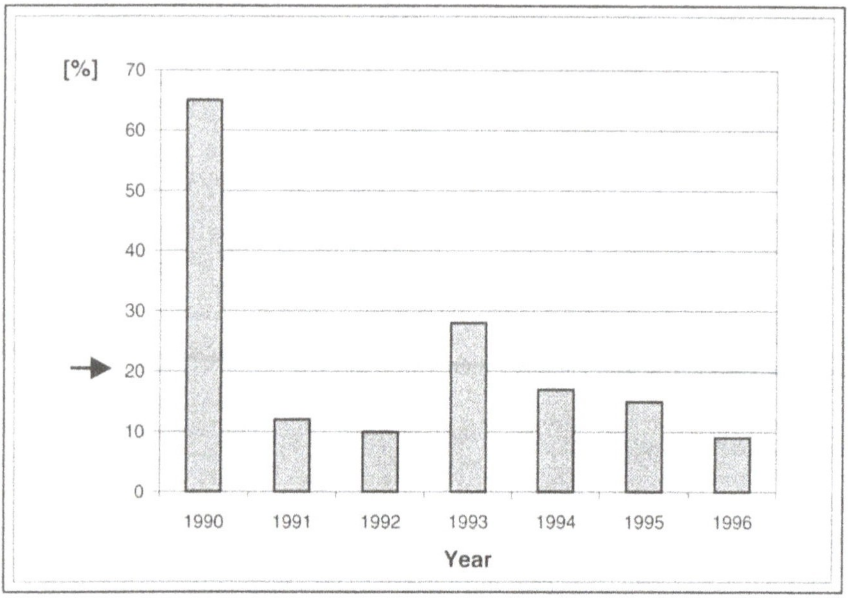

Fig. 1. Proportion of annual salvage felling in relation to the annual allowable cut, taking as an example the management of the municipal forests of Vallorbe (Canton of Vaud). The total volume harvested amounts to 6,000 $m^3 yr^{-1}$.

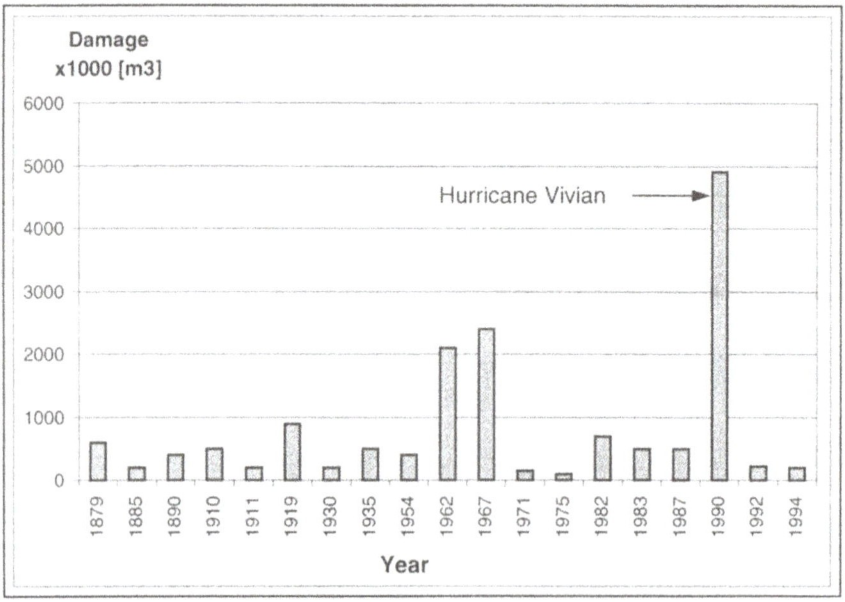

Fig. 2. Salvage felling in Swiss forests from 1879 to 1994 (only major events due to wind and snow, according to Holenstein 1994)

Our experience has shown that in Swiss temperate forests, these phenomena are more likely to be "disastrous" for the forestry service involved and for the timber market (and hence the owners), than for the forest ecosystem itself. The reason is that they result in an exceptional concentration of timber which it is difficult to log and market in a sector which is not accustomed to reacting swiftly. The criteria for measuring a forest disaster are closely related to the management activities arising from it, namely:

- the volume of timber harvested (following a disaster) and placed on the market,
- the number of seedlings needed for restocking, and
- the extent of areas newly planted and restored.

It is difficult to determine the evolution over time of certain recurrent scourges which attack our forests. Whether it be bark beetle or a thunderstorm, forest managers have always been used to taking natural biotic and abiotic factors into account in their daily work. The situation is worrying, however, and it becomes disastrous as soon as the impact of natural hazards on normal forestry routines exceeds a certain percentage of annual harvest. Locally, over the medium term, the proportion of salvage felling has stabilised at around 20% (Figure 1). In the regular, typical pattern of salvage felling – the kind which is part of the routine in the Jura – a forest manager will find half the trees toppled (uprooted) and half of them standing and withered (often as a result of an attack by bark beetle); snapped trees are even rarer than trees which have been struck by lightning.

Although the amount of salvage felling attributable to thunderstorms and weight of snow (Figure 2) in Swiss forests over the last thirty years has increased, it does not necessarily follow that the frequency or the extent of these natural hazards has increased substantially. However, the risk of forest disasters has increased, as the standing timber volume in Swiss forests has increased from 120 m^3 ha^{-1} to 360 m^3 ha^{-1} within the last 150 years. The 1962, 1967 and 1990 hurricanes are certainly remembered as being extreme, but they were also isolated incidents which have left their mark primarily where forests have been managed properly, with all windthrown timber being systematically logged.

3. The raging elements

3.1 Former studies and definitions

The salvage felling of recent decades, caused by extreme weather conditions, and especially by the hurricanes of 1962, 1967, 1987 and 1990, has been the subject of several studies (e.g. Holenstein 1994; Lätt 1991; Schiesser at al. 1997). Possible correlations have been sought, in particular between the nature of the damage and site factors such as slope, relief and soil class, structure of the tree stand, stand age, species mixture, crown closure, the number of years which have passed since the last cutting and the presence of root rot (Schmid-Haas and Bachofen 1991; Wangler 1997). To a certain extent, these studies have made it possible to characterise the endangered stands in Swiss forests which were affected by the hurricanes of 1990, 1967 and 1962, and which will continue to suffer from the resulting damage for a long time (Scherrer and Schmidtke 1997). These same studies also describe the combination of meteorological conditions which caused each of these storms. However, they do not establish any link between these conditions and the nature of the damage to the forests.

The following observations are an attempt to reconstruct the combination of climatic factors which has led to a particular pattern of damage. The combination of the following factors:

- precipitation
- winds
- temperature
- seasonal state of development (phenology)

as they prevailed at each site in the weeks before the disaster they may, at least in part, determine both the extent and the nature of the damage. As we do not have data for windgusts, we have used daily average wind speed as a variable which we expect to give us a representation of windgusts. Our experience of extreme meteorological events is also used to comment on the relevance of these data.

Fig. 3. Types of damage caused to trees by winds and thunderstorms (adapted from Schmid-Haas and Bachofen 1991)

Damage may be classed as follows (Fig. 3, according to Schmid-Haas und Bachofen 1991):

- snapping of treetop
- fracture of stem base
- breaking of rootstock
- uprooting.

Other kinds of damage may be added to these:

- loosening of roots (tilted trees)
- lightning strike
- breakage due to snow or glaze
- ring shake (internal loosening of annual rings).

Incipient damage caused by meteorological incidents and which later becomes apparent in forest stands, such as through an increase in bark beetle populations, is not taken into account here.

3.2 Combinations of climatic factors

All wind-related forest disasters of the past decades show a typical combination of climatic factors in the preceding weeks. While a certain proportion of the damage will be caused by extremes of just one factor, it is generally the combination of several factors which determines its severity. The following three examples illustrate this hypothesis which, however, still requires verification:

A) The maximum wind speeds ever observed on the Swiss Plateau were those of 28 and 29 January 1995, with registered speeds up to 160 km hr^{-1}. This led to some damage but to a lesser extent than Hurricane Vivian, which hit Switzerland in February 1990 with lower daily average wind speeds but some long lasting gusts.

B) Hurricane Vivian occurred on 27 and 28 February 1990 (Scherrer and Schmidtke 1997), with maximum wind speeds reaching 159 km hr^{-1} in Zurich (Holenstein 1994) and daily average wind speeds of up to 39 km hr^{-1} recorded for instance at La Chaux-de-Fonds. Exceptional daily average wind speeds had, indeed, been measured previously, with peaks of 26.5 on 24 January and 30.1 km hr^{-1} on 14 February, but only after periods of low rainfall (Fig. 4). The fact that forest soils had been waterlogged by mid-February downpours partly explains the impact Vivian had on the forests.

C) The succession of thunderstorms from 20 February to 28 March 1967, and especially that of 13 March, with daily average wind speeds of up to 30 km hr^{-1}, recorded at La Chaux-de-Fonds, were particularly destructive for our forests. Higher daily average wind speeds had in fact been measured in the preceding months, with a peak of 38.5 km hr^{-1} on 21 February 1967, at a time when rainfall had just started again after three weeks without rain (Fig. 5). The fact that forest soils were waterlogged after a long period of rainfall from 18

February to 13 March 1967, together with the occurrence of destabilising gusts, partly explains the impact of the storm of 13 March on our forests.

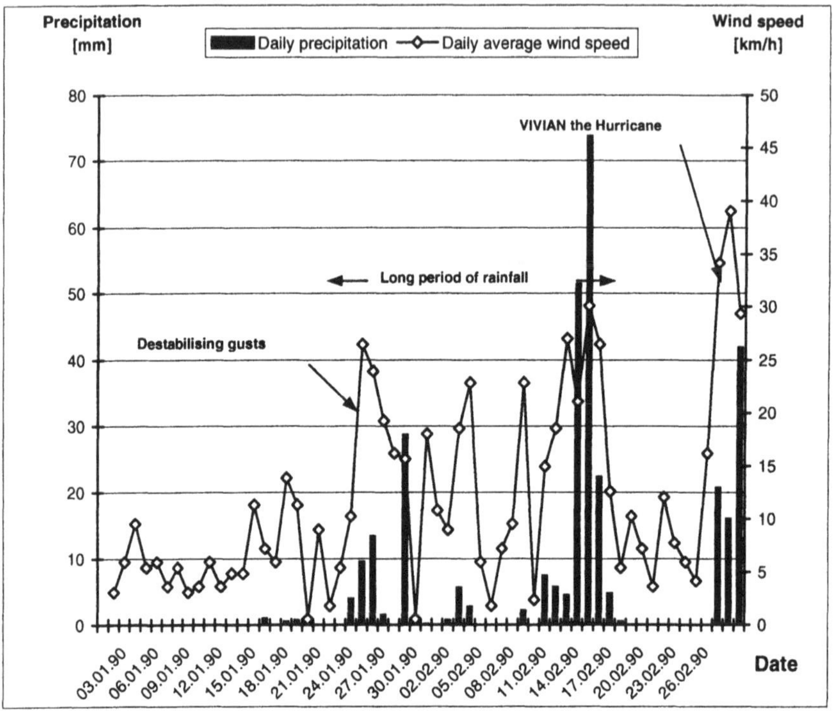

Fig. 4. Precipitation and daily average windspeed during the 60 days before Hurricane Vivian on 27 and 28 February 1990. La Chaux-de-Fonds (Bantle 1989).

Examples B and C above, illustrated by Figures 4 and 5, prompt the following questions:

1. Are daily average wind speeds of any relevance for windthrow or are only gusts relevant, or the combination of both? Despite the fact that this question is very important it remains difficult to answer, since many climatological stations in Switzerland register wind speeds at fixed time intervals and do not measure gusts.
2. What is the critical time interval between a period of rain and wind-gusting if the former is to have a significant impact on the type and extent of damage?

3.3 Combinations of meteorological conditions and damage types

The combination of the above climatic factors seems to have a decisive influence on the nature of the damage. As I have said, "forest disaster" is a subjective concept, which depends primarily on the disturbance it causes to a forestry enterprise. Timber logging, necessitated by natural disturbance, in any case entails economic losses for the owner. The seriousness of these losses, however, depends on the type and the extent of the damage.

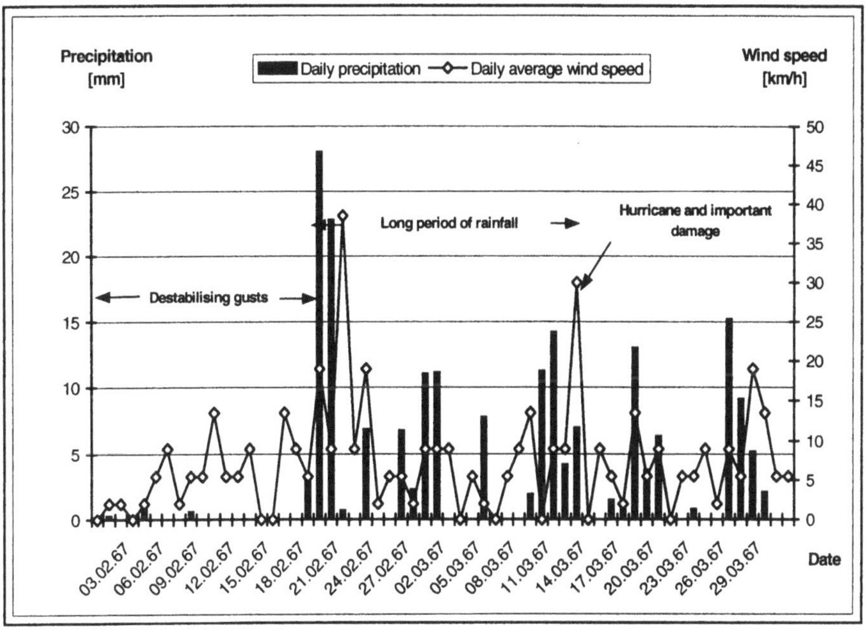

Fig. 5. Precipitation and daily average windspeed during the 30 days before the hurricane of 13 March 1967. La Chaux-de-Fonds (Bantle 1989).

Of the different types of damage, windthrow is the least damaging from an economic point of view. Although the tree is toppled over, once cut from its rootstock, it can normally be processed as usual. The same applies to loosened or tilted trees, even if they are leaning against neighbouring ones. On the other hand, all types of fracture (of the crown, the stem base or the rootstock), snapping by snow or glaze, ring shake and the fragmentation of timber as a result of being struck by lightning, will lead to a total economic loss. With fractured timber, sawmill-quality logs have to be downgraded into industrial or energy wood, depending on the species.

The nature of the damage depends primarily on the type of stand and its health, and on all the factors characterising a site, but it also depends on the climatic factors in the period before the incident. After a long spell of rain, a hurricane may leave a large proportion of toppled and loosened trees and a smaller number of broken ones. In soil which is frozen or desiccated, roots are held in a vice-like grip and the same gust will cause a greater proportion of broken timber and twisted trunks. It should be noted, however, that these patterns, which seem logically defensible, have not been confirmed by a study carried out on the nature of damage to *Picea abies* on the Swiss Plateau, which compared some 802 trees damaged by the 1967 hurricane with 121 trees felled by Hurricane Vivian in 1990. In this analysis, only 5% of the *Picea abies* were uprooted, whereas 95% of them were snapped (Schmid-Haas und Bachofen 1991).

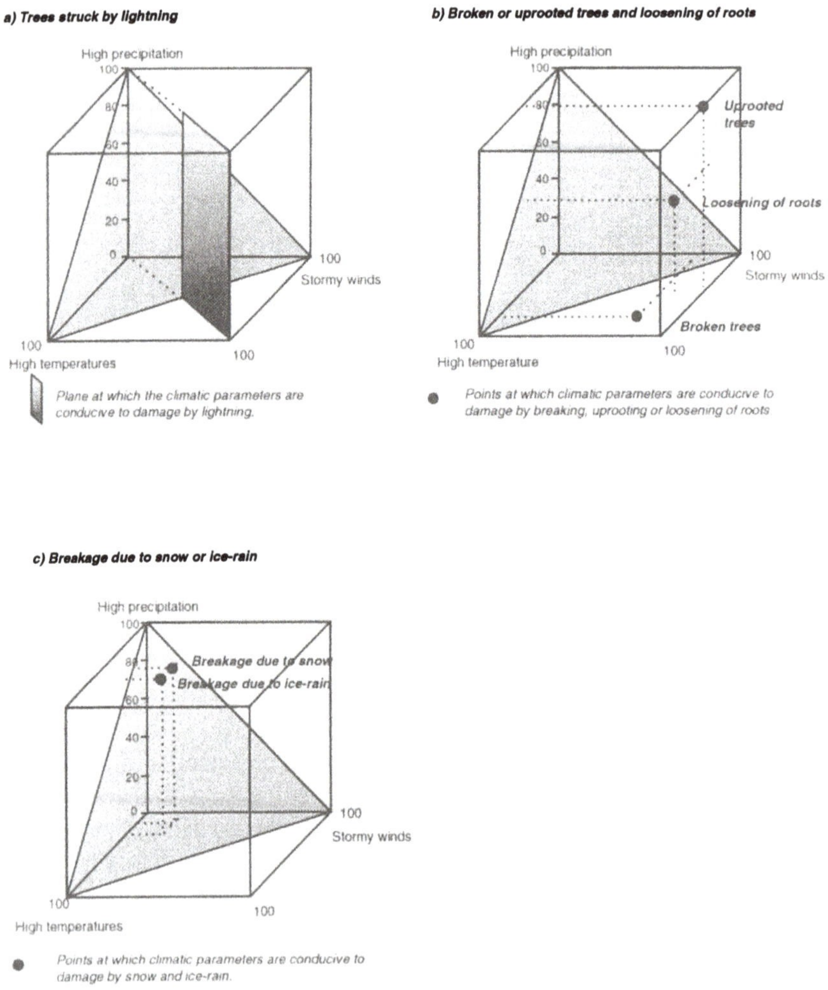

Fig. 6. There is a name for "forest disasters"... Is it possible to group the hazards affecting our forests according to the combination of climatic factors which are characteristic of them?

4. A hypothesis and some suggestions

The combination of climatic factors – rainfall, temperature and wind – together with the seasonal state of development as defined by phenological indicators (Defila 1991), may entail events which are damaging to the forest. Owners and managers know them well and use very precise terms for labelling them (see Figs. 6a, 6b and 6c). In this connection it is vital to pay special attention to phenology, since a stand with fully-developed leaves offers much more resistance to wind.

These observations lead us to propose a balanced appraisal of the most important meteorological factors which are likely to lead to a forest disaster. There has to be some critical level which combines the wind, rainfall and temperature factors, and which indicates a high risk to the forest. This risk would be greater for all the scenarios above this critical level than the maximum values for each of the factors taken in isolation (see Fig. 7). In a matrix of correlations, exceptional values for rainfall, temperature and wind speed should be assessed at least two weeks before the day of the forest disaster, and linked to the phenological data.

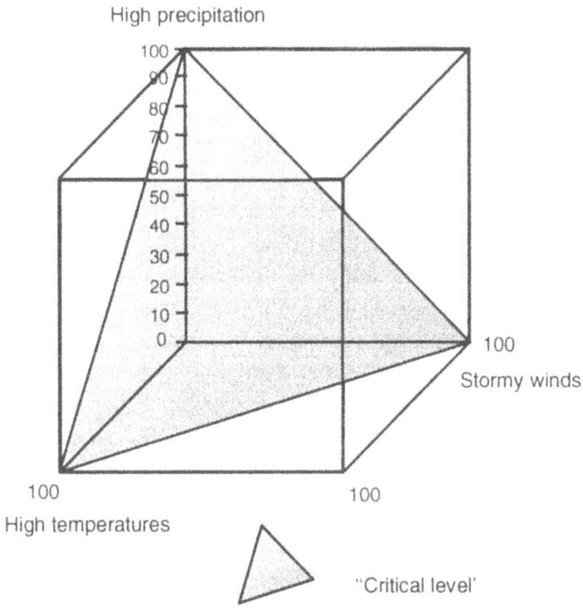

Fig. 7. A critical level resulting from exceptional values for rainfall, temperature and wind speed. For all combinations above that level, forest disasters might occur.

5. Some conclusions drawn by the forest manager

Disasters are traumatic because they destroy material property which we value highly. Up to a point, for nature there is no such thing as a disaster, and extreme phenomena are often opportunities for a new beginning in the evolving process of an ecosystem. But a combination of a few extremes can lead to a point of no return, beyond which natural cycles are no longer able to recover. Can we qualify and quantify this threshold? Is it possible to anticipate disasters by recognising, in advance, trends which signal danger?

In the face of so-called forest disasters, it is essential to promote an integral approach which will make it possible, if necessary, to recognise the combination of conditions which may induce potential danger instead of looking out for one single factor which could be responsible for an exceptional event. In all cases, such a procedure should also include all relevant site factors and, obviously, the characteristics of the stands in question.

Salvage felling as a result of the storms of recent decades concerns all those who work in forestry and, using the criterion of stand stability, puts to the test silviculture as it is practised in Swiss forests. There is little doubt that well-graded and multiform stands, composed of species which are suitable for the site and, if possible, of local provenance, generally have a better chance of withstanding climatic extremes or particularly adverse combinations of atmospheric factors. There is, however, no magic formula: In the municipal forests of Vallorbe, a typical Jura stand of *Fagus sylvatica* with a natural admixture of *Picea abies* and *Abies alba* (Div. 401) was particularly badly damaged by Hurricane Vivian on 28 February 1990 because the conifers, with their fully-needled crowns, leant against the *Fagus sylvatica* which would have resisted the wind quite well, as they were leafless at the time.

Incidentally, it is not impossible that a retrospective explanation may, at least partially, be found for certain symptoms of the forest decline of the 1980s: since the late 1970s a repeated combination of high summer temperatures, prolonged periods of drought and gales weakened the roots of many *Picea abies* and *Abies alba*. The subjective and not very spectacular phenomenon of leaning trees could partly explain why our forests looked such a sorry sight for a decade. Fifteen years on, the most questionable trees have been harvested, either through continuous salvage felling or in the context of periodic cuttings. The "forest disaster" of the 1980s has gradually been superseded by a combination of more harmonious climatic factors.

Acknowledgements. Thanks are due to Martine Rebetez for providing the meteorological data and many suggestions and to Karine Pythoud for drawing the figures and illustrations.

6. References

Bantle H (1989) Programmdokumentation Klima-Datenbank am RZ-ETH Zürich. Swiss Meteorological Institute, Zurich

Defila C (1991) Pflanzenphänologie der Schweiz. Diss. Uni Zürich. In: Veröffentlichungen der Schweiz. Meteorologischen Anstalt, Nr. 50, pp 235

Holenstein B (1994) Dégâts provoqués par la tempête de 1990 dans les forêts en Suisse. Cahier de l'environnement N° 218. OFEFP, Berne, pp 41

Lätt N (1991) Zum Zusammenhang zwischen Kronenschäden und Windfallholzanteil. Schweiz Z Forstwes 142:109–131

Schiesser H-H, Waldvogel A, Schmid W, Willemse S (1997) Klimatologie der Stürme und Sturmsysteme anhand von Radar- und Schadendaten. Vdf Publishers, Zurich, pp 134

Schmid-Haas P, Bachofen H (1991) Die Sturmgefährdung von Einzelbäumen und Beständen. Schweiz Z Forstwes 142:477–504

Scherrer H-U, Schmidtke H (1997) Sturmschäden im Wald. Projektschlussbericht im Rahmen des Nationalen Forschungsprogrammes "Klimaänderungen und Naturkatastrophen", NFP 31. Vdf, Hochschulverlag, Zürich, pp 38

Wangler F (1997) Die Sturmgefährdung der Wälder in Südwestdeutschland – Eine waldbauliche Auswertung der Sturmkatastrophe 1967. Diss. Univ. Freiburg, pp 226

Worldwide positions of alpine treelines and their causes

Christian Körner

Abstract. On a global scale, climate-driven high-altitude treelines occur at surprisingly similar growing-season temperatures (means from 5.5 to 7.5 °C), whereas season length varies between 2.5 and 12 months, and many other climatic constraints which have been suggested as contributing to alpine treeline formation show large regional variations. In order to resolve this apparent discrepancy, which has hindered attempts to explain treelines, I suggest separating "modulative" (regional) from "fundamental" (global) forcing factors. Most of the literature relates to modulative factors on treeline position (i.e. the fine tuning – such as by winter drought damage on saplings, which is absent in the tropics). As the underlying and unifying global determinant (with less precise predictive potential at the local scale), I suggest direct impacts of low temperature on meristem activity. Hence, I hypothesize a sink, rather than a carbon source (photosynthesis) limitation of tree growth at high elevations. I suggest upright trees become thermally constrained by their own life form, because of the close thermal coupling of their shoots to the atmosphere and through self-shading of their root zone. Alpine vegetation (including small tree seedlings) escapes close thermal coupling and cold soil by adopting compact life forms, by radiant canopy warming and by a high soil heat flux. The physiological limitations are likely to be the same.

Key words. Timberline, altitude, climate, temperature, growth, developement, photosynthesis

1. Introduction

A substantial literature exists on the alpine treeline phenomenon, dating back to the last century (see reviews by Brockmann-Jerosch 1919, Daubenmire 1954, Wardle 1974, Tranquillini 1979, Grace 1989, Ohsawa 1990, Miehe and Miehe 1994). Some of the studies reviewed by these authors have brought us nearer, some further away from a global perspective, and much of the current knowledge reflects experiences from selected regions of the temperate zone, largely the central Rocky Mountains and the central Alps. There is no conclusive explanation for the climatically driven alpine treeline and some authors believe that none should be expected. In this brief paper I summarize those facts which suggest a common climatic control of alpine treelines and I develop ideas which may lead to a better functional understanding of this important vegetation boundary. A more detailed discussion and a more complete assessment of the relevant literature will be published elsewhere (Körner 1999).

2. The altitudinal position of treelines

Any definition of a vegetation boundary is a convention (Fig. 1). Here I define alpine treeline (a synonym for alpine forest line) as a line connecting the upper ends of patches of closed forest (individuals >3 m), a boundary which is above the limit of continuous forest ('timberline', often even more difficult to define) and below the line of tree species occurrence (sometimes associated with the upper end of a 'krummholz' belt). Fig. 2 illustrates the global altitudinal distribution of treeline and snowline as a function of latitude. Treeline altitudes increase from subarctic to subtropical latitudes by 130 m per degree of latitude, with the altitudinal advance accelerated in the subtropics and slower in the subarctic–temperate latitudes (75 m per degree). The pan-tropical plateau of treeline altitudes parallels trends in snowline and seems to reflect the physical consequences of increased cloudiness and precipitation at equatorial latitudes (e.g. Ohsawa 1990).

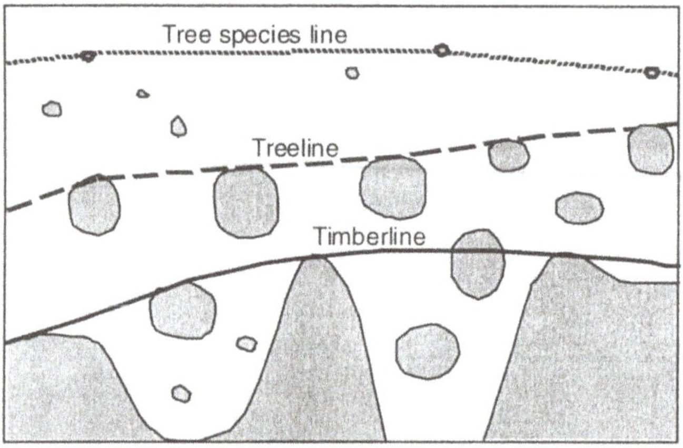

Fig. 1. The definition of vegetation boundaries such as the altitudinal limit of forests is a matter of convention. This figure illustrates the three most commonly used high-altitude boundary lines of forests.

In the absence of human interference or wildfires the treeline occurs at thermally similar altitudes across the globe (see below), independently of local radiation regimes, wind, soil conditions or extreme temperature events. It is thus plausible to assume that the treeline is, above all, a temperature-driven boundary. The question then is which element of the temperature climate is critical:

- air or soil temperature, or both ?
- the absolute minimum temperature ?
- the temperature of the warmest month?
- seasonal mean temperature?
- duration of periods with temperatures above a certain threshold?
- length of the snow-free period?
- the sum of temperatures above a certain threshold?
- yet others?

Absolute temperature minima can easily be ruled out as treeline determinants. There are no indications that the treeline is controlled by frost tolerance (Tranquillini 1979, Larcher 1985, Sakai and Larcher 1987). This does not mean that treeline trees do not experience sporadic freezing damage. Together with other similar sporadic or regional phenomena (e.g. winter desiccation, mechanical damage, pathogens etc.) such events contribute to the modulation of the treeline position but cannot explain its more fundamental cause, the sole topic of this paper (see below).

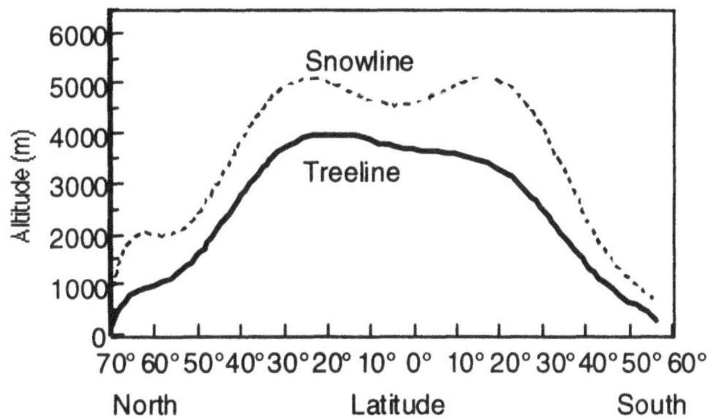

Fig. 2. A schematic presentation of the latitudinal variation of treeline positions across the globe. Original data for this diagram were derived from Hermes (1955), Wardle (1974) and personal observation, and include 150 treeline and 120 snowline positions. The snowline provides a physics-driven reference, a thermal boundary which is connecting points above which the ground remains snow covered throughout most of the year and precipitation falls as snow (an approximation of the elevation of the 0°C isotherm of the warmest month). Note the steep increase of treeline altitude across the subarctic–subtropical transition and the parallel pan-tropical plateauing of treeline and snowline altitudes (for details see Körner (1999).

The mean air temperature of the warmest month has little predictive value on a global scale. It may vary between 5.6 and 13 °C according to the literature (upper line in Fig. 3) and only in the Rocky Mountains and the Alps does it appear to fit the frequently suggested 10°C (July) isotherm (for references see first paragraph). There is also no physiological explanation as to why the mean temperature of a certain month should determine the existence of trees. Because the length of the growing season varies from 2.5 months at sub-arctic and alpine treelines to 12 months in the tropics, the predictive value of season length is also poor. Hence, other aspects of the thermal climate, such as sums of temperatures or number of days above a certain threshold temperature (5°C has been shown to be useful) or means of growing season temperature for either air or the root zone appear ecologically more relevant. It is not yet clear which of these has both greatest predictive and explanatory power. However, it is important to note that frost, season length and the temperature of the warmest month have no or little predictive value at a global scale. A survey of treelines for which whole-season climatic data are available suggests that a seasonal mean air temperature somewhere between 5 and 7.5°C is the most consistent predictor of the altitudinal position of treelines on a worldwide basis (lower line in Fig. 3).

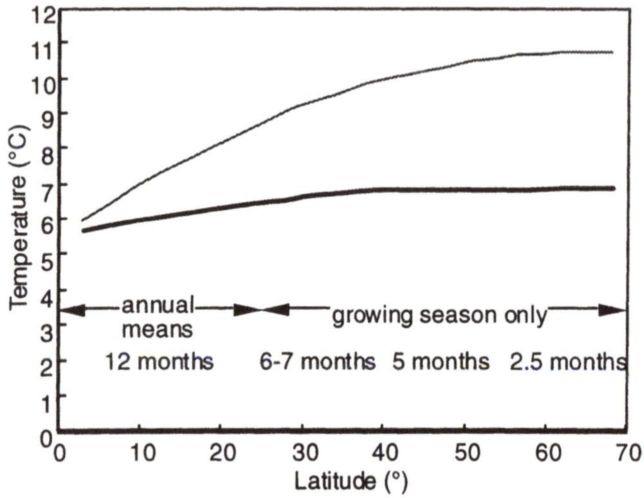

Fig. 3. Relationship of alpine treeline position to the warmest month temperature (upper thin line) and the mean air temperature during the whole growing season (lower line), based on an extensive literature review (simplified from Körner 1999).

3. Possible causes for the alpine treeline

What are the possible causes of forest limitation at this common thermal boundary? In answering this question, one must account for the possibility of autocorrelation of temperature with other components of the climate, a latitudinal bias of means because of different day/night length and non-linearity of biological temperature responses, and memory effects on treeline position (reflecting historical rather than current climate, see the discussion in Körner 1999). In view of the above discussion about freezing resistance and season length, three possible explanations for a thermal boundary seem plausible: (1) winter drought, (2) limitations of photosynthesis and the carbon balance, (3) direct constraints for assimilate investments (tissue formation).

The first phenomenon is of only regional relevance and may, similar to frost, modulate treeline position, but cannot be considered a fundamental driver at a global scale (Körner 1994, 1999). Treelines exist worldwide at seasonal mean temperatures comparable to those found in the temperate zone (lower line in Fig. 3), where winter drought has been suggested to be important, but maritime, subtropical and tropical treelines cannot be explained by 'winter' drought. According to current knowledge there is also no plausible reason to believe that trees at treeline suffer from insufficient photosynthetic carbon acquisition: optimal thermal acclimation of photosynthesis is a well-documented phenomenon in treeline trees (e.g. Tranquillini 1979) and, as mentioned above, the duration of the photosynthetically active period cannot be considered a worldwide determinant, because alpine treelines occur at rather different season lengths. Reports of a recent stimulation of tree growth near treeline by atmospheric CO_2-enrichment are not conclusive because they either relate to responses of heavily damaged stems (Graybill and Idso 1993; the damage history alone could lead to the stimulation) or to trees growing below the treeline (Nicolussi et al. 1995), and neither moisture effects nor responses to soluble nitrogen deposition or altered forest management can be ruled out. A global survey of tree-ring responses in boreal treelines did not reveal any stimulation over the last decades, but some qualitative changes (Briffa et al. 1998).

Hence, phenomena associated with the growth process *per se* need to be considered. Similar to conclusions for alpine plant growth in general (Körner and Larcher 1988) it is hypothesized here that beyond the alpine treeline temperatures are too low for trees to develop new tissue, irrespective of the carbon supply status, an inference previously cited by Däniker (1923). A thermal limit to such formative processes was also suggested by Higgins and Spomer (1976). However, why should such constraints be restricted to trees? There are also hundreds of non-arboreal species which can cope with the climate at much higher altitudes.

There is a fundamental, morphology-driven difference between the climate of trees and low stature vegetation. First, the tree life-form causes self-shading of its root zone and an uplift of energy exchange several meters above the ground, resulting in a reduction of soil heat-flux to almost zero (Figs. 4, 5). As a consequence, a life-form, evolutionarily selected for light competition screens solar radiation so that its own roots may not receive sufficient warmth to continue cell division and cell differentiation. It seems that the only chance for a tree to survive above treeline altitudes is to (teleologically speaking) separate itself from other trees and allow solar radiation to reach the ground. Such wider tree spacing is seen at all low-temperature treelines worldwide, including the arctic lowland ones. Tall isolated tree individuals are often found several hundred meters above treeline (e.g. Miehe and Miehe 1994). Their roots spread into surrounding low-stature vegetation and profit from radiative soil warming as illustrated in Fig. 4 and 5. As soon as trees form clusters and develop into larger groups, they risk cutting their own thermal lifeline (e.g. Holtmeier and Broll 1992).

A second problem is that trees have their shoot apices exposed to the fully convective atmosphere and thus are experiencing lower temperatures than prostrate vegetation. Critically low temperatures may restrict meristematic activity every night (Körner and Pelaez Menendez-Riedl 1989). James et al. (1994) observed that shoot extension growth in *Pinus sylvestris* at the Scottish treeline is closely related to temperature, and a variety of growth chamber studies suggest such a thermal limit for growth, independent of photosynthesis (see reviews by Tranquillini 1979, Körner 1999). In this respect upright trees once more contrast with low-stature vegetation, which either has meristems embedded in the surface-soil and profits from the heat stored during the day, or is favoured by the well-known higher temperatures within a narrow prostrate canopy as a result of reduced atmospheric coupling (Grace and Norton 1990, Squeo et al. 1991).

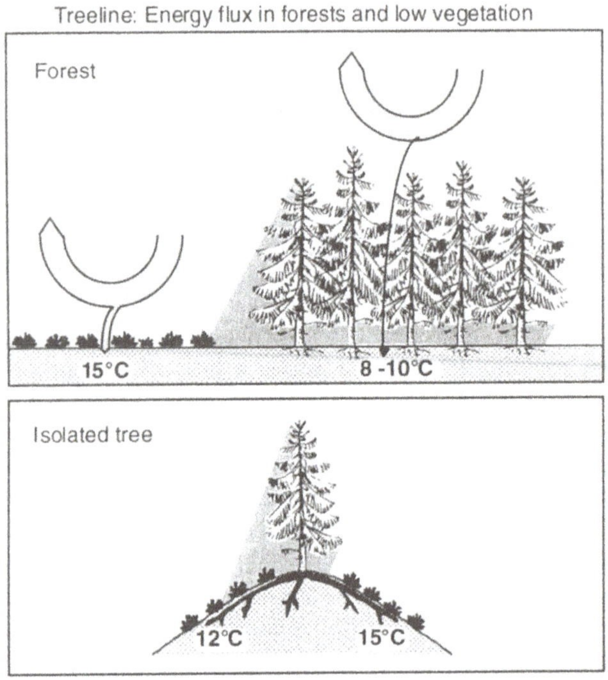

Fig. 4. A schematic illustration of the negative effect of a closed tree canopy on root zone temperature at the treeline. Examples of root zone temperatures for a midsummer situation in the Alps (mean soil temperatures from own field measurements). Arrows symbolize the major daytime fluxes of energy (radiation input versus sensible plus latent heat output and soil heat flux). Note the warmer root zone around isolated trees (see also Fig. 5).

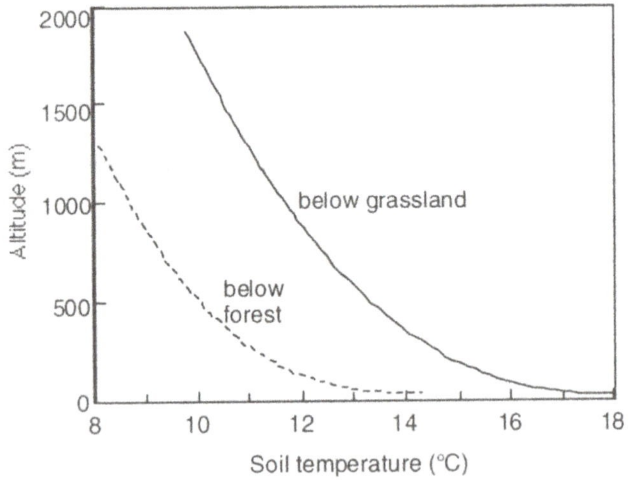

Fig. 5 The altitudinal variatio.n of mid-season -20 cm soil temperature under low stature vegetation (grassland) or under closed *Nothofagus* forests. Data are from an elevational transect in the southern island of New Zealand (from Körner et al. 1986).

In other words, I suggest that the tree life-form is unable to grow and survive in closed groups (i.e. patches of 'forest') above treeline altitudes, because both shoot and root apices are thermally limited in their carbon investment activity. Several studies have shown that shoot and root growth stop at temperatures below 7°C, temperatures which would only moderately constrain photosynthesis (Tranquillini 1979, Körner 1999; see also Körner and Larcher 1988 for the discrepancy between growth and gas-exchange limitations by low temperature). There is clear climatological evidence that high-altitude soils under trees are always colder than in nearby open terrain. Alpine vegetation several hundred meters above treeline often experiences soil temperatures much warmer than those measured under a closed tree canopy in the upper mountain forest (Fig. 4, 5).

4. Conclusion

I propose an investment limitation rather than a carbon gain limitation for treeline trees. This is the only plausible explanation for why trees in montane forests show only modest reductions in annual increments of tree-ring width up to 100 to 200 m below treeline. The dramatic decline of ring width, often restricted to the last few tens of meters of altitude, can neither be explained by changes in growing season length nor photosynthetic rate. Because frost and other more locally important phenomena can be ruled out as global drivers of treeline, developmental and formative processes are the most likely candidates which determine and limit tree growth at high altitudes. The life form specific restriction of soil heat flux seems to strongly contribute to the abrupt transition from forest to tree-less vegetation.

As a first approximation, I suggest in accordance with earlier suppositions (e.g. Walter and Medina 1969) that natural climatic treelines should occur where mean growing season temperatures drop below 6.5 ± 1°C. However, as mentioned above, means are only a first approximation and need to be replaced by better descriptors such as frequency distributions of temperatures or thermal sums. Soil temperatures measured under trees near treeline appear to have the greatest predictive power, but have rarely been studied (see review by Körner 1999). A global survey of treeline soil temperatures is now under way[3].

5. Future Research

In order to substantiate these conclusions, future research would have to focus on three areas: (1) document the thermal life conditions of trees at climatic alpine treelines around the world with specific emphasis on root zone and shoot apex

[3] note added in press: A first set of full season treeline root zone temperatures from 2 sites in Kirgistan (ca 2900m) and 3 sites in the Swiss Central Alps (ca 2300m) reveals a surprisingly narrow range of means between 6.3 and 7.3 °C (median 6.0 to 7.4 °C, -10 cm soil depth under dense tree canopies) with growing periods between 128 and 181 days duration (as indicated by sharp temperature transitions associated with spring snowmelt or autumnal freezing; unpublished data which became available after this text was completed). These numbers are in line with the above prediction, but data from the subtropics and tropics are not yet available.

temperatures. (2) Assess the seasonal variation of the carbon supply status of trees at and below treeline by monitoring potentially mobile pools of C-compounds in all organs. (3) Study the temperature dependency of the division and differentiation of cells in meristems of treeline trees. Such studies should adopt a comparative approach and include different taxa of treeline forming species and treelines from contrasting climatic zones.

6. References

Briffa KR, Schweingruber FH, Jones PD, Osborn TJ, Shiyatov SG, Vaganov EA (1998). Reduced sensitivity of recent tree-growth to temperature at high northern latitudes. Nature 39:1678 – 682

Brockmann-Jerosch H (1919) Baumgrenze und Klimacharakter. Pflanzengeographische Kommission der Schweiz. Naturforschenden Gesellschaft, Beiträge zur geobotanischen Landesaufnahme 6, Rascher & Cie., Zürich.

Däniker A (1923) Biologische Studien über Baum- und Waldgrenze, insbesondere über die klimatischen Ursachen und deren Zusammenhänge. Vierteljahresschrift Naturf Ges Zür 68:1–102

Daubenmire R (1954) Alpine timberlines in the Americas and their interpretation. Butler Univ Bot Stud 2:119–136

Grace PJ (1989) Tree lines. Phil Trans R Soc London, Ser B 324:233–245

Grace J, Norton DA (1990) Climate and growth of Pinus sylvestris at its upper altitudinal limit in Scotland: Evidence from tree growth-rings. J Ecol 78:601–610

Graybill DA, Idso SB (1993) Detecting the aerial fertilization effect of atmospheric CO2 enrichment in tree-ring chronologies. Global Biogeochem Cycles 7:81–95

Hermes K (1955) Die Lage der oberen Waldgrenze in den Gebirgen der Erde und ihr Abstand zur Schneegrenze. Kölner geographische Arbeiten Heft 5 , Geographisches Institut, Universität Köln

Higgins PD, Spomer GG (1976) Soil temperature effects on root respiration and the ecology of alpine and subalpine plants. Bot Gaz 137:110–120

James JC, Grace J, Hoad SP (1994) Growth and photosynthesis of Pinus sylvestris at its altitudinal limit in Scotland. J Ecol 82:297–306

Holtmeier FK, Broll G (1992) The influence of tree islands and microtopography on pedoecological conditions in the forest–alpine tundra ecotone on Niwot Ridge, Colorado Front Range, U.S.A. Arct Alp Res 24:216–228

Körner C, Pelaez Menendez-Riedl S (1989) The significance of developmental aspects in plant growth analysis. In: Lambers H, Cambridge ML, Konings H, Pons TL (eds) Causes and consequences of variation in growth rate and productivity of higher plants. SPB Acad Publ, The Hague, The Netherlands pp 141–157

Körner Ch (1994) Biomass fractionation in plants: a reconsideration of definitions based on plant functions. In: Roy J, Garnier E (eds) A whole plant perspective on carbon–nitrogen interactions. SPB Acad Publ, The Hague, The Netherlands pp 173–185

Körner Ch (1999) Alpine plant life. Springer Verlag, Heidelberg, in press

Körner Ch, Bannister P, Mark AF (1986) Altitudinal variation in stomatal conductance, nitrogen content and leaf anatomy in different plant life forms in New Zealand. Oecologia 69:577–588

Körner Ch, Larcher W (1988) Plant life in cold climates. In: Long SF, Woodward FI (eds) Plants and temperature. Symp Soc Exp Biol 42: 25–57. The Company of Biol Ltd, Cambridge

Larcher W (1985) Winter stress in high mountains. In: Turner H, Tranquillini W (eds) Establishment and tending of subalpine forest: Research and management. Proc 3rd IUFRO Workshop p 1.07-00, 1984. Ber Eidg Anst forstl Versuchswes 270:11–20

Miehe G, Miehe S (1994) Zur oberen Waldgrenze in tropischen Gebirgen. Phytocoenologia 24:53–110

Nicolussi K, Bortenschlager S, Körner Ch (1995) Increase in tree-ring width in subalpine *Pinus cembra* from the central Alps that may be CO_2-related. Trees 9:181–189

Ohsawa M (1990) An interpretation of latitudinal patterns of forest limits in south and east Asian mountains. J Ecol 78:326–339

Sakai A, Larcher W (1987) Frost survival of plants. Responses and adaptation to freezing stress. Ecological Studies 62, Springer, Berlin, Heidelberg, NY

Squeo A, Rada F, Azocar A, Goldstein G (1991) Freezing tolerance and avoidance in high tropical Andean plants: Is it equally represented in species with different plant height? Oecologia 86:378–382

Tranquillini W (1979) Physiological ecology of the alpine timberline. Tree existence at high altitudes with special references to the European Alps. Ecological Studies 31, Springer, Berlin, Heidelberg, New York

Walter H, Medina E (1969) Die Bodentemperatur als ausschlaggebender Faktor für die Gliederung der subalpinen und alpinen Stufe in den Anden Venezuelas. Ber Dtsch Bot Ges 82:275–281

Wardle P (1974) Alpine timberlines. In: Ives JD, Barry RG (eds) Arctic and alpine environments. Methuen, London, pp 371–402

Vascular plant species richness in relation to altitudinal and slope gradients in mountain landscapes of central Norway

Jarle I. Holten

Abstract. Local plant ecological investigations in the central Norwegian mountains in 1992–1997 have shown some interesting features regarding the variability of vascular plant species richness along altitudinal gradients. The material reveals two peaks of vascular plant species richness with increasing elevation, a lowland peak at 0–400 m a.s.l. and a peak at the timberline area (upper part of the northern boreal zone), around the inflection line. Mountains with highly acidic bedrock have a vegetation discontinuity around the transition between discontinuous and continuous permafrost (1500 m in the Dovrefjell area), with a change from dwarf shrubs to more graminoid life forms. The angle of slope is decisive for soil-forming processes. The instability of steep slopes prevents the formation and accumulation of organic top-soils. The data show a high, positive correlation between the slope of habitat plots and the richness of vascular plant species, in both the forested and the alpine zones. A working hypothesis is put forward that, due to high substratum instability, steep terrain encourages high species richness due to the greater 'openness' of habitats and the higher pH of the top-soils. It is suggested that this effect of local topography on species richness is strongest around the inflection line.

1. Introduction

A literature review (Rahbek 1995) of altitudinal patterns of species richness in various taxonomic groups shows a great variation between the latter and geographical regions. For the high alpine zone (> 3000 m) in central Europe, Grabherr et al. (1995) described a characteristic negative exponential decrease in species richness towards higher elevations. Holten & Wilmann (1996) and co-workers found a maximum in species richness around the timberline (800–1150 m asl). Recent investigations by the author in the continental Dovrefjell area support Grabherr et al.'s (1995) negative exponential pattern of decrease.

This paper considers the following questions concerning causal relationships between landscape processes and biodiversity:

- What patterns of altitudinal decrease of species richness can be recognised among vascular plants in the central and southern Scandes?
- Do the life forms of vascular plants change towards higher elevations?
- What connection can we find between terrain slope and species richness?

2. Material and Methods

To study the altitudinal variation of species richness, diversity data were collected by 'walking up' altitudinal transects on mountain slopes during the period 1990–1997. Total floras (florulae), consisting of lists of vascular plants, were made separately for each 50 m or 100 m altitudinal interval and their abundance was noted on a scale from 1 to 4 (1 – rare, 2 – sporadic, 3 – common, 4 – dominant).

About one hour was spent on each 100 m belt. The area investigated within each altitudinal belt varied due to local variations in steepness and other changes in the topography. Based on a 20 m broad corridor investigated within each altitudinal belt of 100 m, and a total length varying from 300 to 800 m (mainly depending on the steepness), the 'unit area' for each flora list varied from approximately 60 to 160 ha.

Data for studying the relationship between slope and species richness were taken from Holten & Wilmann (1996) and their co-workers. Spearman rank correlation analysis was used to quantify this relationship.

3. Results

3.1 Altitudinal gradient

Preliminary results from investigations of vascular plant species richness in a vertical coast–inland transect in western-central Norway (own observations 1990–1997), indicate two maxima of species richness (Figure 1): a general lowland maximum from sea level to about 300 m, and a maximum from about 200 m below to 200 m above the timberline. They also indicate that the upper 'peak' of maximum species richness rises as one moves inland. The optimum altitude seems to be 300–400 m on coastal slopes, around 700 m in the western valley districts, and about 1000–1100 m close to the main watershed of the Scandes mountain range in the Oppdal–Dovrefjell area. The upper peak of species richness seems to be situated just below the current *Betula* timberline (Figure 2).

The EU 'Mountain Biodiversity Project' (Holten & Wilmann 1996) undertook biodiversity studies in the coast–inland transect described above. It concentrated on a mountainous area (20 x 12 km) in the suboceanic to subcontinental valleys of western-central Norway. Holten & Wilmann (1996) showed a surprisingly high vascular plant species richness at relatively high elevations in this area. The zone of maximum species richness seemed to be between 800 and 1100 m asl. A comparable graph for the Scottish site of the project, in the Cairngorms, differs completely, showing an almost negatively exponential decrease from 550 to 1050 m asl.

From 1992 to 1997, the author collected data on the altitudinal variation of species richness along south-facing mountain slopes in a coast–inland gradient in central Norway, from the extreme coast with its mild, humid climate to the Dovrefjell plateau in the central part of the Scandes mountain range, which experiences a dry climate with cold winters. Eight mountains were investigated, from the lowest possible to the highest possible altitude (Figure 2). These mountains represent three different macroclimates (see top of bars in Figure 2), oceanic (M, H, R), suboceanic to subcontinental (F, G) and continental (Si, Sn, N). Extremely steep topography with many inaccessible slopes and summits has so far prevented the area between 40 and 80 km from the coast from being investigated.

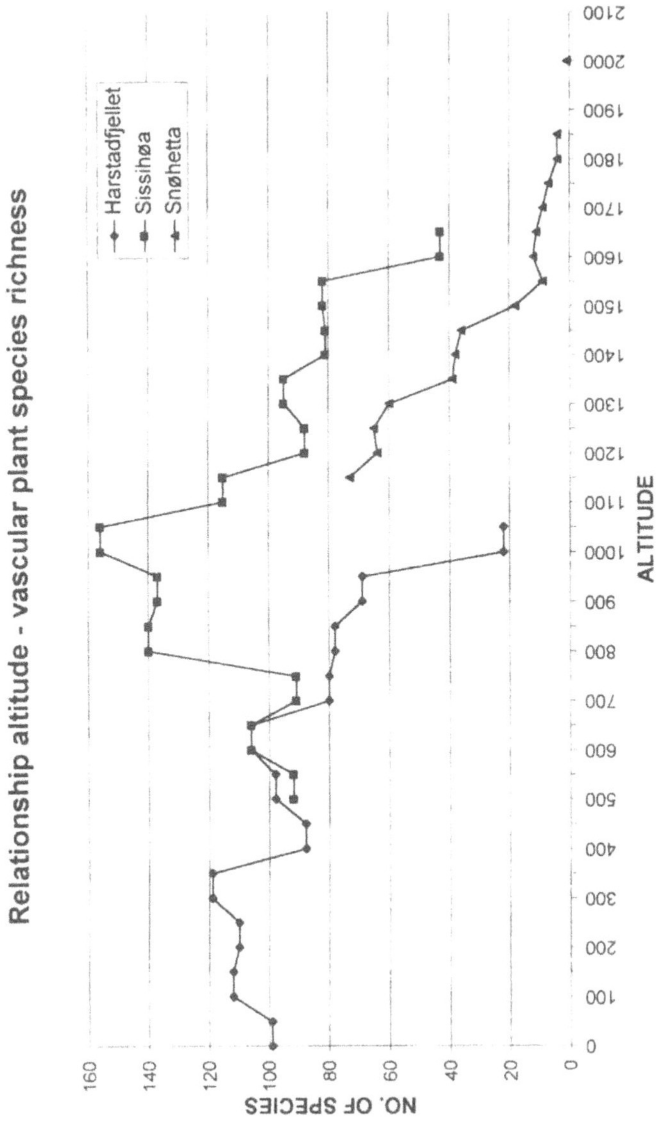

Fig. 1. The pattern in the decrease of vascular plant species richness towards higher levels on three mountains in central Norway: The oceanic Harstadfjellet (1004 m) has hard, acidic bedrock, the continental Sissihøa (1621 m) has calcareous bedrock, and the continental Snøhetta (2286 m) has hard, acidic bedrock.

As this is a pilot study to form working hypotheses and draw preliminary conclusions about the factors governing vascular plant species richness along altitudinal transects, only three (the broad bars in Figure 2) of the eight mountains were selected for special study: These were the oceanic Harstadfjellet with hard, acidic bedrock, the continental Sissihøa with calcareous bedrock, and the continental Snøhetta with hard, acidic bedrock.

The altitudinal variation in species richness on these three mountains differed (Figure 1). Harstadfjellet shows a gradual decrease in species richness up to the summit plateau, from about 70 species in the 900–1000 m belt to 18 within the 4 vertical metres spanning the summit plateau (1000–1004 m). Sissihøa (1621 m) shows a marked peak in species richness between 1000 and 1100 m, which is close to the timberline; its summit area is rounded and has a fairly high species richness, totalling 43 species belonging in the upper part of the middle alpine zone. Snøhetta (2286 m) shows a negative exponential decrease in species richness from 1150 to 2005 m. At the lower limit of the discontinuous permafrost zone (Ødegård et al. 1996), there is a marked discontinuity in species richness at 1500 m, decreasing by 50% through 50 vertical metres, from 36 species at 1500 m to 18 at 1550 m.

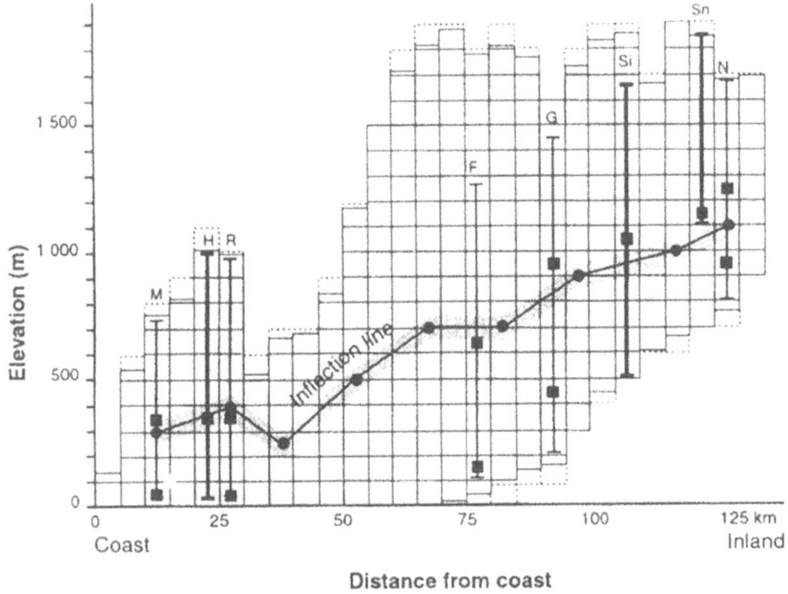

Fig. 2. The coast–inland altitudinal transect showing the position of the eight south-facing mountain slopes. Broad bars: the mountains of Harstadfjellet (H), Sissihøa (Si) and Snøhetta (Sn), (Figure 1) selected for more thorough treatment in this paper; narrow bars: the mountains of Mælen (M), Reinsfjellet (R), Fahlelia (F), Gråurda (G) and Nord Knutshø (N), where lists of vascular plants were made. The inflection line (based on Holten 1986) joins points identified on maps as marking the significant break of slope on the hillside concerned. Peaks of species richness are shown by squares.

3.2 Slope gradient

Slope information was derived from data collected from 1130 plant community plots studied during the 1993–95 EU Mountain Biodiversity Project (Holten & Wilmann 1996). A Spearman rank correlation analysis of slope information related to vascular plant species richness gave a coefficient of $r = 0.70$ ($p < 0.001$), (see Figure 2). The wet, more or less mire-like habitats, oligotrophic forests and heaths, are concentrated in the lower and left parts of the diagram (habitat series M1-5, H1-5, PS1-3 in Figure 3), all of which have almost flat (0–5^{g}) to gently sloping plots (5–15^{g}). These habitats are characterised by soils having a rich organic component, such as peat or strongly podzolised soils with thick raw humus. The species richness is very low in these habitats, varying from 8 to about 17, with a maximum in M5 of 19 species per plot.

The habitats on moderate slopes (15–25^{g}) are much richer in herbs and graminoids, typified by the *Betula* habitat series HB1-2 and B1-3 with intermediate species richness (12–22 species). Eutrophic deciduous forest vegetation is the predominant habitat in steep terrain. Only two of the habitats studied are in this 'steep' category, H3 (mesic, eutrophic heath) and MS1 (eutrophic herb snow patch), the latter having the second highest species richness recorded (29.4 species). No information about species richness is available for slopes in excess of 45^{g}.

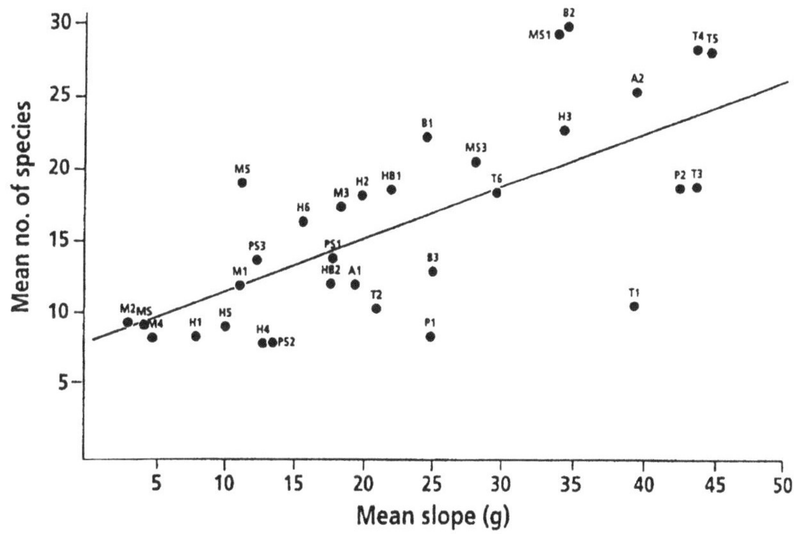

Fig. 3. The relationship between vascular plant species richness and the slope (inclination) of plant community plots. The slope figures in the diagram are based on the median values of slope figures within each community type.

4. Discussion

4.1 Altitudinal gradient

The increase in the vascular plant species richness observed at Harstadfjellet (H) and Sissihøa (Si), reaching a maximum in the upper part of the wooded zone, could have various causes (Figures 1 and 2). The same general pattern is also observed on the coastal mountains of Mælen (M) and Reinsfjellet (R), the suboceanic to subcontinental mountains of Fahlelia (F) and Gråurda (G), and the continental Nord Knutshø (N). A lowland maximum species richness is also observed on these five mountains. Based on topographical maps (1:50 000), Holten (1986) defined an inflection line in an altitudinal transect described from the same coast–inland transect (Figure 2). This represents the transitional area between the predominantly concave topography of the lowlands and the predominantly convex topography of the uplands. It may also be defined as the point along an altitudinal gradient where the slope of the terrain changes from being steeper upwards to becoming less and less steep (= convex terrain). In other words, the inflection line marks the area in the altitudinal gradient that generally has the steepest terrain. The inflection line can be at the same time interpreted as a kind of border area between two hydrological regimes, an upland area with predominantly downward-moving soil water (podzolisation), giving acidic soils, and a lowland area with lateral- or upward-moving soil water, giving less podzolised or 'brown forest' soils with a higher nutritional status. However, the entire altitudinal transect has a mixture of convex and concave topography giving a variety of soil types; hence, the inflection line is regarded as a generalisation for a wide area. The inflection line is thought to be important for both soil-forming processes and the formation of local climatic regimes, including a much windier and cooler climate above the line and a less windy and locally warmer climate below. It is interesting to see that the upper peak of species richness in Figure 2 is closely associated with the position of the inflection line. However, the data so far available are unable to satisfy the demands of a statistical test to show whether this correlation is significant. The following working hypothesis to explain the peak of species richness around the inflection line is put forward:

The area around the inflection line mainly consists of a mixture of boreal–alpine and lowland habitats containing species that are widespread to dominant in the area, as well as many additional, less common, species that are on the border of their range, resulting in a local maximum of species richness.

There are alternative working hypotheses to explain the upper peak of species richness observed in seven of the eight transects. A steep south-facing slope may have an altitudinally fairly wide thermal belt with a favourable climate for lowland plants (Geiger 1976), resulting in a wide altitudinal range of constant, or even increasing, species richness towards higher levels. Alternatively, the grazing pressure by domestic animals, mainly sheep, has operated for centuries, mainly in the northern boreal and low alpine zones, keeping the landscape open and more favourable with regard to the light conditions, thereby creating niches for many vascular plant species that are adapted to, or even stimulated by, grazing.

The pattern of the decrease at Snøhetta differs from that in the other transects (Figure 1). The Snøhetta transect starts only about 50 m above the *Betula* timberline and species richness decreases more or less evenly up to the limit of vascular plant life at about 2000 m. This pattern may be regarded as being negatively exponential. A similar pattern was described by Grabherr et al. (1995) for the nival zone in the Alps. Such a pattern is probably characteristic for high levels of mountains like Snøhetta that are composed of hard, siliceous bedrock.

4.2 Slope gradient

Many authors have pointed out the importance of slope forces and soil instability for high plant species richness in mountainous landscapes. Chapin and Körner (1995) gave the following summary of the importance of slope forces or gravity for species richness: "In cold-dominated ecosystems the balance between the formation of a soil organic mat and disturbance results in an inverse relationship between soil carbon and species diversity... In the alpine zone, gravity (1) prevents water accumulation that would reduce decomposition and cause organic accumulation and (2) disrupts the soil organic mat as freeze–thaw action displaces the soil surface down-slope, opening space for many colonizing species. Such slope effects are found in both arctic and alpine regions, so that within each region the greatest species diversity is found on slopes steep enough to minimize soil organic accumulation" (see also Walker 1995).

The results (see Fig. 3) from the slope and species richness studies in central Norway strongly support the conclusions of Chapin and Koerner (1995) and Walker (1995). The habitats rich in carbon (peaty and podzolised soils) have flat (0–5^{g}), gentle (5–15^{g}) or moderate inclination (15–30^{g}). They are typically species-poor, containing oligotrophic species with a 'competitor' life strategy (Grime 1979). These habitats are very 'resistant' to invasion by other, more mesotrophic to eutrophic, species. With a long-term (100 to 1000 years) stable climate, podzolic soils will probably 'invade' and 'establish' the steeper slopes. However, such areas will easily become disrupted after thick mats of humus have built up, and will slide downwards. This kind of cyclical succession is probably characteristic for steep, acidic slopes in cool, oceanic climates. Changes in the species composition of such 'organic' habitats will only take place following physical disturbance caused by gravity or slope forces. Lapin and Barnes (1995) also recorded a maximum of species richness (alpha diversity) in the most steeply sloping parts of moist, nutrient-rich ecosystems in a temperate forest landscape in Michigan.

Where the angle of slope exceeds about 40^{g} (see Fig. 3), depending on the type of bedrock, slope forces will become active, giving high substratum instability and resulting in open scree vegetation or no vegetation at all (Blikra 1994). The habitats on very steep slopes are represented in the wooded zone by *Ulmus* forests (T 4-5), tall herb/fern *Betula/Alnus* forest (A2), calcareous *Pinus sylvestris* forest (P2), and continental meadows (T1, T3), and in the alpine zones by two types: meadow snow beds (MS1) and mesic, eutrophic heath (H3). It is suggested that the physical disturbance created by the slope forces through landslides, avalanches, etc., stimulates species with a somewhat ruderal life strategy in the sense of Grime (1979), frequently herbs and graminoids. Substratum instability will stimulate heterogeneity of micro-habitats on mountain slopes, from quite open communities where landslides have recently occurred to mesotrophic to eutrophic forest some tens of metres to one side. On a landscape scale, such heterogeneity will provide higher species richness, perhaps through cyclical changes determined by steepness and frequency of instability events. Beyond a slope of 45^g, substratum instability becomes very high, perhaps so high that species richness will be abruptly reduced when the inclination increases still more.

The relationship between species richness and slope angle will probably become non-linear after the curve peaks at 40 to 50^g. The 'critical slope' where species richness decreases again with greater steepness will probably vary from one type of bedrock to another, and will depend on the climatic regime. The 'critical slope of instability' is probably steeper in wet, oceanic climates where podzolisation and peat formation can take place on steeper slopes than in dry, continental climates with less cover of higher vegetation or, in other words, a higher cover of open mineral soil having less inherent cohesion.

If the results embodied in Figure 2 are compared with those in Figure 3, it is striking that the altitudinal area having the highest angle of slope (steepness) is, indeed, the area around the inflection line. The highest frequency of plant communities with the highest species richness may be expected in this area, as Figure 3 shows. Hence, increased substratum instability around the inflection line will stimulate ecosystem heterogeneity on various scales (mainly population, community and landscape scales), leading to higher species richness.

5. Conclusions

1) Preliminary results of a study of the altitudinal distribution of species richness of vascular plants in central Norway show two different patterns of decrease towards higher elevations:
   a) a negatively exponential decrease on high (often > 2000 m) mountains with acidic bedrock and alpine relief, and
   b) plant diversity studies in all eight altitudinal transects in central Norway show two peaks of species richness, a weaker peak in the lowlands, and a stronger one around the inflection line.
2) The peak of species richness observed in the area around the inflection line (the upper part of the northern boreal zone) is interpreted as being a result of steep terrain and substratum instability causing higher ecosystem hetero-

geneity and mosaic communities, probably accentuated in some areas by the high intensity of grazing by domestic animals.

3) The transition from sporadic to discontinuous permafrost at about 1500 m in the Snøhetta area coincides fairly well with the transition from the low alpine dwarf shrub zone to the middle alpine graminoid zone.
4) The slope of the terrain is decisive for soil-forming processes and many attributes of the vegetation, including species richness. Paludification and podzolisation on flat (0–5^g) and moderate slopes (15–30^g) create extensive mats of organic soils in the long run, reducing species richness.

6. References

Blikra LH (1994) Postglacial colluvium in Western Norway, sedimentology, geomorphology and palaeoclimatic record. – Dissertation, Univ. of Bergen

Chapin FS, Körner C (eds) (1995) Arctic and alpine biodiversity: patterns, causes and ecosystem consequences. Ecological studies 113, Springer-Verlag, Berlin

Geiger R (1976) The climate near the ground. Harvard University Press, Harvard

Grabherr G, Gottfried M, Gruber A, Pauli H (1995) Patterns and Current Changes in Alpine Diversity. Ecological Studies 113:167–181

Grime P (1979) Plant strategies and vegetation processes. John Wiley & Sons, Chichester

Holten JI (1986) Autecological and phytogeographical investigations along a coast–inland transect in Central Norway. University of Trondheim, Institute of Botany. Dr. philos. thesis. pp 342

Holten J.I, Wilmann B (1996) Habitat and associated species richness along altitudinal and slope gradients in Grødalen, Western Central Norway. In: Hill et al. (eds) Effects of Rapid Climatic Change on Plant Biodiversity in Boreal and Montane Ecosystems. Final Report. Institute of Terrestrial Ecology. Report to CEC DG XII. ENVIRONMENT 1991–1994, Contract No. EV5V-CT92-0090. pp 11–29

Lapin M, Barnes BV (1995) Using the landscape ecosystem approach to assess species and ecosystem diversity. Conserv Biol 9:1148–1158

Rahbek C (1995) The elevational gradient of species richness: a uniform pattern? Ecography 18:200–205

Walker MD (1995) Patterns and causes of arctic plant community diversity. Pages 3–20 in Chapin & Koerner (eds): Arctic and alpine biodiversity. Ecological Studies 113. Springer Verlag, Berlin, Heidelberg.

Ødegård RS, Hoelzle M, Vedel Johansen K, Sollid J L (1996) Permafrost mapping and prospecting in southern Norway. Norsk geogr Tidsskr 50:41–53

Environmental information from stable isotopes in tree rings of *Fagus sylvatica*

Matthias Saurer, Rolf Siegwolf, Silvio Borella and Fritz Schweingruber

Abstract. Stable isotopes in biological systems provide a valuable tool to infer environmental information from the present and the past. From isotope fractionation models, it is known that the $^{13}C/^{12}C$ ratio of plant material is dependent on the stomatal limitation of photosynthesis and on the intercellular CO_2-concentration, c_i. As these physiological parameters are dependent on micro-climatic conditions, the isotope ratios are also influenced by climate. We present data from cellulose of *Fagus sylvatica* covering the period from 1935 to 1990 from several sites in Switzerland and demonstrate that increased $^{13}C/^{12}C$ ratios are correlated with warm and dry conditions during the growing season. Further, the calculated c_i-values for *F. sylvatica* increase by about 20 ppm during the last 50 years which is significantly less than the increase of the atmospheric CO_2-concentration, c_a, during the same period. Accordingly, there is a strongly increasing trend for c_a-c_i whereas c_i/c_a is decreasing. The $^{18}O/^{16}O$ ratio in plant material is determined (i) by the oxygen isotope variations in precipitation which in turn are dependent on temperature, and (ii) by the ^{18}O enrichment in leaf water due to transpiration which is influenced mainly by the relative humidity of the atmosphere. From our tree-ring data, we conclude that $^{18}O/^{16}O$ variations in precipitation are preserved in tree-ring cellulose, whereas the signal for leaf-water enrichment is strongly dampened.

1. Introduction

Many elements of biological importance occur naturally with two or more stable isotopes. These include ^{12}C and ^{13}C for carbon, ^{16}O, ^{17}O and ^{18}O for oxygen, ^{1}H and ^{2}H (D) for hydrogen, whereby the major isotope is present with about 99% abundance or more. With the development of isotope-ratio mass-spectrometers, it has been found that there are variations in the isotope ratios between different parts of the environment. Such isotope differences are caused by physical, chemical and biological processes which usually discriminate against the heavier isotope. For instance, the $^{13}C/^{12}C$ isotope ratio in the carbon pool in the atmosphere (i.e. CO_2) is higher than the $^{13}C/^{12}C$ isotope ratio in the organic carbon pool in plants. As isotope discrimination during photosynthesis is dependent on the micro-climatic conditions in the surroundings of the plant, the isotope ratio of organic material contains valuable information about the environment.

The isotope signal is not only stored in leaf material, but is also reflected in stem wood, which is the basis for the use of stable isotopes in tree rings as a palaeoclimatic tool. An understanding of the fractionation processes during the uptake of isotopes by plants is a prerequisite for the interpretation of isotope variations. The respective theory is developed in the next two sections, focusing mainly on carbon and oxygen isotopes. We then present tree-ring data from several sites in Switzerland and evaluate the usefulness of isotopes in tree rings as a tool to infer climatic and physiological parameters.

2. Carbon isotopes

As natural isotope variations are small, it is convenient to express them as per mil-deviations from an internationally agreed standard in the δ-notation, i.e. $\delta^{13}C=[(^{13}C/^{12}C)_{sample}/(^{13}C/^{12}C)_{standard}-1]*1000‰$ (in an analogous way, $\delta^{18}O$ is defined as $[(^{18}O/^{16}O)_{sample}/(^{18}O/^{16}O)_{standard}-1]*1000‰$). The only source of carbon for plants is CO_2 and thus the isotopic composition of atmospheric CO_2 is the first value to be known to interpret the carbon isotope composition in the plant. Today, atmospheric CO_2 has a $\delta^{13}C$-value of about -8‰. This value has varied in the past, steadily decreasing during the last century due to the input of fossil CO_2 (from oil and gas combustion) with a more negative $\delta^{13}C$-value. The changes of atmospheric $\delta^{13}C$ through time are relatively well known from ice core data (Friedli et al. 1986) and more recently from atmospheric measurements (Keeling et al. 1989). In the organic matter of C_3-plants, the $\delta^{13}C$-values range from about -23‰ to -32‰, i.e. a depletion of about 20‰ compared to atmospheric CO_2 is observed. This depletion during photosynthesis is caused by fractionation due to the diffusion of CO_2 through the stomata and biochemical fractionation due to carboxylation, i.e. the fixation of CO_2 by the rubisco enzyme. The relative limitations of photosynthesis by these two factors determines the isotope fractionation. A model integrating the above processes was developed by Farquhar et al. (1982) who found $\delta^{13}C$ in plant material, $\delta^{13}C_{plant}$, to be linearly related to the CO_2 concentration in intercellular spaces, c_i:

$$\delta^{13}C_{plant} = \delta^{13}C_{atm} - a - (b - a)c_i/c_a \qquad (1)$$

where c_a is the CO_2-concentration in the atmosphere, $\delta^{13}C_{atm}$ is the $\delta^{13}C$ of atmospheric CO_2, and a (=4.4‰) and b (=27‰) are the isotope fractionations caused by diffusion through the stomata and by carboxylation, respectively.

From this model, two potential applications for tree-ring studies appear feasible. First, climatic information can be expected to be stored in the $\delta^{13}C$ of organic material. This assumption is based on the fact that dry and warm conditions will increase the stomatal limitation imposed on photosynthesis and will result in low c_i-values. According to eq. 1, we will find relatively high $\delta^{13}C_{plant}$-values in this scenario. Indeed, $\delta^{13}C$-variations in tree rings are related to the precipitation amount and temperature (Leavitt and Long 1988; Robertson et al. 1997; Saurer et al. 1997b).

Second, attempts can be made to reconstruct long-term changes in gas-exchange parameters. There is a question as to how much the rising CO_2-concentration in the atmosphere during the last two centuries has influenced the photosynthesis and the biomass production of terrestrial plants (Mooney et al. 1991). To know whether there have been changes in c_i would be a very valuable contribution to this discussion. Eq. 1 can be solved for c_i (eq. 2) and thus c_i can be estimated from the measurement of $\delta^{13}C_{plant}$. Accordingly, c_i (as well as c_i/c_a) can be reconstructed through time from tree-ring $\delta^{13}C$ data provided that the concentration and the $\delta^{13}C$ of the atmospheric CO_2 are known for the time period under consideration.

$$c_i/c_a = (\delta^{13}C_{atm} - \delta^{13}C_{plant} - a)/(b-a) \qquad (2)$$

3. Oxygen isotopes

$\delta^{18}O$-variations are a much-used tool in palaeoclimate research (Lowe and Walker 1997). In particular, this method is well-established for water stored in ice sheets and carbonates in marine sediments, although relatively few studies deal with $\delta^{18}O$ in organic matter. The basis of the above applications is that the $\delta^{18}O$ of precipitation (as well as the δD) is correlated with surface air temperature (Dansgaard 1964; Rozanski et al. 1992), a relationship which is caused by evaporation and condensation processes in the global water cycle altering the isotopic composition. Dansgaard (1964) reports an average change in $\delta^{18}O$ of 0.7‰ per degree Celsius for a range of polar and coastal stations. This temperature signal can in principal be expected to be stored in the $\delta^{18}O$ of tree rings, but several processes altering the isotopic composition during uptake of the oxygen by the plant have to be considered (Fig. 1).

(i) The relative uptake of surface water and of soil water by the roots of a tree may be important. In response to a strong seasonal variation in $\delta^{18}O$ of precipitation (Siegenthaler and Oeschger 1980), there is also a seasonal variation in surface water, whereas the $\delta^{18}O$ of groundwater represents a long-term average. Therefore, the $\delta^{18}O$ of surface water during the vegetation period (summer) can be about 3‰ higher than groundwater.

(ii) A remarkable ^{18}O enrichment takes place in the leaf due to transpiration (up to about 20‰). This enrichment has been shown to be dependent mainly on the relative humidity in the atmosphere, whereby dry conditions result in high $\delta^{18}O$ values of the leaf water. This process can be readily described by a physical model (Dongman et al. 1974).

(iii) For a variety of plants, a difference of about 27‰ between the $\delta^{18}O$ of cellulose and water at the site of synthesis has been found (DeNiro and Epstein 1979). This biochemical fractionation appears to be due to the isotopic exchange of carbonyl oxygen atoms of organic molecules with water during hydration reactions (Sternberg et al. 1986). It is thus the oxygen of the water (and not the oxygen of CO_2 or O_2) which determines the oxygen isotope composition of the organic material.

(iv) Sucrose transported from the leaves to the stem exchanges part of the oxygen atoms with stem water during cellulose build-up (Hill et al. 1995). Therefore, the $\delta^{18}O$ enrichment from the leaf water, which is reflected in the organic matter of the leaf, may be partly lost in the cellulose of the tree rings.

In summary, the $\delta^{18}O$ of precipitation appears to be the main factor influencing the oxygen isotope composition in tree rings, but the relative humidity of the atmosphere can also be important via its effect on leaf-water enrichment (Burk and Stuiver 1981; Edwards et al. 1985). A rapid increase of studies in this field can be expected due to the development of on-line techniques (Werner et al. 1996). Further, the combined analysis of $\delta^{18}O$ and $\delta^{13}C$ may be particularly promising (Lipp et al. 1996). This approach may allow the separation of the signals of source water $\delta^{18}O$ (i.e. temperature) and evaporative enrichment of leaf water (i.e. humidity) and thus result in a more quantitative analysis of climatic signals.

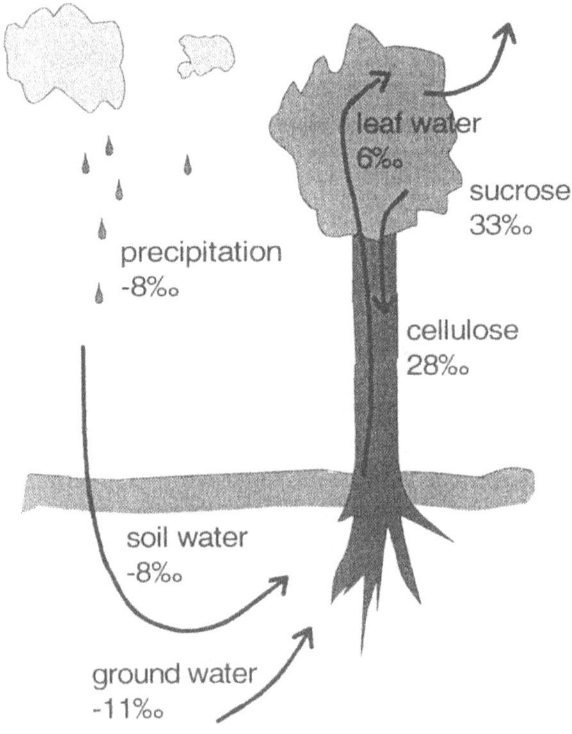

Fig. 1. The path of oxygen atoms of water into plants and typical $\delta^{18}O$ values (versus VSMOW). The indicated values of $\delta^{18}O$ in precipitation and soil water are representative for summer in Switzerland, the values for leaf water and glucose are estimated according to the theory in section 3. The value for the cellulose is the mean of our data for *Fagus sylvatica*. From Saurer et al. (1997a)

4. The influence of site conditions

Climate reconstruction using tree-ring width variations has been most successful at sites where trees grow near climatically controlled limits of their distribution

(Schweingruber 1988). At these sites, one dominant climatic factor is determining tree growth and, accordingly, a high degree of common variance in ring-width variation among different trees is observed. Such sites are found, for instance, in high-latitude and high-elevation regions or in the arid southwestern part of the USA (Fritts 1976). However, in more temperate climates the influence of climate variables on tree growth is more complex and no simple climate–ring width relationships can be found. The common variance of ring-width variations is usually lower than in more stressed environments.

The use of stable isotopes in tree rings may be particularly promising in these more temperate climates. The larger amount of work for the isotope analysis compared to ring-width measurements could be counterbalanced by the possibility of expanding climate reconstructions into temperate or maritime regions (Robertson et al. 1997). The influence of site conditions on the climate–isotope relationship can be best evaluated by comparing sites with contrasting conditions. In this study, we realise this approach by investigating *Fagus sylvatica* trees on dry and humid sites.

5. Materials and methods

The study area is the Swiss Central Plateau, with all sites being located within 30 kilometres of Bern: Twann (47°06'N 7°10'E), Burgdorf (47°02'N 7°36'E), Krauchthal (47°01'N 7°34'E), and Hub (47°00'N 7°32'E). At three of the four locations, dry and humid sites close together have been selected. The terms dry and humid refer to the soil moisture conditions due to different expositions and micro-climates, which are also reflected in the plant community (Saurer et al. 1997b). Accordingly, there are four dry sites (Twann-Dry = T-Dry; Burgdorf-Dry = B-Dry; Krauchthal-Dry = K-Dry; Hub-Dry = H-Dry) and three humid sites (Twann-Humid = T-Humid; Burgdorf-Humid = B-Humid; Krauchthal-Humid = K-Humid). At each site, four to six co-dominant trees have been sampled by taking two 0.5 cm cores. Approximately the last 50 years were prepared for isotope analysis by dividing into three-year-ring groups (1-year samples at Krauchthal).

The samples were milled and the cellulose was extracted. Tree-ring width was analysed on separate cores. The $\delta^{13}C$-values of the cellulose samples from T-Dry/Humid, K-Dry/Humid and H-Dry were determined with an off-line technique which is based on combustion in quartz tubes with excess cupric oxide and the subsequent measurement of the produced CO_2 in an isotope-ratio-mass-spectrometer (Finnigan, Germany). The $\delta^{13}C$-values of the samples from B-Dry/Humid were determined with an on-line technique using an elemental analyser coupled to the mass-spectrometer (Saurer et al. 1997b). The $^{18}O/^{16}O$ ratio (for the sites T-Dry, K-Dry/Humid) was measured with a nickel-pyrolysis technique (Saurer et al. 1997a).

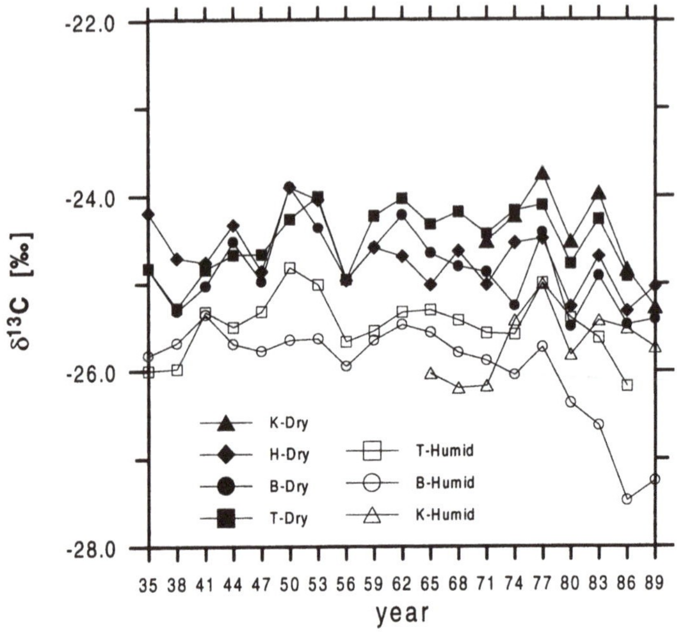

Fig. 2. $\delta^{13}C$ tree-ring series of all sites. Each curve represents the mean of four to six *Fagus sylvatica* from one site. Dry sites are indicated by full symbols and humid sites by open symbols. From Saurer et al. (1997b)

6. Results and discussion

6.1 Carbon isotopes

In Figure 2, the mean site $\delta^{13}C$-curves as calculated from four to six *Fagus sylvatica* are shown for the period 1934–1989. All the curves from the dry sites have more positive values than the humid sites, in agreement with the theory presented above. As drought stress is greater at the dry sites, this results in a higher average $\delta^{13}C$. The short term $\delta^{13}C$ variations for the different dry sites are highly correlated with each other, whereas the curves for the humid sites show less coherence. In particular, the curve from the B-Humid site is affected by a decreasing trend starting in 1977, but this may be a local phenomenon as tree-ring width strongly increases in about the same period at this site (probably caused by felling some of the trees). Correlation analysis with climatic data revealed that the time period May–June has the greatest influence on the $\delta^{13}C$ variations (Table 1).

There is a negative correlation with precipitation amount and a positive correlation with temperature. This means that warm, dry conditions result in relatively high values of $\delta^{13}C$ in the cellulose of tree rings. The correlation with climate is improved when first differences instead of the original values are considered, indicating that it is mainly the short-term variations that can be explained by climatic variability. The tree-ring width variations at these sites, in contrast, are very poorly related to climate. This confirms the finding that $\delta^{13}C$ variations in tree rings can be used for climatic reconstruction in a relatively broad range of site conditions (Saurer et al. 1997b; Robertson et al. 1997), whereas tree-ring width variations are best used on sites where one factor dominates growth (Schweingruber 1988).

		T-Dry (n=18)	B-Dry (n=19)	H-Dry (n=19)	K-Dry (n=7)	T-Humid (n=18)	B-Humid (n=19)	K-Humid (n=9)
original	P	-0.72 **	-0.65 **	-0.47 *	-0.04	-0.67 **	-0.29	-0.51
values	T	0.44	0.35	0.26	0.55	0.19	0.09	0.28
first	P	-0.78 **	-0.61 **	-0.70 **	-0.19	-0.57 *	-0.54 *	-0.22
differences	T	0.53 *	0.31	0.48 *	0.05	0.00	0.06	0.12

Table 1. Correlation coefficients r between climatic data and $\delta^{13}C$ data (original values) and between first differences of climatic data and first differences of $\delta^{13}C$ data for precipitation amount (=P) and temperature (=T; mean of May, June and July using data from the meteorological station in Bern). The first differences are defined as the differences between consecutive values. Significance levels $p<0.05$ *, $p<0.01$ **

We estimated long-term trends of c_i by applying eq. 2 as follows. First, we calculated a mean dry and a mean humid $\delta^{13}C$ curve by pooling the $\delta^{13}C$ curves from all dry and humid sites (but neglected the 1977–89 period from B-Humid). In the next step, we eliminated climatic influences by using the regression equations discussed above. This reduced the short term $\delta^{13}C$ variations to some degree. In a third step, we corrected for the declining trend in the $\delta^{13}C$ of atmospheric CO_2 using a spline fit through the data from Friedli et al. (1986) and Keeling et al. (1989). Finally, the ca. 1.3‰ difference between cellulose and bulk wood had to be considered for the inference of c_i-values from $\delta^{13}C$ (Marshall and Monserud 1996). The corrected values were then inserted in eq. 2 to calculate c_i-values from 1934 to 1989 (Fig. 3a). The c_i-curve estimated for the humid sites is about 20 ppm higher than the respective curve for the dry site throughout the whole period. This means that the stomatal limitation to photosynthesis for *Fagus sylvatica* studied in this area is greater for the dry sites. Further, an increasing trend is found for both dry and humid conditions: c_i increases from 205 ppm to about 225 ppm for the dry sites and from 215 ppm to 235 ppm for the humid sites, in contrast to some evidence suggesting that c_i may be held constant for this range of c_a (Francey and Farquhar 1982). The increase of about 20 ppm over the last 50 years is smaller than the increase of c_a of about 45 ppm in the same period.

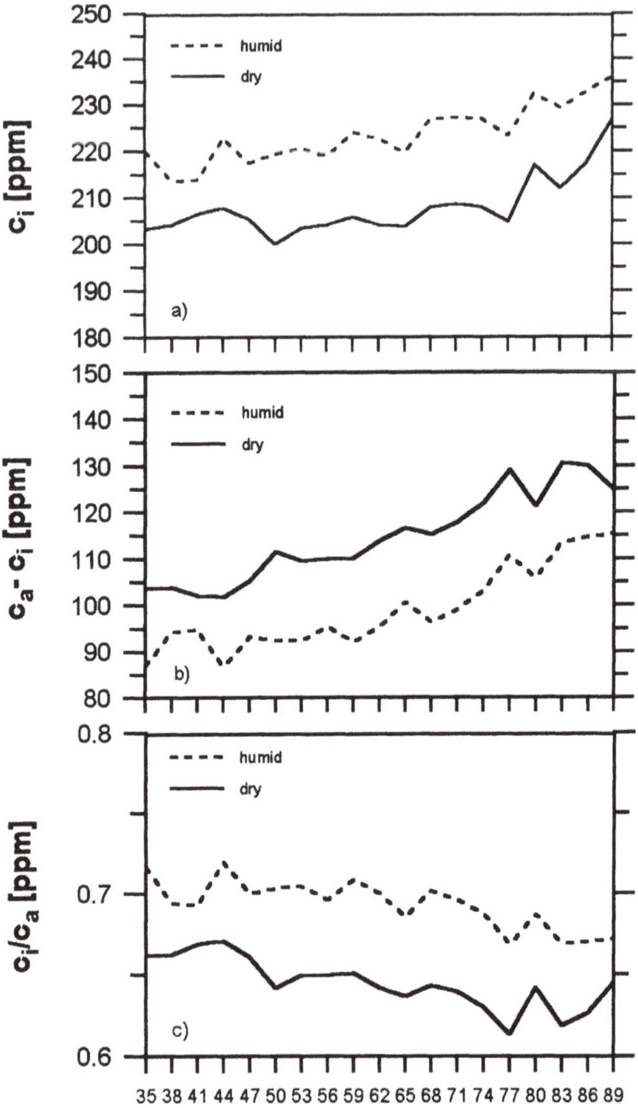

Fig. 3. a) Time course of c_i (intercellular CO_2-concentration) as calculated from the tree-ring $\delta^{13}C$ data of cellulose (*Fagus sylvatica*). The dashed line represents the trees from all humid sites, and the solid line represents all dry sites. See text for details b) Time course of c_a-c_i (where c_a is the atmospheric CO_2-concentration) c) Time course of c_i/c_a

Therefore, we find a strongly increasing trend for c_a-c_i whereas c_i/c_a is decreasing (Figs 3b, c). As the net photosynthesis is given by $A=g(c_a-c_i)$ where g is the leaf conductance to CO_2, the increase of c_a-c_i of about 20% (as estimated from Fig. 3b) could indicate that A has significantly increased in this period. This interpretation, however, is based on the assumption that g was held constant (which we do not know). More likely, an acclimation to rising c_a has occurred, accompanied by decreasing stomatal conductance (Bowes 1993). Some caution is necessary in the interpretation of the calculated c_i-values because age-related changes could interfere with the results, i.e. the changes in c_i could be caused by an age trend affecting photosynthesis rather than by the change in atmospheric CO_2 (Marshall and Monserud 1996). This question cannot be fully answered here although it may not be so important here as we discarded the first 20 rings (avoiding the well-known "juvenile effect" in $\delta^{13}C$, Francey and Farquhar 1982). Furthermore, we found that trees with strongly differing ages at the beginning of the investigation period (ranging from about 20 to 90 years) nevertheless showed a similar shift in c_i.

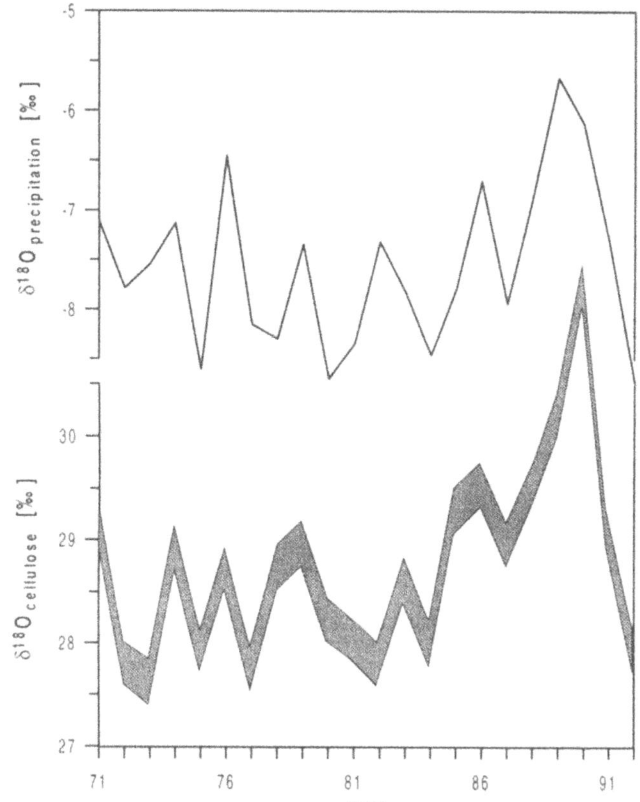

Fig. 4. Upper curve: mean $\delta^{18}O$ of precipitation from May to August at Bern. Lower curve: mean $\delta^{18}O$-curve of the cellulose of *Fagus sylvatica* from K-Humid with the measurement precision indicated as the width of the band. From Saurer et al. (1997a)

6.2 Oxygen isotopes

Monthly values of the $\delta^{18}O$ of precipitation have been determined in Bern since 1971. This provides a very valuable data set for comparison with tree-ring measurements in this area. In Figure 4, we present the $\delta^{18}O$ tree-ring series from K-Humid (which were determined with a one-year resolution), together with the $\delta^{18}O$ series of precipitation representing the mean of the months May, June, July and August. The two curves are quite similar and the correlation coefficient between the $\delta^{18}O$ of precipitation and cellulose is r=0.74. The four-month period from May to August was found to yield the best correlation for K-Humid, whereas for K-Dry the correlation was highest with June alone (r=0.68).

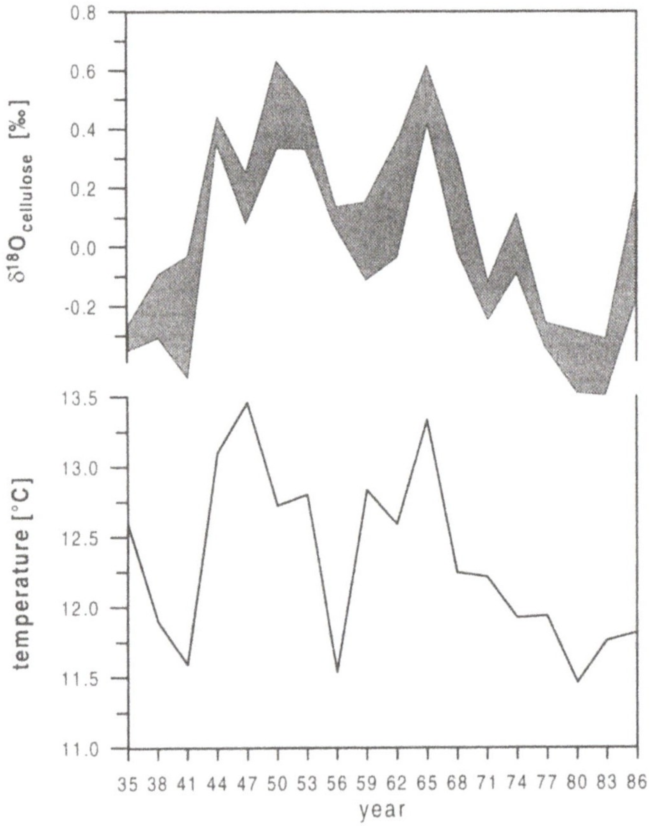

Fig. 5. Upper curve: mean $\delta^{18}O$-curve of the trees from T-Dry (anomalies) with standard deviation (width of the band). Lower curve: mean temperature of the months April/May/June at the Bern weather station. From Saurer et al. (1997a)

These results clearly indicate that the $\delta^{18}O$-signal of precipitation is recorded in tree rings, although the slopes of the regression lines are significantly below 1 (0.76±0.15‰ for K-Humid, 0.31±0.08‰ for K-Dry). The period influencing the isotopes is relatively short for K-Dry (June) compared to K-Humid (May to August) which may be due to a shorter growing season at the dry location or due to different soil drainage conditions. Further, changing relative humidity of the atmosphere might interfere via its effect on leaf $\delta^{18}O$-enrichment, and indeed, a correlation between $\delta^{18}O$ in the tree rings at K-Humid and the daytime (1300 h) humidity in Bern has been found. The slope of the respective correlation (0.13‰/%) is only about one third of the theoretical value as calculated with the model from Dongman et al. (1974). This means that the influence of leaf-water enrichment is strongly dampened in the cellulose of the stem.

The fact that there is a correlation between $\delta^{18}O$ in tree rings and $\delta^{18}O$ in precipitation suggests that a correlation also exists between tree-ring $\delta^{18}O$ and temperature (due to the well-established relationship between $\delta^{18}O$ in precipitation and temperature). No such correlation was however found for the sites K-Dry and K-Humid. This may be partly due to the relatively short period investigated (1971–92). On the other hand, we found the expected correlation for the T-Dry site (1935–86 with 3 year resolution, see Fig. 5). The $\delta^{18}O$-data in Fig. 5 represent the mean of three trees and are given as anomalies, i.e., we calculated the deviations from the mean for each tree before calculating the mean. Using anomalies reduces the variation between the trees because the slight offset is eliminated. From Fig. 5, it is evident that it is mainly the long-term variations (>10 years) in temperature that are reflected in the $\delta^{18}O$ of tree-ring cellulose (best correlation for the mean temperature of April to June). Interestingly, the slope of the regression line (0.33‰/°C±0.09‰/°C) appears to be smaller than the value of 0.7‰/°C given by Dansgaard (1964) for precipitation.

7. Conclusions

The causes for $^{13}C/^{12}C$ variations in plants are fairly well understood and carbon isotope analysis of tree rings appears to be a reliable method for climate reconstruction. Information can be gained mainly about the occurrence and frequency of warm and dry conditions during the growing season. The site conditions are probably not as important for the climate–isotope relationship as they are for the climate–tree-ring width relationship. Further, $^{13}C/^{12}C$ is a unique means to detect past changes in physiological parameters related to photosynthesis in response to rising atmospheric CO_2. In particular, estimates of c_i can be made by applying the Farquhar-model for isotope fractionation. Relatively few studies exist which deal with oxygen isotope variations in tree rings. Our data from *Fagus sylvatica* from Swiss sites show that $^{18}O/^{16}O$ variations in precipitation are reflected in tree rings. Accordingly, temperature changes can be reconstructed. The oxygen isotope analysis of tree rings may be one of the great under-exploited areas of climate research.

Acknowledgements. This study was supported by the Swiss National Foundation (NFP-31) and the Swiss Federal Institute for Forest, Snow and Landscape Research (WSL, Birmensdorf).

8. References

Bowes G (1993) Facing the inevitable: Plants and increasing CO_2. Ann Rev Plant Physiol Plant Molec Biol 44:309–332

Burk RL, Stuiver M (1981) Oxygen isotope ratios in tree cellulose reflect mean annual temperature and humidity. Science 211:1417–1419

Dansgaard W (1964) Stable isotopes in precipitation. Tellus 16B:436

DeNiro MJ, Epstein S (1979) Relationship between oxygen isotope ratios of terrestrial plant cellulose, carbon dioxide and water. Science 204:51–53

Dongmann G, Nürnberg HW, Förstel H, Wagener K (1974) On the enrichment of $H_2{}^{18}O$ in the leaves of transpiring plants. Radiation Environm Biophys 11:41–52

Edwards TWD, Aravena RO, Fritz P, Morgan AV (1985) Interpreting paleoclimate from ^{18}O and ^{2}H in plant cellulose: comparison with evidence from fossil insects and relict permafrost in southwestern Ontario. Can J Earth Sci 22:1720–26

Farquhar GD, O'Leary MH, Berry JA (1982) On the relationship between carbon isotope discrimination and the intercellular carbon dioxide concentration in leaves. Aust J Plant Physiol 9:121–137

Francey RJ, Farquhar GD (1982) An explanation of $^{13}C/^{12}C$ variations in tree rings. Nature (London) 297:28–31

Friedli H, Loetscher H, Oeschger H, Siegenthaler U, Stauffer B (1986) Ice record of the $^{13}C/^{12}C$ ratio of atmospheric CO_2 in the past two centuries. Nature (London) 324:237–238

Fritts HC (1976) Tree rings and climate. Academic Press, San Diego

Hill SA, Waterhouse JS, Field EM, Switsur VR, Ap Rees T (1995) Rapid recycling of triose phosphates in oak stem tissue. Plant Cell Environ 18:931–936

Keeling CD, Bacastow RB, Carter AF, Piper SC, Whorf TP, Heimann M, Mook WG, Roeloffzen H (1989) A three-dimensional model of atmospheric CO_2 transport based on observed winds: 1. Analysis of observational data. Geophys Monogr Am Geophys Union 55:165–236

Leavitt SW, Long A (1988) Stable carbon isotope chronologies from trees in the southwestern United States. Global Biogeochem Cycles 2:189–198

Lipp J, Trimborn P, Edwards T, Waisel Y, Yakir D (1996) Climatic effects on the $\delta^{18}O$ and $\delta^{13}C$ of cellulose in the desert tree *Tamarix jordanis*. Geochim Cosmochim Acta 60:3305–3309

Lowe JJ, Walker MJC (1997) Reconstructing Quaternary Environments. Longman, Essex

Marshall JD, Monserud RA (1996) Homeostatic gas-exchange parameters inferred from $^{13}C/^{12}C$ in tree rings of conifers. Oecologia 105:13–21

Mooney HA, Drake BG, Luxmoore RJ, Oechel WC, Pitelka LF (1991) Predicting ecosystem responses to elevated CO_2 concentrations. Bioscience 41:96–104

Robertson I, Switsur VR, Carter AHC, Barker AC, Waterhouse JS, Briffa KR, Jones PD (1997) Signal strength and climate relationships in $^{13}C/^{12}C$ ratios of tree ring cellulose from oak in east England. J Geophys Res 102 (D16):19507–19516

Rozanski K, Araguas-Araguas L, Gonfiantini R (1992) Long-term trends of oxygen-18 isotope composition of precipitation and climate. Science 258:981–985

Saurer M, Borella S, Leuenberger M (1997a) • ^{18}O of tree rings of beech (*Fagus silvatica*) as a record of $\delta^{18}O$ of the growing season precipitation. Tellus 49B:80–92

Saurer M, Borella S, Schweingruber F, Siegwolf R (1997b) Stable carbon isotopes in tree rings of beech: Climatic versus site-related influences. Trees 11:291–297

Schweingruber F (1988) Basics and applications of dendrochronology. Reidel, Dordrecht, Boston, Lancaster, Tokyo

Siegenthaler U, Oeschger H (1980) Correlation of ^{18}O in precipitation with temperature and altitude. Nature (London) 285:314–317

Sternberg LSL, DeNiro MJ, Savidge RA (1986) Oxygen isotope exchange between metabolites and water during biochemical reactions leading to cellulose synthesis. Plant Physiol 82:423–427

Werner RA, Kornexl BE, Rossmann A, Schmidt HL (1996) On-line determination of $\delta^{18}O$-values of organic substances. Analytica Chimica Acta 319:159–164

Simulated impacts of mean vs. intra-snnual climate changes on forests

Rüdiger Grote, Gerd Bürger and Felicitas Suckow

Abstract. This paper investigates the impacts of climatic variations on coniferous ecosystems using a physiologically-based model. The investigation is based on the assumption that the overall reaction of a forest to a new combination of climatic influences can only be assessed from the bottom upwards (given that the underlying processes are sufficiently understood). Thus, a model is used that considers direct and indirect effects of environmental changes.

For the investigation of responses to different kinds of climatic variations, we used a physiology-based model of the soil–vegetation complex with linked balances of carbon, water and nitrogen. The model was developed to reproduce detailed physiological, soil-physical and chemical measurements in *Pinus sylvestris* L. plantations in eastern Germany. A special feature of the model is that allocation and litterfall processes are modelled as a function of plant carbon requirements rather than being fixed ratios or empirical functions of specific environmental conditions. Thus, seasonal tree responses, e.g. increased root growth or foliage mortality in response to drought, are also considered.

The model is initialised with measured stand and soil conditions. Tree development is simulated under 24 different climatic scenarios. The scenarios are defined by systematically modifying observed climatic conditions with respect to either mean temperature or precipitation (+/-2 K and +/-30 %, respectively), and superimposing a set of redistributed climatic parameters within the year (seasonal increase or decrease as above without changing the average conditions). The actual simulation of the corresponding weather uses a climate disaggregation scheme.

The results of the simulation are: (1) The seasonal shifts in temperature conditions have a greater impact on water and carbon balance than the changes in annual averages at the investigated sites; (2) precipitation changes are most effective if they occur in the dry season; (3) stem growth responds more than net primary production; and (4) the intensity of the impacts is reversibly related to the soil water capacity at the sites. We conclude that potential changes in the seasonal distribution of climatic variables need to be considered if climate impacts on forests are being investigated.

1. Introduction

A number of process-based models have been developed and tested during the last 15 years that enable the assessment of forest responses to changing environmental conditions (e.g. Mohren et al. 1993; Running 1994; Zhang et al. 1994). Furthermore, some models have been applied to local and regional case studies based on specific climate change scenarios (Aber et al. 1995; Bowes and Sedjo 1993; Pan and Raynal 1995). Because the computation of these calculations is expensive, the studies investigate only one (or very few) scenarios derived from Global Circulation Models (GCMs). However, current climate projections by GCMs are rather uncertain for local predictions, particularly with respect to the ecologically important aspects of climate. Thus, it seems appropriate to investigate the potential forest responses to a broad range of climatic conditions rather than the response to some specific climate changes.

If the responses of forests to a broad range of conditions are to be assessed, it is necessary to consider a general decrease or increase of some climatic variables as well as climatic changes in specific seasons (Fiedler and Wenk 1973; Perry et al. 1994). This is of particular importance because a difference between temperature increases in summer and winter has been already reported for various regions (Houghton et al. 1995). Nevertheless, we are not aware of any sensitivity analysis that is concerned about climatic changes in particular seasons. A possible explanation is that the dimension of a seasonal change is even more difficult to estimate than the change in annual average conditions. This argument increases the scientific caution needed with predictions of forest development. However, it cannot be used against the application of scenarios in a risk analysis. Another reason lies in the lack of sensitivity of the models. Although many models contain a detailed description of photosynthesis, other important physiological responses are often treated independently from climatic conditions. Thus, a response of phenological development or a change in the carbon allocation pattern cannot be considered and may lead to substantial errors in the estimation of tree development (King 1993; Kramer 1995).

In the current study, we investigated the responses of a *Pinus sylvestris* L. forest in eastern Germany to a variety of climatic variations. The scenarios used for the simulations differ in their annual average conditions of temperature and precipitation as well as in the seasonal distribution of these climatic variables. The impact analysis is done with the model FORSANA, a physiology-based simulation model of the soil–vegetation complex, which calculates carbon, water and nitrogen cycles of the stand with a daily time step. It was developed to reproduce detailed physiological, soil-physical and chemical measurements under actual environmental conditions at different sites over several decades (Grote et al. 1995, 1996). Thus, it is assumed to be also suitable for scenario assessments within a relatively wide range of climatic conditions.

2. Methods

2.1 Climate Scenarios

The climatic input is given by the meteorological core variables of minimum, average, and maximum temperature, precipitation, average humidity, and sunshine duration, at Potsdam, Germany. The climate is relatively dry (approximately 550 mm annual precipitation) with a slight maximum of precipitation in summer. The average temperature is 8°C. Each of the following scenarios consists of 100 years of simulated daily weather that is representative for a specific climate. The model we used for the simulation is described at the end of this section.

The baseline climate is taken from the annual average conditions as observed in the period 1961–90. These conditions were then systematically modified along the following lines: We defined a group of temperature scenarios and precipitation scenarios, each of which contains three subgroups which represent, for temperature, a cold, mean, and warm climate, and, for precipitation, a dry, mean, and wet climate; all of these climatic shifts uniformly affect the whole annual cycle. Each of the groups contains three temperature and five precipitation climates that reflect a specific redistribution of climatic conditions within the year that is superimposed on the given annual conditions. For temperature, we use two "seasons" (winter and summer), while for precipitation this is redistributed using the four traditional seasons. Each group furthermore contains a control climate. This defines a total of $3 \times (3 + 5) = 24$ scenarios whose characteristics are summarised in the following tables.

2.1.1 Precipitation (P-)Scenarios

current	ΔP=(0,0,0,0)%
spring	ΔP=(-30,-30,-30,-30)%
summer	ΔP=(+30,+30,+30,+30)%

Table 1. Definitions of the precipitation scenarios with different annual precipitation

current	ΔP=(0,0,0,0)%
spring	$\Delta P=(+30,-p_2,-p_3,-p_4)\%$
summer	$\Delta P=(-p_1,+30,-p_3,-p_4)\%$
autumn	$\Delta P=(-p_1,-p_2,+30,-p_4)\%$
winter	$\Delta P=(-p_1,-p_2,-p_3,+30)\%$

Table 2. Precipitation scenarios with different seasonal rainfall distribution

The ΔP-vector indicates the four seasons, starting with spring; "wet", e.g., means a scenario that shows a 30% increase in precipitation for all seasons. The intra-annual redistribution of precipitation is obtained according to specific rules. For example, the vector $(-p_1,-p_2,+30,-p_4)\%$ describes a 30% increase of autumn (third season) precipitation and a corresponding decrease $(-p_i)$ for the other seasons, such that the annual precipitation sum is preserved (Fig. 1). Note that the various p_i's are different according to the different seasonal contributions to the annual precipitation sum.

2.1.2 Temperature (T-)Scenarios

cold	ΔT=(-2.0,-2.0)K
mean	ΔT=(0.0,0.0)K
warm	ΔT=(+2.0,+2.0)K

Table 3. Definitions of the temperature scenarios with different annual temperature

mean	ΔT=(0.0,0.0)K
summer	ΔT=(+2.0,-2.0)K
winter	ΔT=(-2.0,+2.0)K

Table 4. Definitions of the temperature scenarios with different intra-annual temperature distribution

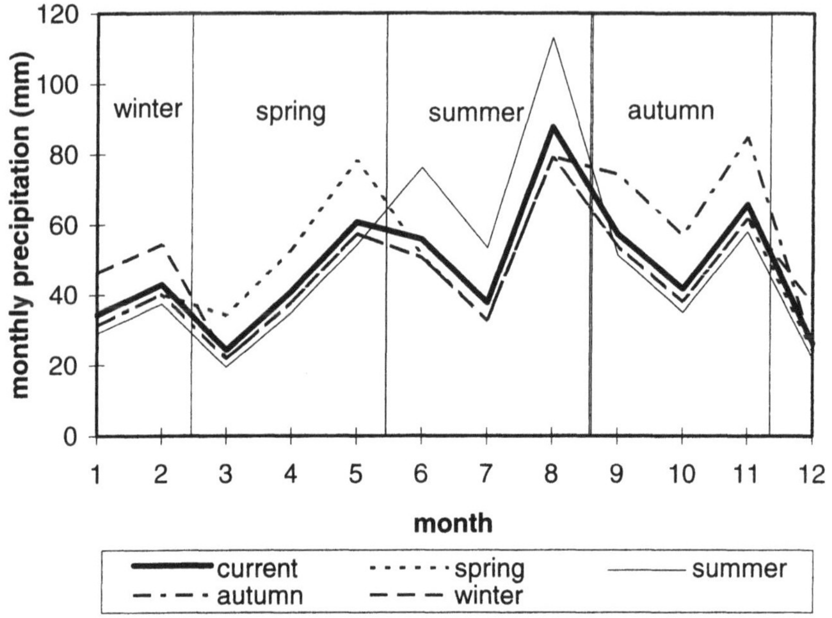

Fig. 1. Example of precipitation distribution in different scenarios (current = current distribution of precipitation at Potsdam in the year 1952, other seasons = distribution of precipitation in the same year in the investigated scenarios).

So far, the above scenarios are just simple definitions of various annual cycles. An important question is now how to turn these annual cycles into actual weather. Here we apply the disaggregation scheme C2W (Bürger 1997). C2W is capable of generating realistic weather records for the above six variables from their long-term means; long-term variability is parameterised by those means. Furthermore, the scheme allows the disaggregation of anomalies from the mean, such as monthly or seasonal temperature anomalies. The resulting weather converges, in a statistical sense, to the input aggregations. Thus, the expected value of the generated weather equals the input value. The daily variability, which C2W models, is as close to the observations as a single-parameter (the mean) disaggregator allows.

A feature of C2W is that a precipitation increase is realised as an increase in both frequency and amount. This is caused by a parametric regression based on observations between daily mean and variability of precipitation, which is used by C2W. A shift of these relations, caused by a possible climate change, cannot be modelled in this way.

2.2 Model Description

FORSANA uses a wide range of ecological relations, extending from basic physiological processes with small time steps to stand disturbances which occur only once in several years. Only the daily processes are briefly described because stand development is not considered in the simulations presented. A complete model documentation can be found in Grote and Suckow (in press). An overview of the linkages between the processes is given in Figure 2.

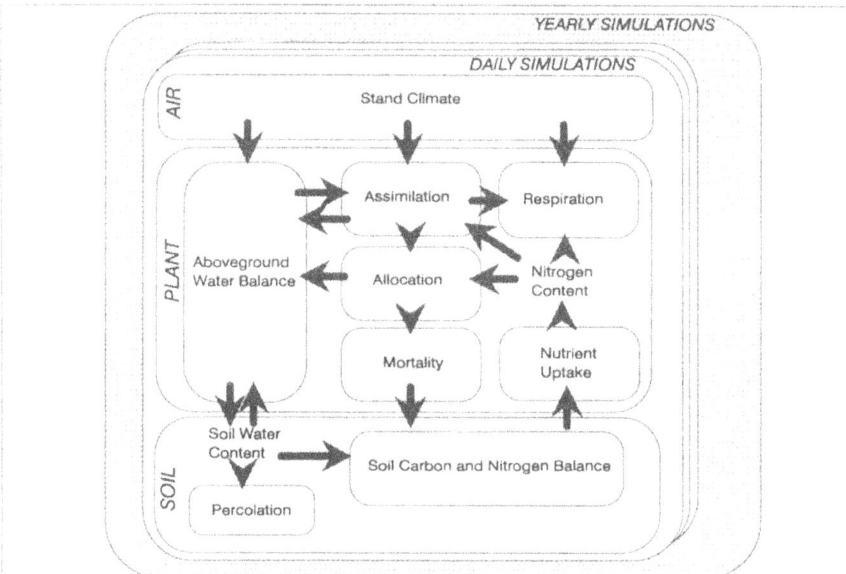

Fig. 2. Model overview of interactions between major processes within and between stand climate, vegetation and soil.

The dependency of photosynthesis on light is represented by a saturation curve (Goudriaan 1982). The temperature influence is described with an optimum function (Bossel 1994), and the response to foliage nitrogen concentration, which is explicitly calculated from the nitrogen uptake by roots and canopy, is linear over a wide range (Aber et al. 1996). A shortage in water supply reduces carbon gain, which is firstly determined without any stomata-induced limitations. A drought stress factor is therefore calculated from the relationship between transpiration demand and potentially available water. The transpiration demand is calculated according to the Penman–Monteith equation (Monteith 1965), and the available water is a function of relative soil water content, fine root biomass and sapwood water storage in the tree (Grote et al. 1997). Maintenance respiration in each tree compartment depends on compartment mass, temperature and nutrient content (Penning de Vries et al. 1989).

The distribution of carbohydrates to the reserve pool, fine root and sapwood compartments is simulated by assuming an organ-specific growth demand, which is calculated from optimum and actual biomass values (Grote et al. 1997). Optimum weights are determined according to their relation to foliage mass. For fine roots, this relationship depends on the supply of water and nitrogen relative to the respective demand. The seasonal dynamics of allocation results from the temperature-induced flushing of foliage and the different rates of mortality in each compartment.

The total amount of mortality in each compartment is calculated from a base biomass and a compartment-specific mortality function. This function considers base longevity parameters, which are modified under environmental stress. With respect to foliage mortality, the total amount is calculated in advance from empirical relations to total foliage biomass. The daily rate of litterfall, however, is determined from the ratio between gross photosynthesis and total maintenance respiration. If this ratio decreases, the mortality is increased. Thus, an increasing litterfall at the end of the vegetation period is simulated without any direct dependency on the day of the year. The base biomass for fine roots is the actual fine root biomass and the mortality is enhanced with decreasing soil water content relative to maximum soil water content (Grote et al., 1997).

The soil module includes detailed calculations of the mineralisation processes and water movement (Kartschall et al. 1990). Nitrogen concentration within the plant can thus be calculated in close relationship to water uptake. The calculations of water and nitrogen cycles consider the competitive effects of the ground vegetation in relation to the rooting intensity and soil conditions in each horizon (Grote et al. 1997).

2.3 Initial Conditions

For the initial stand conditions, one of the *Pinus sylvestris* stands was chosen for which the model had been developed for, and for which the seasonal dynamics and stand development under current conditions were successfully simulated (Grote et al. submitted). This stand was 60 years old in 1994, with an average height of 20.1m and a diameter at breast height of 21.6 cm. Total stand volume is 339 m^3 ha^{-1}, and leaf area index (one sided) was approximately 3. Ground vegetation is dominated by *Avenella flexuosa*, which reaches a peak biomass of more than 1000 kg ha^{-1} in the late summer.

As it was intended only to investigate the effects on the physiological processes of the vegetation in this simulation study, stand development was neglected. The simulation was carried out over the whole scenario length of 100 years, but the initial stand conditions were restored at each year of the simulation period. Thus, the stand conditions in each year were the same, despite the modifying direct influences of stress conditions on the mortality in specific compartments (e.g. a decrease in fine root biomass under drought conditions). The results represent the average response of the current forest ecosystem to a change in climatic conditions.

Two different soils, a loamy brown soil (type 19), and a sandy podzol (type 31), were used to analyse the importance of the soil-water holding capacity in relation to vegetation responses to a changed climate. These soils were derived from the BGR soil map of Brandenburg, Germany, and initialised assuming a rooting depth of 1 m. Within this depth the available soil water under water saturation (maximum amount of water between wilting point and field capacity) was 190 mm (type 19) and 60 mm (type 31), respectively.

3. Results and Discussion

As outlined in the model description, temperature and precipitation affect a large number of processes in the soil and vegetation. Therefore, we focus on the presentation and discussion of some central processes within the water and carbon cycles. An overview of these variables and their linkages to other processes in the system is given in Figure 3.

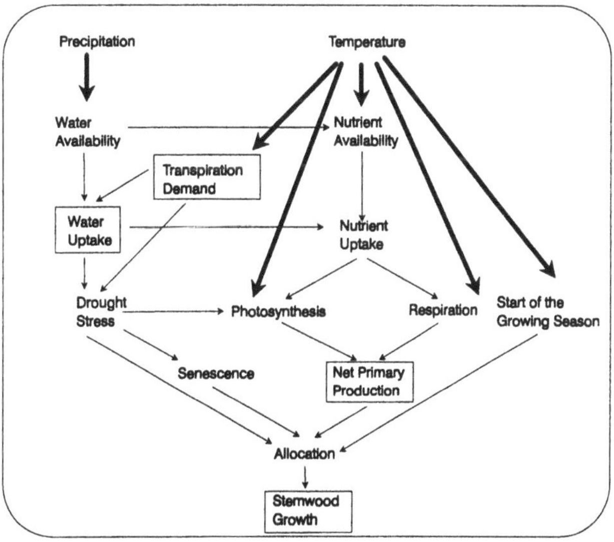

Fig. 3. Overview of the direct (thick arrows) and indirect (thin arrows) effects of precipitation and temperature as represented in FORSANA. Analysed processes are given in rectangles. Note that the feedback of growth on other physiological processes is not shown so as to keep the scheme easier to understand.

3.1 Temperature Scenarios

In general, the most important effect on transpiration demand is the annual average temperature. A higher temperature always results in a higher evaporation demand and thus increases transpiration demand. Thus, the scenarios with the increased summer temperature in particular yielded the highest evaporation demand (Figure 4). This can be explained partly by the larger leaf area index in summer, because potential evaporation depends on canopy resistance, which decreases with increasing LAI (Monteith 1965). Additionally, non-linear relations with other climatic factors such as vapour pressure deficit of the air may play a certain role but are not quantified here.

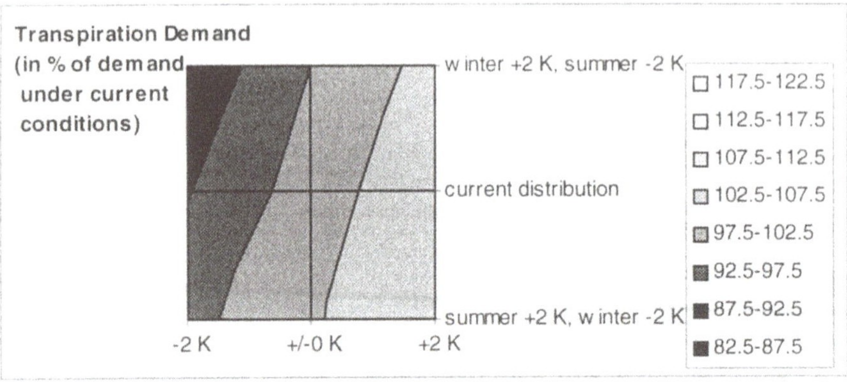

Fig. 4. Transpiration demand in several temperature scenarios relative to the transpiration demand under current climate conditions (+/-0, -2 and +2 denote the scenarios with a respective change average temperature; an increase in winter temperature and a corresponding decrease of temperature in summer is shown at the top of the figure, whereas the result of the opposite temperature scenarios are given at the bottom; K stands for Kelvin).

Transpiration follows the same pattern with respect to the annual average temperature, but shows a contrasting response to the seasonal weighting of temperature compared to potential evaporation. This is because actual water uptake during the winter is limited almost solely by potential evaporation (except in frozen soils), whereas in the summer, water uptake is limited by water availability. Thus, only a temperature increase in the summer increases the drought stress for the plants. The degree of stress increase, however, depends strongly on the water holding capacity of the soils (Figs. 5a and 5b).

Consequently, net primary production is decreased under increased summer temperatures because of the increasing drought stress, whereas the increase in respiration plays only a minor role (results not shown). However, if only the average temperature is increased and the temperature distribution remains unchanged, negative and positive effects are largely balanced and the overall change in assimilation compared to the current conditions is only small (Figs. 6a and 6b). In this case, the earlier start of the vegetation period and the positive effect on photosynthesis (especially during the early summer) increase the net carbon gain if the temperature increases. When the temperature increase is concentrated in the winter and the drought stress effect is absent, the positive effects dominate the overall tree response.

Stemwood growth responds in much the same way as net primary production to a change in temperature. However, the response is more intense, because drought stress increases the share of carbon that is used for fine root production, and thus additionally decreases the carbon available for stemwood growth.

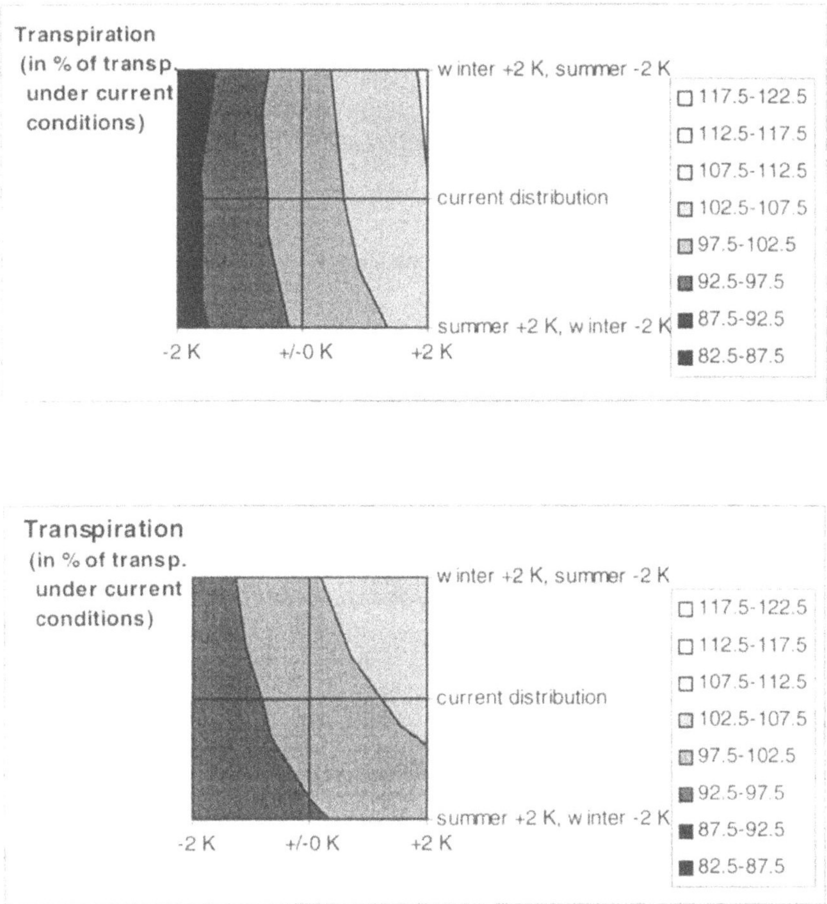

Fig. 5. Actual transpiration in several temperature scenarios relative to the transpiration under current climate conditions. A) With a soil water holding capacity of 190 mm (upper figure), B) With a soil water holding capacity of 60 mm (lower figure). For further explanations see Figure 4.

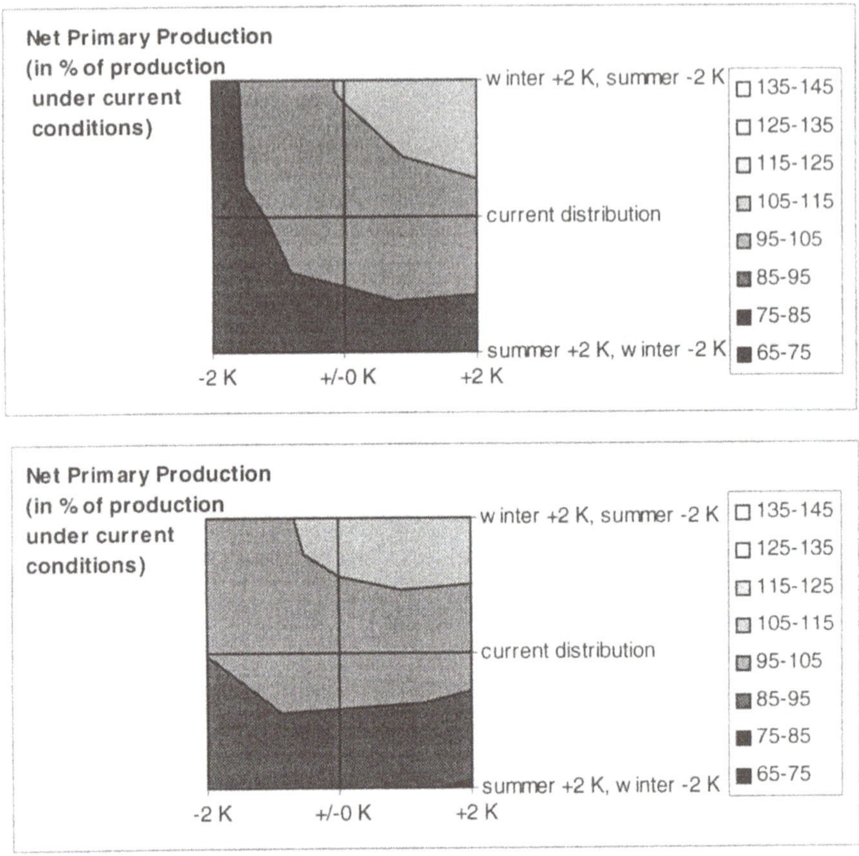

Fig. 6. Actual net primary production in several temperature scenarios relative to the production under current climate conditions. A) With a soil water holding capacity of 190 mm (upper figure), B) With a soil water holding capacity of 60 mm (lower figure). For further explanations see Figure 4.

3.2 Precipitation Scenarios

The relative changes of net primary production and stemwood growth with a changed precipitation are shown in Figures 7a and 7b. Without changing the rainfall distribution, a 30% decrease in precipitation leads to a decrease of stemwood production of 10 and 20% at sites with a large and a small soil water capacity, respectively. An increase in rainfall of the same amount, however, yields only a 5–10% increase in stemwood production. According to the same mechanisms discussed for the temperature effect on drought stress, the response of net primary production is significantly smaller than that for wood growth.

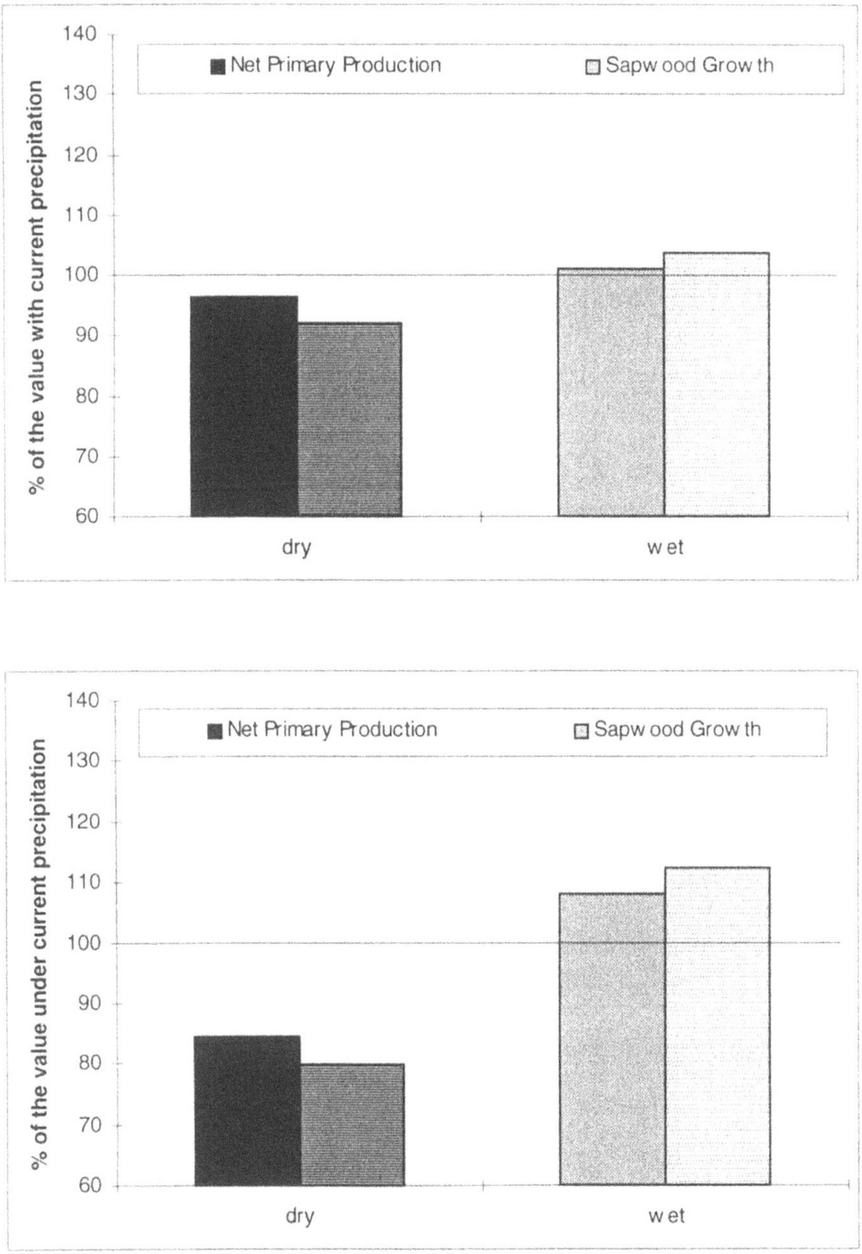

Fig. 7. Actual net primary production and stemwood growth in two precipitation scenarios (dry = 380, wet = 715 mm) relative to the respective values under current precipitation (550 mm). A) With a soil water holding capacity of 190 mm (upper figure), B) With a soil water holding capacity of 60 mm (lower figure).

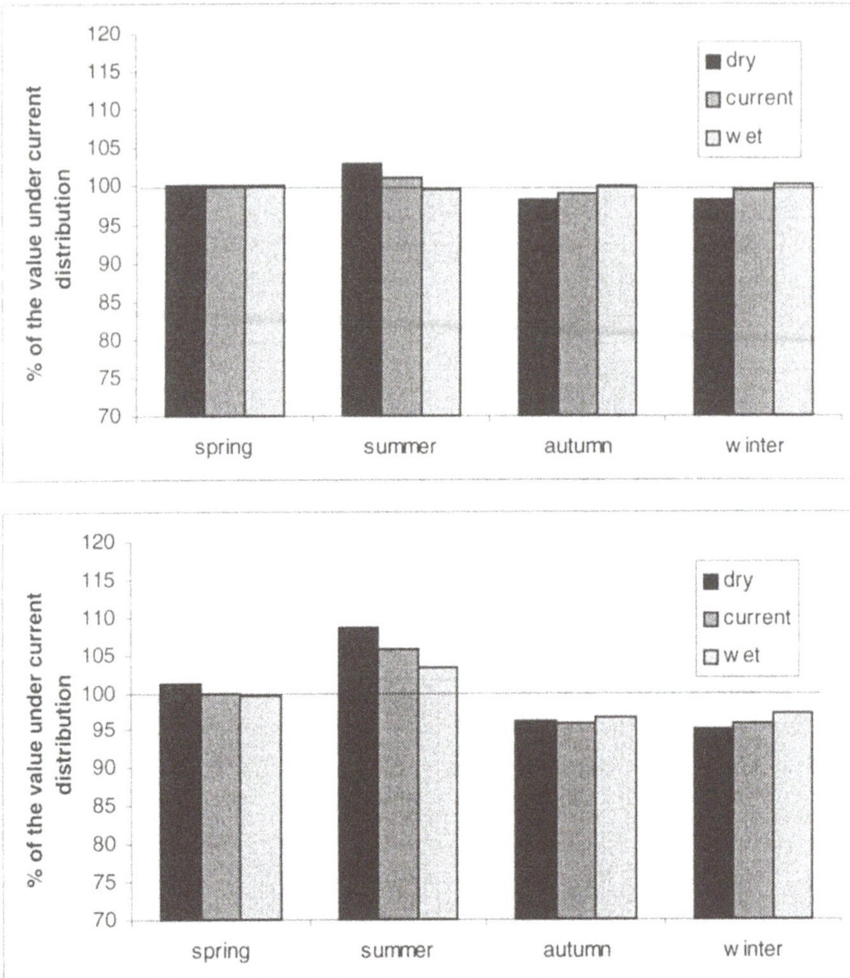

Fig. 8. Transpiration in 12 precipitation scenarios relative to the transpiration under the current amount and distribution of precipitation (dry = 380, current = 550, wet = 715 mm). The seasons on the x-axis denote the time period with increased precipitation relative to the current distribution. A) With a soil water holding capacity of 190 mm (upper figure). B) With a soil water holding capacity of 60 mm (lower figure).

If the rainfall in summer is increased in relation to the other seasons, this additional water is largely taken up by the vegetation (Figs. 8a and 8b) and decreases the drought stress in this season. The differences to the current climate are smaller if the total amount of precipitation or the soil water holding capacity is large, because under these conditions the drought stress in summer is already less intense. If the precipitation increases in autumn or winter without a change in total precipitation, less water is available for transpiration during the vegetation period, which increases the drought stress during this period (Figs. 9a and 9b). Thus, stemwood growth is decreased under these conditions, particularly if the soil-water holding capacity is small (Figs. 10a and 10b).

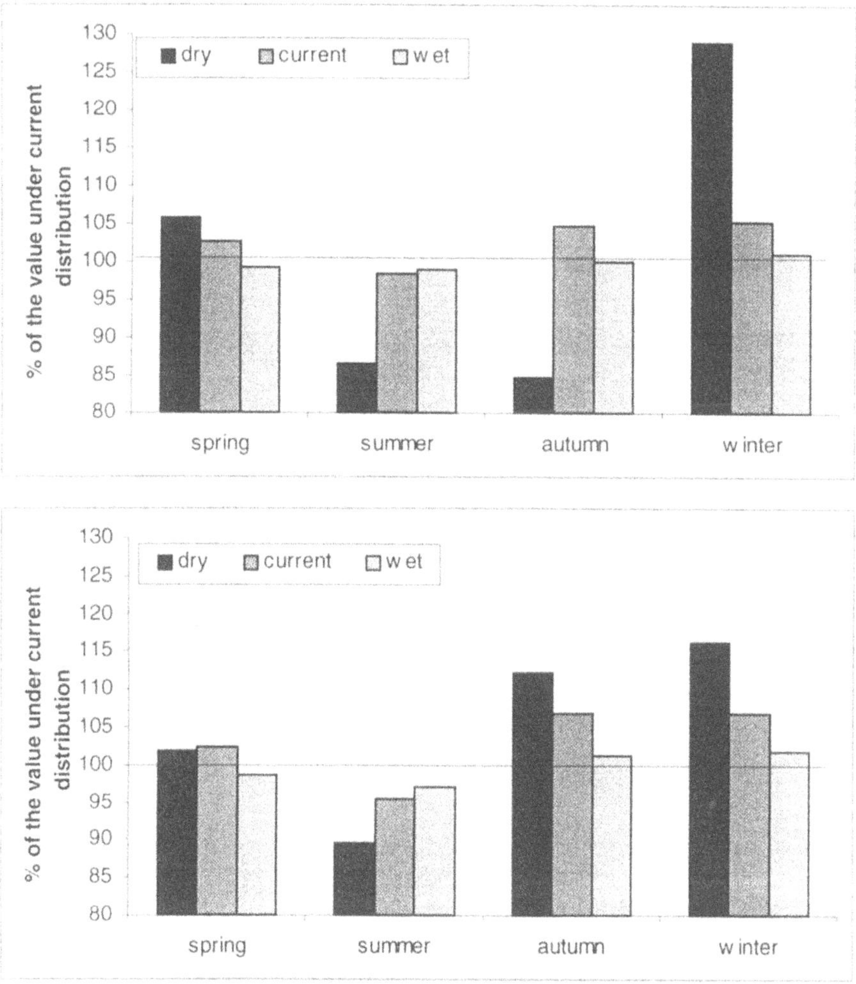

Fig. 9. Percolation in 12 precipitation scenarios relative to the transpiration under the current amount and distribution of precipitation. For further explanations see figure 8.

It may be surprising that the relatively large changes in precipitation lead only to moderate responses of tree growth, in particular if conditions get more favourable. However, additional water can only be used for transpiration (and thus mitigation of drought stress) if the evaporation demand is high and the trees are physiologically capable of additional water uptake. Thus, the effect also depends on the tree species and its morphological properties at the particular site (e.g. the amount of fine roots, rooting depth, leaf area index and hydraulic conductivity). However, the tree species investigated as well as the initial stand conditions are already adjusted to a poor water supply (e.g. relatively high root / shoot ratio). The possibility for the trees to adjust to wetter conditions through changed allocation is limited because the response period in this investigation is only one year (after this period the initial conditions are restored again). It is assumed that the growth increase would be considerably higher if a long-term adjustment to wetter conditions (which would lead, for example to higher foliage biomass relative to root biomass) were taken into account.

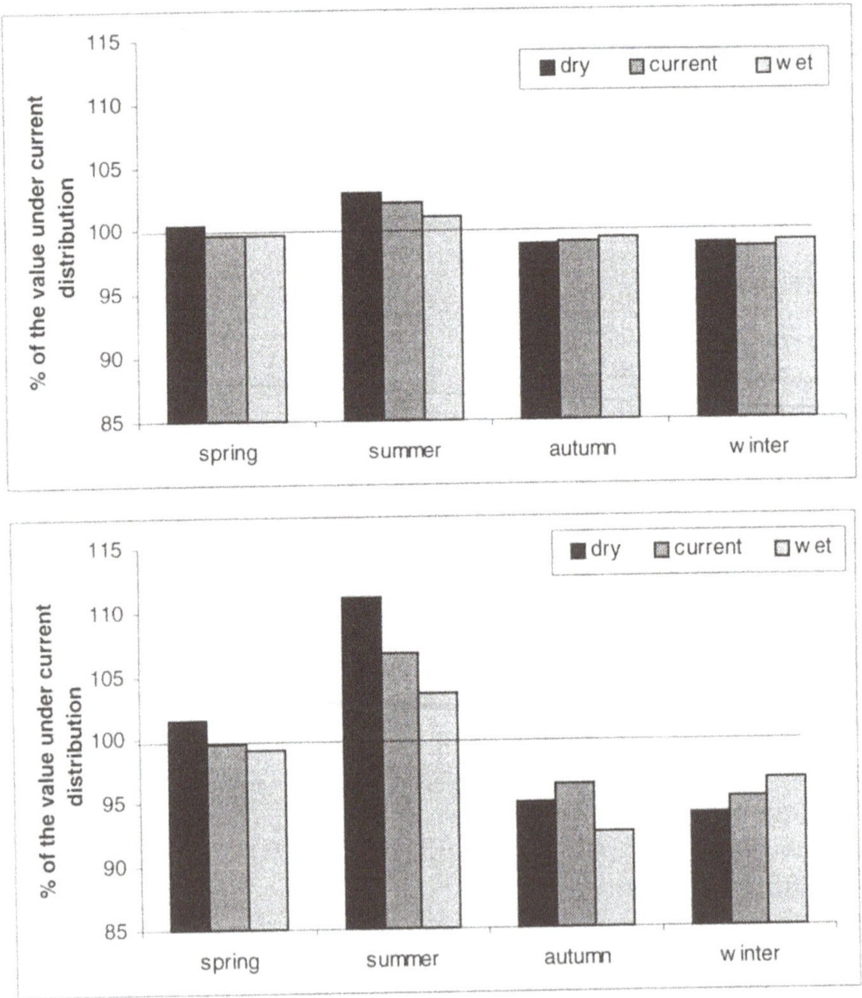

Fig. 10. Stemwood production in 12 precipitation scenarios relative to the transpiration under the current amount and distribution of precipitation. For further explanations see Figure 8.

4. Conclusion

The simulation study indicates that variations in the seasonal distribution of climatic variables affect key processes in forest ecosystems similarly to changes in average conditions. It is therefore necessary to assess systematically the possible range of new climatic conditions. The result of the investigation is a risk analysis for a particular site rather than a prediction of future development, which is actually impossible as long as the development of environmental conditions at a local or regional scale is unknown.

Furthermore, such a sensitivity study indicates the suitability of the applied model for the range of investigated conditions. In this respect a model is assumed to be unsuitable if it gives unrealistic results under any kind of current environmental conditions.

Another important result of the analysis is that the overall response of a forest ecosystem to a changing climate depends very much on the site conditions. This is clearly demonstrated by the different growth reactions to different precipitation scenarios at sites which vary in water holding capacity. Additionally, it is also expected that stand conditions and morphological adjustments of the rooting system and the canopy have to be considered if the growth response at a particular site is to be judged.

For more detailed analyses, additional scenarios can be developed, including a climate that is assumed as most likely for the particular site. With such scenarios, the effect of a simultaneous change of temperature and precipitation can be investigated. Furthermore, it would be interesting to assess the effects of a change in the amount of precipitation relative to a shift in precipitation frequency. If the amount of precipitation per event decreases without a change in total precipitation, less water percolates through the soil during strong rainfall events or wet seasons and the precipitation is used more effectively by vegetation. Concerning the specific site conditions, the analyses can be enlarged to assess the growth responses of different tree species and thus the risk for the forest owner under an uncertain climatic development.

The underlying model assumptions need to be further evaluated to make the risk analysis more reliable and to include additional tree species. This can be done, for example, by comparing the observations of bud burst in years with different spring temperatures, and the start of foliage growth in the simulation. Possibly, new mechanisms should also be included to account for dormancy requirements or frost damage (Hänninen 1990). Furthermore, more research is needed to enlarge the range of evaluation of the allocation responses, which is currently based on investigations at only three sites (Grote et al. 1997). Finally, natural (tree mortality) and anthropogenic (thinning) adjustment responses at the stand level need to be included if the simulation is to give a realistic long-term response to a specific scenario of changing boundary conditions.

5. References

Aber JD, Ollinger SV, Federer CA, Reich PB, Goulden ML, Kicklighter DW, Melillo JM, Lathrop RGJ (1995) Predicting the effects of climate change on water yield and forest production in the northeastern United States. Clim Res 5:207–222

Aber JD, Reich PB, Goulden L (1996) Extrapolating leaf CO_2 exchange to the canopy: a generalized model of forest photosynthesis compared with measurements by eddy correlation. Oecologia 106:257–265

Bossel H (1994) TREEDYN3 Forest Simulation Model. Forschungszentrum Waldökosysteme, B/35. Universität Göttingen, Göttingen, pp 118

Bowes MD, Sedjo RA (1993). Impacts and responses to climate change in forests of the MINK region. Climatic Change 24:63–82

Bürger G (1997) On the disaggregation of climatological means and anomalies. Clim Res 8:183–194

Fiedler F, Wenk G (1973) Der jahreszeitliche Ablauf des Dickenzuwachses von Fichten und Kiefern und seine Abhängigkeit von meteorologischen Faktoren. Wissenschaftliche Zeitschrift der Technischen Universität Dresden 22:531–535

Goudriaan J (1982) Potential production processes. In: Penning de Vries FWT, van Laar HH (eds) Simulation of plant growth and crop production. PUDOC, Wageningen, pp 98–113

Grote R, Erhard M, Suckow F (1997) Evaluation of the physiologically-based forest growth model FORSANA. PIK Report 32, Potsdam Institute for Climate Impact Research, pp 64

Grote R, Erhard M, Flechsig M (1995) Investigation and simulation of pollution effects on pine forests in Eastern Germany. In: Ebel A, Moussiopoulos N. (eds) Observation and Simulation of Air Pollution: Results from SANA and EUMAC. Air Pollution III. Computational Mechanics Publications, Southampton, pp 43–51

Grote R, Suckow F, Bellmann K (1996) A physiology-based growth model for simulations under changing conditions of climate, nitrogen availability, air pollution and ground cover competition. Conference on Effects of Environmental Factores on Tree and Stand Growth. TU Dresden, Berggießhübel, Germany, pp 68–77

Grote R, Suckow F (in press) FORSANA – Model Documentation. PIK Report, pp 110

Grote R, Suckow F, Bellmann K (submitted) Modelling of of carbon-, nitrogen-, and water balances in pine stands under changing air pollution and deposition. Nutrients in Ecosystems

Hänninen H (1990) Modeling dormancy release in trees from cool and temperate regions. In: Dixon RK, Meldahl RS, Ruark GA, Warren WG (eds) Process modeling of forest growth responses to environmental stress. Timber Press, Portland, Oregon, pp 159–165

Houghton JT, Filho LGM, Callander BA, Harris N, Kattenberg A, Maskell K (eds) (1995) Climate Change 1995 – The Science of Climate Change. Cambridge University Press, Cambridge, UK, pp 572

Kartschall T, Döring P, Suckow F (1990) Simulation of Nitrogen, Water and Temperature Dynamics in Soil. Syst. Anal. Model Simul 7:33–40

King DA (1993) A model analysis of the influence of root and foliage allocation on forest production and competition between trees. Tree Physiol 12:119–135

Kramer K (1995) Modelling comparison to evaluate the importance of phenology for the effects of climate change on growth of temperate-zone deciduous trees. Clim Res 5:119–130

Mohren GMJ, Bartelink HH, Jorritsma ITM, Kramer K (1993) A process-based growth model (FORGRO) for analysis of forest dynamics in relation to environmental factors. In: Broekmeijer M, Vos W., Koop HGJM (eds), European Forest Reserves. Proc. of the European Forest Reserves Workshop, 6–8 May 1992, The Netherlands. Pudoc, Wageningen, 273–280

Monteith JL (1965) Evaporation and environment. In: Fogg GE (ed) The State and Movement of Water in Living Organisms. Symp Soc Exp Biol Academic Press, London, 205–234

Pan Y, Raynal DJ (1995) Predicting growth of plantation conifers in the Adirondack Mountains in response to climate change. Can J For Res 25:48–56.

Penning de Vries FWT, Jansen DM, ten Berge HFM, Bakema A (1989) Simulation of Ecophysiological Processes of Growth in Several Annual Crops. Simulation Monographs, 29. PUDOC, Wageningen, The Netherlands.

Perry MA, Mitchell RJ, Zutter BR, Glover GR, Gjierstad DH (1994) Seasonal variation in competitive effect on water stress and pine responses. Can J For Res 24:1440–1449

Running SW (1994) Testing Forest-BGC ecosystem process simulations across a climatic gradient in Oregon. Ecol Appl 4:238–247

Zhang Y, Reed DD, Cattelino PJ, Gale MR, Jones EA, Liechty HO, Mroz GD (1994) A process-based growth model for young red pine. For Ecol Manage 69:21–40

Sensitivity analysis of a forest gap model concerning current and future climate variability

Petra Lasch, Felicitas Suckow, Gerd Bürger, Marcus Lindner

Abstract. The ability of a forest gap model to simulate the effects of climate variability and extreme events depends on the temporal resolution of the weather data that are used and the internal processing of these data for growth, regeneration and mortality. The climatological driving forces of most current gap models are based on monthly means of weather data and their standard deviations, and long-term monthly means are used for calculating yearly aggregated response functions for ecological processes. In this study, the results of sensitivity analyses using the forest gap model FORSKA_P and involving climate data of different resolutions, from long-term monthly means to daily time series, including extreme events, are presented for the current climate and for a climate change scenario. The model was applied at two sites with differing soil conditions in the federal state of Brandenburg, Germany.
The sensitivity of the model concerning climate variations and different climate input resolutions is analysed and evaluated. The climate variability used for the model investigations affected the behaviour of the model substantially.

1. Introduction

Climate change impact investigations in forests have been carried out during recent years using succession models, e.g. FORCLIM (Bugmann 1997), FORSKA (Prentice et al. 1993; Lindner et al. 1997), and various other succession models (Solomon 1986; Pastor and Post 1988; Kienast 1991; Botkin and Nisbet 1992; Prentice et al. 1993; Kräuchi 1995). These studies used different climate change scenarios and different application methods. Assumptions about the increase or decrease of long-term monthly means of air temperature and sums of precipitation under elevated atmospheric CO_2 were obtained from GCM runs and IPCC scenarios (Kienast 1991; Gerstengarbe and Werner 1996; Lauenroth 1996; Werner and Gerstengarbe 1997). Transient and steady-state scenarios were derived in different ways from current time series of climatic data.

A preliminary study on the sensitivity of the forest gap model FORSKA to climatic data input (Lindner et al. 1996) has shown that the temporal resolution of the climatic data is important for simulating forest species composition realistically. In this study, an adapted version of the succession model FORSKA (Prentice et al. 1993; Lasch and Lindner 1995b) was applied. This model originally used long-term monthly means to calculate the climate conditions and response functions in the model. The results of our earlier studies indicated that FORSKA simulated more realistic species composition when using climatic data that included observed short-term climatic variability as compared to using long-term monthly means. This was due to a more realistic description of the yearly aggregated drought stress index, which was calculated as a function of the ratio of actual to potential evapotranspiration.

These investigations are extended here by using FORSKA_P, a new version of FORSKA, which was applied at two sites in Brandenburg (Germany). The study is concerned with the response of this model version when calibrated for two sites, both under current climate and under a climate scenario. The model uses input data of different temporal resolution. The selected sites differ in their climatic conditions. In addition, two soil types at each site were assumed. Long-term time series of monthly weather characteristics, a weather generator for producing monthly time series, and a weather generator for daily time series were used for current climate and the climate scenario. Both the applied steady-state climate scenario and the new weather generator C2W were produced by the Climate Research Department at PIK.

The investigation falls into two parts. Firstly, analyses of the behaviour of the model in response to climate input data of different temporal resolution are carried out for current climate. Secondly, analogous simulation experiments for climate change conditions are analysed. Based on the model's behaviour, conclusions are drawn concerning the applicability of the model FORSKA_P in studying the effects of climate change.

2. Methods

2.1 Simulation model

The succession model FORSKA (Prentice et al. 1993) was modified for applications in north-eastern Germany. Previous studies (Lasch and Lindner 1995a; Lindner et al. 1996) showed that modifications to the model and parameters were necessary to simulate a realistic tree species composition in north-eastern Germany and on a transect across Central Europe. The original and modified versions of FORSKA simulate the dynamics of forest vegetation based on site and environmental conditions. The site conditions are characterized by soil physical parameters. Chemical parameters and nutrient availability are not taken into account because the response of trees to site fertility is not modelled. FORSKA uses long-term monthly means of temperature, precipitation and relative sunshine duration as environmental inputs. These values were scaled down to a quasi-daily time step by linear interpolation between mid-months. With this method, extreme drought events hardly ever occur because monthly precipitation is distributed uniformly over the whole month.

The soil water model in FORSKA calculates the daily soil moisture and the potential and actual evapotranspiration. The daily values are aggregated to a drought stress index (DI), defined as:

$$DI = 1. - \frac{\sum_i AET_i}{\sum_i PET_i} , \qquad (1)$$

AET - daily actual evapotranspiration
PET - daily potential evapotranspiration.

This coefficient is calculated over all days i of the growing season, which is defined by temperature thresholds of 5°C (deciduous trees) and 0°C (evergreen trees). The drought stress index DI is used to calculate a species-specific response function which reduces the annual net assimilation (Prentice et al. 1993).

The new version FORSKA_P was derived from the modified version of FORSKA described above. The original soil-water model of FORSKA (a one-layer bucket model) was replaced by a more detailed model to improve the description of the soil water budget. A multi-layer percolation model was implemented to include more information about the soil conditions of the specific site (see also Martin 1992; Kräuchi 1994). It needs only a few well-known soil physical parameters which are available for wide applications. The model calculates the water content in different soil layers. Each layer is described by thickness, field capacity, wilting point and a texture-dependent percolation factor (Glugla 1969; Koitzsch 1977; Grote and Suckow submitted). Percolation only occurs if the water content of the layer in question is greater than field capacity. Water uptake by the trees and soil evaporation are balanced in each layer and are limited by the potential evapotranspiration and the plant-available water above the wilting point. Interception corresponding to the leaf area index and interception evaporation are considered when calculating the infiltration of water into the first soil layer. Precipitation is stored as snow at temperatures of less than 0°C, and the snow pool is emptied slowly and infiltrates into the soil at temperatures above 0°C (Weise and Wendling 1974). The soil-water model of FORSKA_P works on a daily time step and also calculates an annual drought stress index DI as defined by (1) using the daily AET as the sum of water uptake and soil evaporation.

The model FORSKA_P was modified to work with climatic driving forces with three kinds of temporal resolution:

1. long-term monthly means and sums,
2. time series of monthly means and sums, which are internally interpolated to quasi-daily data by linear interpolation between mid-months. The data are derived from observations or a monthly weather generator,
3. daily time series which are given by observations or produced by a daily weather generator.

2.2 Simulation sites and climate characteristics

Two sites in the state of Brandenburg, Germany, were considered in this study. These sites were selected from several examined in a regional impact study of climate change in Brandenburg (Stock and Toth 1996). Angermünde is situated in the north-east of Brandenburg and Potsdam in the centre. The sites have different temperature and precipitation characteristics (Table 1).

Name	Latitude (°N)	Longitude (°E)	Elevation (m)	T_{mean} (°C)	P_{mean} (mm yr^{-1})
Angermünde	53.02	14.0	56	8.3	535
Potsdam	52.23	13.04	81	8.7	596

T_{mean} - long term annual mean of temperature
P_{mean} - long-term mean of annual precipitation sum

Table 1. Location, elevation, and climate data of the simulation sites

To investigate whether the sensitivity of the model FORSKA_P to climatic variability is influenced by soil moisture conditions, two different types of soil conditions (Table 2) were assumed. The soil types differ particularly in their water-holding capacity. The simulations were carried out at each site for both soil types.

Name	Water-holding capacity (mm)	Number of layers	Texture	Soil depth (m)
st230	230	7	loamy sand	1
st130	130	7	Sand	1

Table 2. Characteristics of the selected soil types

2.3 Climate data and scenarios

Current climatic data and a climate change scenario prepared for both sites were used as driving forces for the simulation experiments. Daily weather records are available for both sites. The observation time series for the Potsdam site is more than 100 years long (1893–1992) and is thus suitable for long-term studies. At the Angermünde site, the observation time series only covers the period 1951–1990. This short time series is unsuitable because the model requires long-term weather input data to simulate long-term forest dynamics. Consequently, a sensitivity analysis for the resolution of the climatic input with real weather data was carried out only at the Potsdam site.

The climate scenarios were produced by the Climate Research Department at PIK. The method is based on a statistical modification of measured time series of weather data (Werner and Gerstengarbe 1997). Scenarios were derived according to simulations with the GCM model ECHAM-T21 (Cubasch et al. 1992) for the IPCC scenarios A ('business-as usual'), which projects a temperature increase of about 1.5 K for Europe. In addition, an 'extreme' scenario was derived under the assumption of a temperature increase twice as great as that of scenario A (+3 K). Transient and a steady-state variants with a daily temporal resolution were available for the period 1996 – 2050 for each scenario type. The steady-state scenarios presume an increased temperature and make no assumption about changing precipitation patterns.

Different types of model input data, such as long-term (30 years) monthly and daily means and sums, time series of monthly means and sums, and covariance matrices were derived from the current climate and the climate scenario time series for temperature, precipitation and relative sunshine duration. The covariance matrices were calculated from the time series of monthly means. A multivariate normal random deviate weather generator MWG (Monthly Weather Generator) was used with the covariance matrices to produce monthly time series of the three climate variables temperature, precipitation, and relative sunshine duration, distributed normally around their long-term means (MWG input data). The MWG input data were provided for both the current climate and the climate scenario +3K.

A new weather generator C2W (climate-to-weather disaggregator) was applied (Bürger 1997). C2W is aimed at disaggregating climatological means and anomalies into realistic weather processes, and it can be used to process GCM outputs. C2W generates realistic daily values of six meteorological variables (e.g. average temperature, precipitation, sunshine duration) using long-term climatological means of daily values and specific aggregations such as monthly means. Daily time series of the three required climate input variables were produced (C2W input data) for all climate scenarios using C2W.

2.4 Simulation experiments

Experiments were carried out with FORSKA_P for the Potsdam and Angermünde sites using the current climate and the 'extreme' scenario +3 K. For the Central European region within the next 100 years an increase of temperature of about 2–4 K can be assumed (Gerstengarbe and Werner 1996). The +3 K scenario corresponds to this assumption and seems to be suitable for the comparison of the simulation results. The simulation results were averaged from 50 plots. The species composition did not change when simulations with an increased number of plots were used (Kräuchi 1994; Bugmann et al. 1996). The model run was 600 years for the long-term simulations. The main emphasis was put on the comparison of the species composition. The equilibrium species composition was determined by averaging the yearly output of species biomasses from the last 300 years of a simulation run.

3. Results and Discussion

3.1 Model sensitivity to the resolution of climate input

The first question investigated concerned the sensitivity of the model to different resolutions of weather input data. An analysis with real weather data was therefore undertaken. For the simulation of the dynamic behaviour at the Potsdam site (Figure 1), FORSKA_P was driven by real time series of monthly and daily means and by long-term monthly means. The simulation time was limited to 100 years since this was the approximate extent of the time series of climatic data being used. The simulated species compositions were significantly different between the long-term monthly means and the real daily and monthly time series.

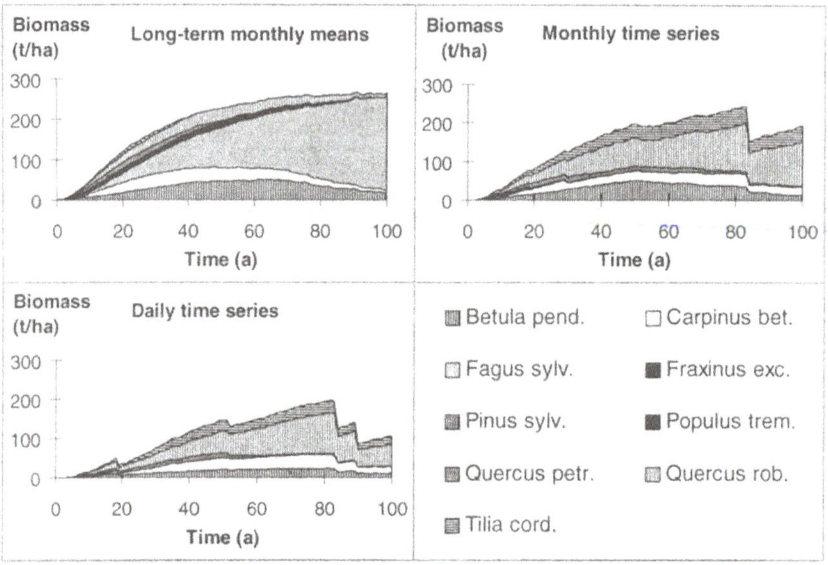

Fig. 1. Simulation of total biomass at the site Potsdam, soil type st130, with long-term monthly means, real time series of monthly means, and time series of real weather data for 100 years

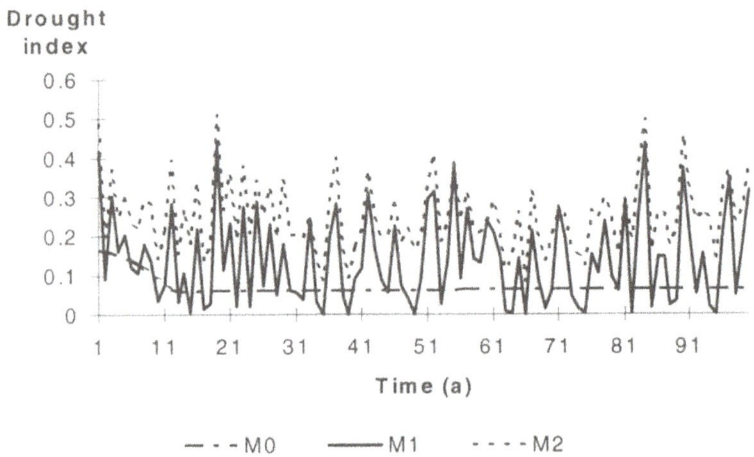

Fig.2. Simulation of drought indices at the site Potsdam, soil type st130, with long-term monthly means (*M0*), real time series of monthly means (*M1*), and time series of real weather data (*M2*) for 100 years

This effect can be explained by the differences in the simulated drought stress indices (Figure 2). In the case of long-term monthly means, a nearly constant low drought stress index was calculated over the whole simulation period.

A statistical analysis of the drought stress simulation is given in Table 3. The drought index varies most strongly when the monthly weather data (M1) are used. It is near or equal to zero in some years. The main reason for this effect is the internal linear interpolation of monthly data to a daily resolution, which is necessary for the soil water model.

In Figure 3 the real and the interpolated daily temperature and precipitation data are compared for two years

Mode	Mean	Minimum	Maximum	Variance	Std. Dev.
M0	0.071	0.059	0.164	0.0055	0.0235
M1	0.141	0.	0.439	0.0133	0.1152
M2	0.251	0.082	0.513	0.0084	0.0917

Table 3. Statistics of drought index simulated over 100 years with long-term monthly means (M0), monthly time series (M1), and daily time series (M2), derived from a daily weather record, for the Potsdam site

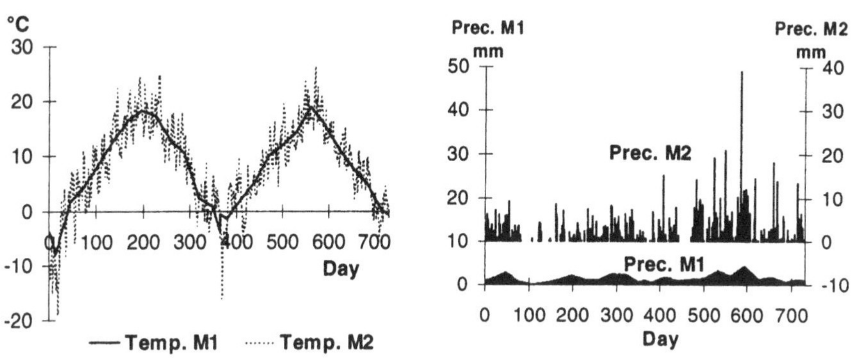

Fig. 3. Comparison of observation values of temperature (left) and precipitation (right) with monthly resolution (*M1*), which are uniformly distributed among all days, and daily resolution (*M2*). The data are shown for 2 years

Extreme temperature and precipitation events do not occur with the interpolated daily data (M1) and thus the simulated drought stress is underestimated in some years. If real daily data are used (M2), a more realistic drought index is calculated, with lower variance but higher mean, minimum, and maximum values (see Figure 2). Higher values of the drought index occur because during the vegetation period the soil becomes dry if days without precipitation occur. In addition, in the case of heavy precipitation events, losses of water by percolation or run-off are possible. Therefore, the drought stress is better reproduced with the daily time series (M2).

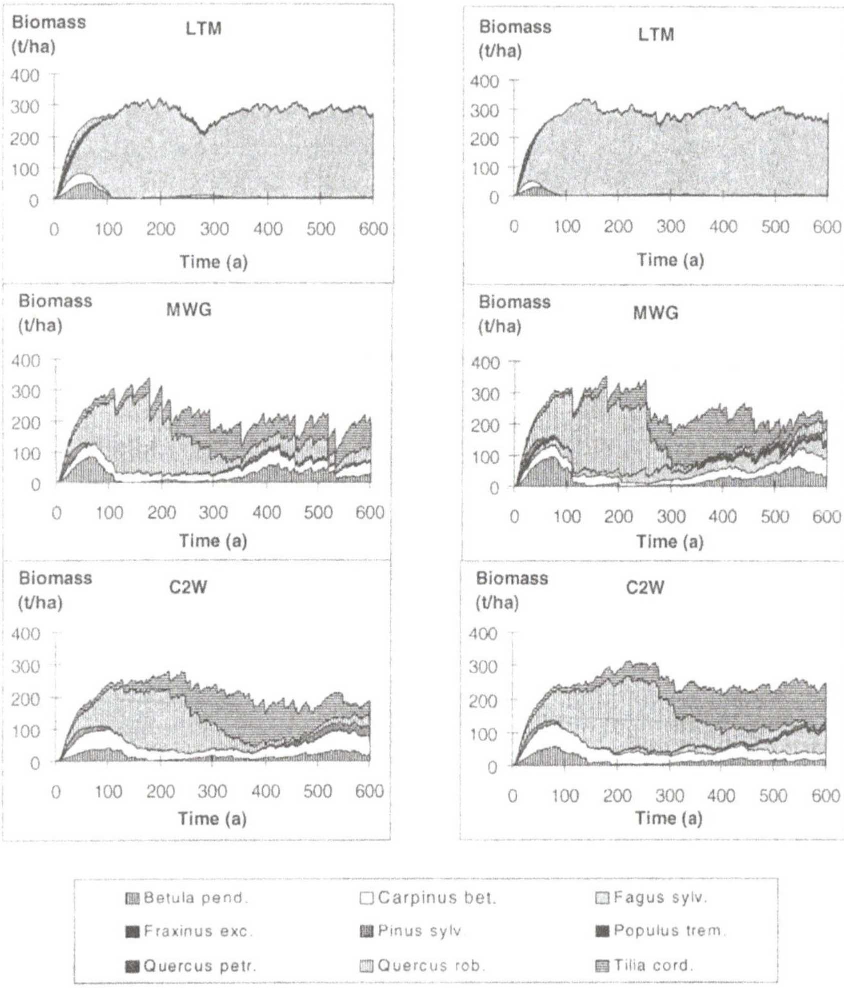

Fig. 4. Simulation of total biomass at the site Potsdam, soil type st130 (left) and st230 (right); for different temporal resolution of climate input data (*LTM*- long-term monthly means, *MWG*- monthly means weather generator, *C2W*- daily weather generator), under current climate

This sensitivity study was continued by the analysis of simulation runs covering 600 years for the Potsdam and Angermünde sites. MWG - and C2W - input data were used for this analysis. The simulation results for the Potsdam site for both soil types are shown in Figure 4.

FORSKA_P simulated similar species compositions for both soil types under the same climate conditions. However, it was only on the loamy sand soil type (st230) that there was some occurrence of *Fagus sylvatica* using MWG - or C2W - input data. This was due to the higher water-holding capacity. Therefore, a lower drought index for this soil type was simulated in comparison with the sandy soil type (st130). The species composition simulated with long-term monthly means differed strongly from those using MWG and C2W input data, independently of the soil type. In the first case the abundance of *F. sylvatica* was overestimated at this site because average site conditions at Potsdam represent conditions just outside the natural range of *F. sylvatica*. In both other cases the simulation of a mixed deciduous species composition was more realistic. The results do not correspond exactly with the expected potential natural vegetation at this site (Krausch 1993) because the model does not account for species-specific responses to site fertility. The variation of the drought index is higher in the case of MWG, leading to unrealistic biomass variations during the simulation period in comparison to biomass simulations with C2W. The differences between simulations with the three different temporal resolutions of weather data inputs become apparent from Figures 5 and 6, which show the equilibrium species composition at the Potsdam and Angermünde sites for both soil types.

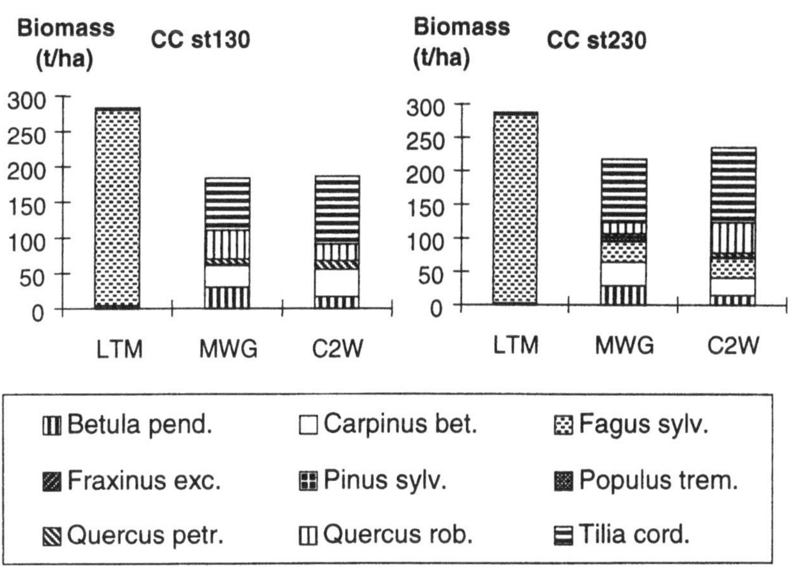

Fig. 5. Equilibrium species composition simulated at the site Potsdam, soil type st130 and st230; for different resolution of climate input data (*LTM*- long-term monthly means, *MWG*- monthly means weather generator, *C2W*- daily weather generator), under current climate (*CC*)

The simulations for the Angermünde site show a similar behaviour of the model to the different resolutions of the weather input data as for Potsdam. The drought indices for both sites and soil types and three types of weather data are given in Table 4.

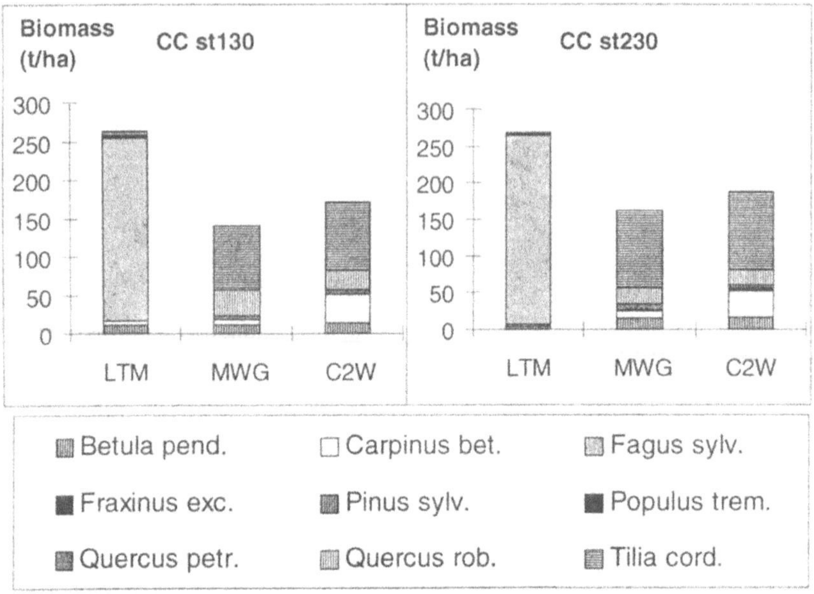

Fig. 6. Equilibrium species composition simulated at the site Angermünde soil type st130 and st230; for different resolution of climate input data (*LTM*- long-term monthly means, *MWG*- monthly means weather generator, *C2W*- daily weather generator), under current climate (*CC*)

Site	Soil type	LTM	MWG	C2W
Potsdam	st130	0.066	0.127	0.183
	st230	0.015	0.069	0.137
Angermünde	st130	0.098	0.151	0.204
	st230	0.079	0.097	0.153

Table 4. Mean values of drought index at the Potsdam and Angermünde sites calculated from simulations with long-term monthly means (LTM), weather inputs from a monthly weather generator (MWG), and daily resolved weather inputs from C2W

The drought indices were lower for the Potsdam site in all cases. The variations in species composition and total biomass at the sites for the same soil type (Figure 5) were caused by differences in the drought index. Similar changes in species composition and total biomass due to different weather data inputs were simulated at both sites.

It can be concluded that a constant weather input over the whole simulation period caused an unrealistic species composition. Furthermore, realistic assumptions about the distribution of monthly temperature and precipitation resulted in a more realistic calculation of the drought stress and therefore in a more plausible simulation of the species composition.

3.2 Model sensitivity concerning climate change scenarios

The same simulation experiment was conducted with the steady-state 'extreme' climate change scenario +3K. Simulation results for the Potsdam site with the sandy soil type (st130) are shown in Figure 7. For all types of climate input data the model simulated a shift in species composition and a decrease of total biomass under the temperature increase of +3K in comparison with the simulations using current climate data (Figure 4).

As with the simulations under current climate, there are remarkable differences in the simulations for the three types of temporal resolution of weather data. The abundance of *Fagus sylvatica* seems to be overestimated with the long-term monthly means. Using monthly and daily time series of weather input data, the simulated species compositions correspond with an expected *Tilia cordata* – *Carpinus betulus* forest with *Quercus* spp.

Large differences in the simulated total biomass occurred between the monthly and the daily weather generators (Figure 7, MWG and C2W). Although the simulated averaged drought index for monthly time series (0.172) was less than the drought index for daily time series (0.23), see Table 5, the total biomass was lower over the whole simulation period.

Site	Soiltype	LTM		MWG		C2W	
		CC	3 K	CC	3 K	CC	3 K
Potsdam	st130	0.066	0.140	0.127	0.172	0.183	0.230
	st230	0.015	0.086	0.069	0.105	0.137	0.171
Angermünde	st130	0.098	0.182	0.151	0.213	0.204	0.253
	st230	0.079	0.175	0.097	0.174	0.153	0.21

Table 5. Mean values of the drought index simulated for weather data input of different temporal resolutions (LTM- long-term monthly means, MWG- monthly weather generator, C2W- daily weather generator under current (CC) and climate change scenario (3K)

This disagreement can be explained by the important influence of the yearly variance and maximum values of the drought index when using the monthly time series. In the left column of Figure 7 the corresponding drought indices were compared for the different resolutions of weather data. The high variance of the drought index caused by the linear interpolation of monthly to daily values (MWG simulation) led to very low total biomasses. There were many years with a very high drought index (ca. 0.4), which was intolerable for most species in the model parameterization that was used. High drought stress indices caused reduced biomass increases and thus high mortality during the simulation. The biomass losses could not be compensated during years with a low drought index. The occurrence of years with very high drought index was largely due to the weather generator MWG, which uses simple assumptions about correlations between temperature, precipitation, and sunshine duration.

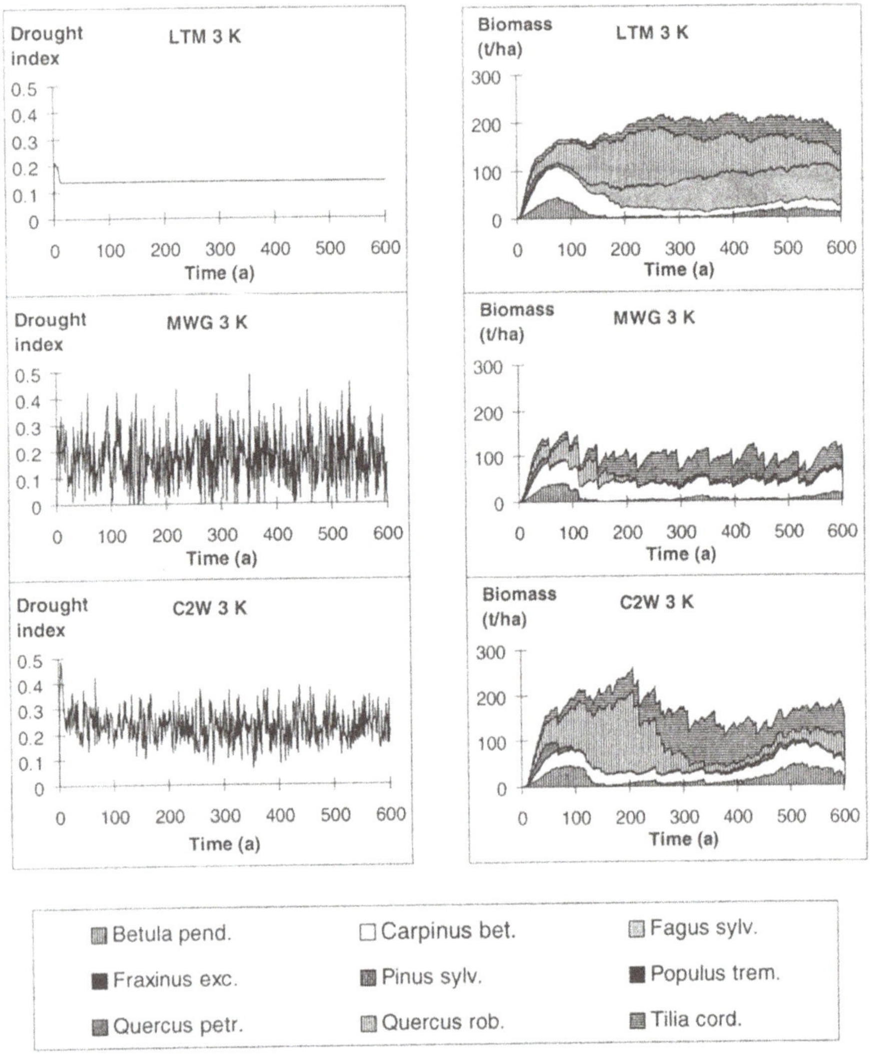

Fig. 7. Simulated total biomasses and drought indices at the Potsdam site with the soil type st130 under the climate change scenario +3K for different temporal resolution of climate input data (*LTM*- long-term monthly means, *MWG*- monthly means weather generator, *C2W*- daily weather generator)

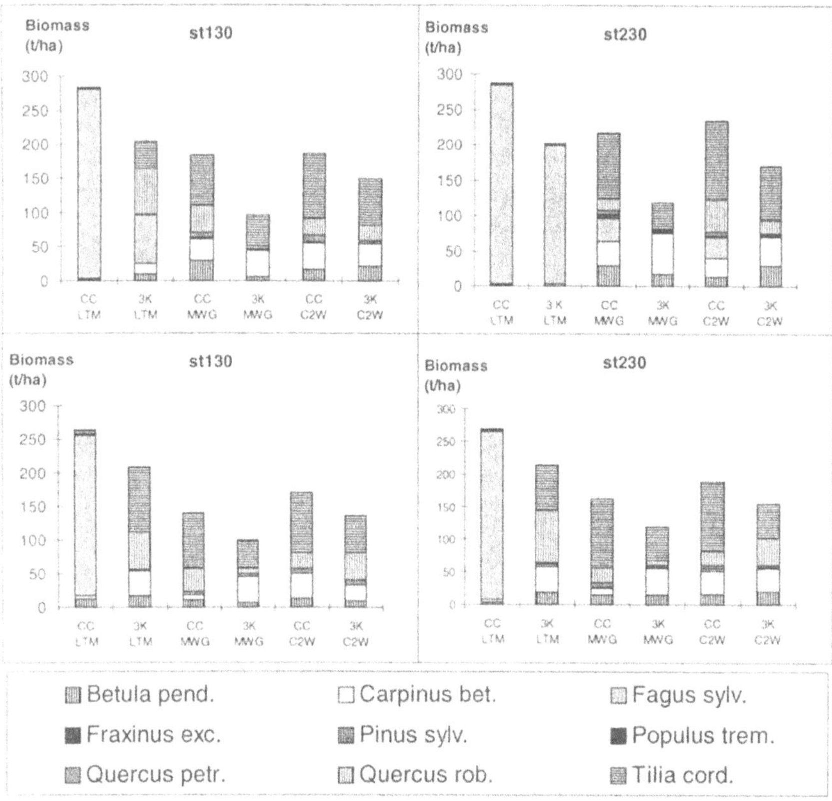

Fig. 8. Equilibrium species composition simulated for the Potsdam (upper row) and Angermünde (lower row) sites with soil type st130 and st230, under current climate (*CC*) and the climate change scenario +3K, using different temporal resolution of climate input data (*LTM*- long-term monthly means, *MWG*- monthly means weather generator, *C2W*- daily weather generator)

Using Table 5, it is possible to compare the averaged drought indices simulated for all sites and soil types. These results can be related to the equilibrium species compositions for all cases given in Figure 8.

The effect of climate change is most severe at the drier site (st130) with less precipitation (Angermünde) (Figure 8, lower row). The differences in total biomass between simulations with monthly and daily time series, discussed above, appear in all cases. In the experiments, the temporal resolution of weather input data was a more important factor in determining the simulated species composition than soil type or climate conditions.

4. Conclusions

These simulation analyses were aimed at testing the sensitivity of the model FORSKA_P concerning climate variability under current and future climate conditions. All in all, the investigations on the behaviour of the model showed a strong sensitivity to the temporal resolution of weather input data. This resulted from the soil-water model used in FORSKA_P, which worked on a daily time step and required the monthly weather input data to be processed into daily weather data. The application of long-term monthly means of meteorological variables for this model is not recommended since the assumptions made about the distribution of climate variables such as temperature and precipitation are too simplistic. In this case the simulated species composition at drier sites or sites with low water-holding capacity did not correspond to the expected natural vegetation. Although the model FORSKA_P was not calibrated to use variable weather input, the simulation results became more realistic if a yearly variability of at least the monthly means or daily values of the meteorological variables was assumed. It was better to drive the model with time series comprising realistic day-to-day variations of the meteorological variables, which included realistic weather events such as long periods without precipitation or heavy rains.

A comparison of the model behaviour when using monthly and daily resolved weather input data indicates that differing drought indices and therefore total biomasses and species composition can be obtained. The choice of the temporal resolution of weather input strongly influences the model behaviour.

The simulated results under current climate differ remarkably from simulations under scenarios of temperature increase in all cases of climate input. This concerns species composition and total biomass. The model response to climate change also depends on climate variability. Many assumptions about climate change scenarios include only changes of mean temperature values or precipitation sums. This study emphasises that climate variability and extreme events may be more important and that they should be taken into account in climate change studies (Riha et al. 1996). The model responses to such events cannot be investigated by a model driven by climate variables of long-term monthly resolution.

The sensitivity of the model's response to climate conditions and their changes depends on the description of the main processes of plant growth and stand development. The model FORSKA_P is limited not only in reproducing differences in site fertility but also in the representation of environmental effects on tree growth. The application of more mechanistic approaches in a succession model would enable better causal interpretations of climate impacts on forests (Friend et al. 1997). The development of a forest gap model which responds realistically to changing environmental conditions is in progress (Bugmann et al. 1997). To improve the predictive power of the model we recommend that detailed information about climate variability and especially about precipitation distribution is used as input to the model.

Acknowledgement. This research was partly supported by the German Federal Ministry of Education and Research. We thank the anonymous reviewers for their comments on earlier drafts of this paper.

5. References

Botkin DB, Nisbet RA (1992) Forest response to climatic change: effects of parameter estimation and choice of weather pattern on the reliability of projections. Climatic Change 20:87–111

Bugmann H (1997) Sensitivity of forests in the European Alps to future climatic change. Clim Res 8:35–44

Bugmann H, Fischlin A, Kienast F (1996) Model convergence and state variable update in forest gap models. Ecol Modelling 89:197–208

Bugmann H, Grote R, Lasch P, Lindner M, Suckow F (1998) A new forest gap model to study the effects of environmental change on forest structure and functioning. In: Mohren GMJ, K Kramer, S Sabate (eds) Impacts of Global Change of Tree Physiology and Forest Ecosystem. Proceedings of the International Conference on Impacts of Global Change on Tree Physiology and Forest Ecosystems, held 26–29 November 1996, Wageningen. Kluwer Academic Publisher, Dordrecht

Bürger G (1997) On the disaggregation of climatological means and anomalies. Clim Res 8:183–194

Cubasch U, Hasselmann K, Höck H, Maier-Reimer E, Mikolajewicz U, Santer BD, Sausen R (1992) Time-dependent greenhouse warming computations with a coupled ocean–atmosphere model. Clim Dyn 8:55–69

Friend AD, Stevens AK, Knox RG, Cannell MGR (1997) A process-based, terrestrial biosphere model of ecosystem dynamics (hybrid v3.0). Ecol Modelling 95:249–287

Gerstengarbe FW, Werner P (1996) Szenarien zur Klimaentwicklung im Land Brandenburg bis zum Jahr 2050. In: Stock M, F Toth (eds) Mögliche Auswirkungen von Klimaänderungen auf das Land Brandenburg. Pilotstudie. Potsdam Institute for Climate Impact Research, Potsdam, pp 28–39

Glugla G (1969) Berechnungsverfahren zur Ermittlung des aktuellen Wassergehaltes und Gravitationswasserabflusses im Boden. Albrecht-Thaer-Archiv 13:371–376

Grote R, Suckow F (submitted) Integrating long-term adaptations into physiological forest growth modeling. I. Effects on water balance and gas exchange. For. Ecol. Manage.

Kienast F (1991) Simulated effects of increasing atmospheric CO2 and changing climate on the successional characteristics of Alpine forest ecosystems. Landsc Ecol 5:225–238

Koitzsch R (1977) Schätzung der Bodenfeuchte aus meteorologischen Daten, Boden- und Pflanzenparametern mit einem Mehrschichtmodell. Z Meteorol 27.5:302–306

Kräuchi N (1994) Modelling forest succession as influenced by a changing environment, Ph.D. Thesis No. 10479, Swiss Federal Institute of Technology Zurich

Kräuchi N (1995) Application of the Model Forsum to the Solling Spruce Site. Ecol Modelling 83:219–228

Krausch H-D (1993) Potentielle natürliche Vegetation. Ökologische Ressourcenplanung Berlin und Umland – Planungsgrundlagen. . In: Umweltbundesamt (ed) FB 90051, UBA-Texte , Berlin, pp 8

Lasch P, Lindner M (1995a) Application of two forest succession models at sites in north east Germany. J Biogeogr 22:485–492

Lasch P, Lindner M (1995b) Wirkung von Klimaveränderungen auf Waldökosysteme. Abschlußbericht zum BMBF-Forschungsvorhaben DLR 01 LK 9109. PIK-Report No.12, Potsdam-Institut für Klimafolgenforschung: 73

Lauenroth WK (1996) Application of patch models to examine regional sensitivity to climate change. Climatic Change 34:155–160

Lindner M, Bugmann H, Lasch P, Flechsig M, Cramer W (1997) Regional impacts of climatic change on forests in the state of Brandenburg, Germany. Agric For Meteorol 84:123–135

Lindner M, Lasch P, Cramer W (1996) Application of a forest succession model to a continentality gradient through Central Europe. Climatic Change 34:191–199

Martin P (1992) EXE: a climatically sensitive model to study climate change and CO2 enrichment effects on forests. Aust J Bot 40:717–735

Pastor J, Post WM (1988) Response of northern forests to CO2-induced climate change. Nature (London) 334:55–58

Prentice IC, Sykes MT, Cramer W (1993) A simulation model for the transient effects of climate change on forest landscapes. Ecol Modelling 65:51–70

Riha SJ, Wilks DS, Simoens P (1996) Impact of Temperature and Precipitation Variability On Crop Model Predictions. Climatic Change 32:293–311

Solomon AM (1986) Transient response to CO_2-induced climate change: simulation modeling experiments in eastern North America. Oecologia 68:567–579

Stock M, Toth F (eds) (1996) Mögliche Auswirkungen von Klimaänderungen auf das Land Brandenburg. Potsdam Institute for Climate Impact Research, Potsdam

Weise K, Wendling U (1974) Zur Berechnung des Bodenfeuchteverlaufs aus meteorologischen und bodenphysikalischen Größen. Archiv für Acker- und Pflanzenbau und Bodenkunde 18:145–154

Werner PC, Gerstengarbe F-W (1997) A proposal for the development of climate scenarios. Clim Res 8:171–182

Simulated effects of bark beetle infestations on stand dynamics in *Picea abies* stands: coupling a patch model and a stand risk model

Manfred J. Lexer and Karl Hönninger

Abstract. Bark beetle infestations in *Picea abies* stands can substantially alter stand structure and subsequently micro-climatic properties within forest stands, which in turn may influence forest vegetation dynamics. Such effects are not considered in conventional forest gap models of the JABOWA/FORET-type. In this study we explore the simulated effects of bark beetle mortality on forest dynamics and species composition by coupling a stand risk model with a spatially explicit forest patch model. The forest model provides input to the risk model which in turn provides estimates of damage probability and intensity for the simulated stand. The simulation experiments for two climate scenarios showed that species composition can differ significantly for periods of up to 300 years if bark beetle mortality is taken into account. Further improvements to gap model mortality algorithms are suggested.

Keywords. patch model, stand risk model, *Picea abies*, mortality, bark beetles

1. Introduction

Since the publication of Botkin et al. (1972) patch models (gap models sensu Shugart1984) have been applied in numerous studies to simulate vegetation development in various ecotones. Conceptually gap models simulate growth, regeneration and death of individual trees stochastically on small forest patches occupying between 0.01 and 0.1 ha. Site resources available for each tree (light, water, nutrients) and their effect on growth and regeneration are considered explicitly. By aggregating the characteristics of many individually simulated forest patches, vegetation characteristics at the ecosystem level can be depicted. Gap models have been subjected to extensive model tests and are considered to be capable of reproducing important features of natural forest ecosystems such as successional behaviour, species composition and diameter distributions. Although not especially designed for the purpose, gap models are a useful tool for studying the possible impacts of a changing climate on forest dynamics and forest vegetation composition.

Prediction of mortality is an essential feature of any vegetation dynamics model (Monserud 1976). Unfortunately mortality is extremely variable and difficult to predict. Lee (1971) distinguished two kinds of mortality in forest models: regular mortality, which results from competition for scarce resources, and irregular mortality, which is the result of external disturbances such as fire or windthrow. In gap models trees die either from an age-independent intrinsic risk of death, or from a growth-related mortality (Botkin 1993). The former assumes that not more than a small fraction (usually 1 or 2 %) of all individuals of a species can reach the species-specific maximum age under optimal growth conditions. Intrinsic mortality is thus independent of stand density. Growth-related mortality assumes that a tree which fails to reach a species-specific minimum diameter increment for a number of successive years has an annual mortality probability of 0.368 (which equals the assumption that only 1 % of such poor-growing individuals will survive the next 10 years) until it is able to increase the diameter increment again (Shugart 1984; Botkin 1993). The rationale for this concept is that a poor-growing tree lacks vigour and, following resource allocation theories, is unable to allocate sufficient amounts of photosynthetic compounds to its defence mechanisms (e.g. Christiansen et al. 1987; Waring and Pitman 1980; Berryman 1976). Therefore the risk of being killed by insects or diseases will be increased. The individual-based probability of death is compared with a uniform random deviate and the tree dies if the random number is less than the corresponding probability of death. The threshold parameters for this mortality algorithm are not calibrated from field measurements but chosen rather arbitrarily. According to the gap model concept, growth-related mortality can be triggered either by stand density or by unfavourable environmental conditions. Growth-related mortality due to unfavourable environmental conditions guarantees that individuals of a species which are close to their physiological limits will be subjected to high mortality rates and will subsequently die. This is meant to characterize the increased risk of disease and insect damage at the edges of the physiological range of a species (Shugart 1984). From these considerations it follows that the growth-related mortality algorithm in gap models characterizes a whole range of possible causes of mortality and represents more or less a "black box". Thus, while the simulated species composition may be plausible, the reasons and pattern of forest succession might not be captured appropriately (Paccala et al. 1993).

Since the middle of the last century *Picea abies* (L.) Karst. has been highly promoted due to its wood quality and suitability for plantation management. Extensive areas in warmer and drier lowlands have been transformed to conifer plantations dominated by *P. abies*. These *Picea* stands are particularly vulnerable to an array of insects and disease organisms. According to climate change scenarios, drought periods are likely to occur more frequently because of altered precipitation patterns and higher temperatures (Houghton et al. 1990, 1996). Thus the well known sensitivity of *P. abies* to drought stress (Schmidt-Vogt 1977) may result in reduced tree vigour which in turn might increase the risk of bark beetle outbreaks (*Ips typographus* L., *Pityogenes chalcographus* L.). Bark beetle infestations can alter stand structure substantially, creating micro-scale environmental conditions which currently are not considered explicitly in forest gap models. If gap models are applied to simulate successional dynamics of such stands then the results might not be realistic.

With these considerations in mind our objectives are:

1. To couple a forest patch model with a stand risk model to account explicitly for bark beetle induced mortality in *Picea abies* stands. To do this, we will introduce an integrated modelling system.

2. To analyse the simulated effects on mortality and species composition under different climate scenarios and subsequently compare the simulation results from model runs with and without explicitly considering bark beetle mortality.

2. Methodology

2.1 The patch model PICUS 1.0

The patch model PICUS 1.0 is used to project forest stand dynamics (Hönninger and Lexer 1997; Lexer and Hönninger in preparation). In PICUS, growth, reproduction and death of individual trees are modelled on 10 x 10 m patches where the position of an individual tree within a patch is unspecified. The 10 x 10 m patch size corresponds approximately to the maximum crown projection area of a dominant overstory tree in Central European forests. The forest consists of any number of patches defined by the user. All patches are arranged on a square grid. Crown cells of 5 m depth account for the third dimension of the simulated forest. The current version of PICUS considers spatial 3D-interactions between forest patches in simulating the radiation regime of the forest as well as in the regeneration sub-model. Thus, unlike in conventional patch models, the simulated forest changes over time as an interactive unit rather than as a series of independent plots. From this point of view PICUS is a descendant of the ZELIG model (Urban 1990). The spatial range of interactions between adjacent patches depends on the characteristics of the simulated stand (tree heights), site characteristics (orientation, slope, latitude) and solar altitude, angle and direction. The spatial structure of the model is schematically presented in Figure 1.

Growth of individual trees is modelled by reducing a size-dependent, species-specific, optimum diameter increment with the combined effect of light, temperature, water supply and site nutrient status (Botkin, 1993). The optimum-growth equation introduced by Moore (1989) is parameterized with data from open grown trees (Hasenauer et al. 1994; Lexer unpublished data). The temperature regime is characterized by the growing degree days above a threshold of 5.5°C (Botkin 1993; Shugart 1984) and the winter minimum temperature, and the soil moisture index SMI is derived from a site-specific water balance (eq. 1).

$$SMI = \frac{AET_{GS}}{PET_{GS}} \tag{1}$$

SMI = soil moisture index

AET_{GS} = actual evapotranspiration in the growing season
PET_{GS} = potential evapotranspiration in the growing season

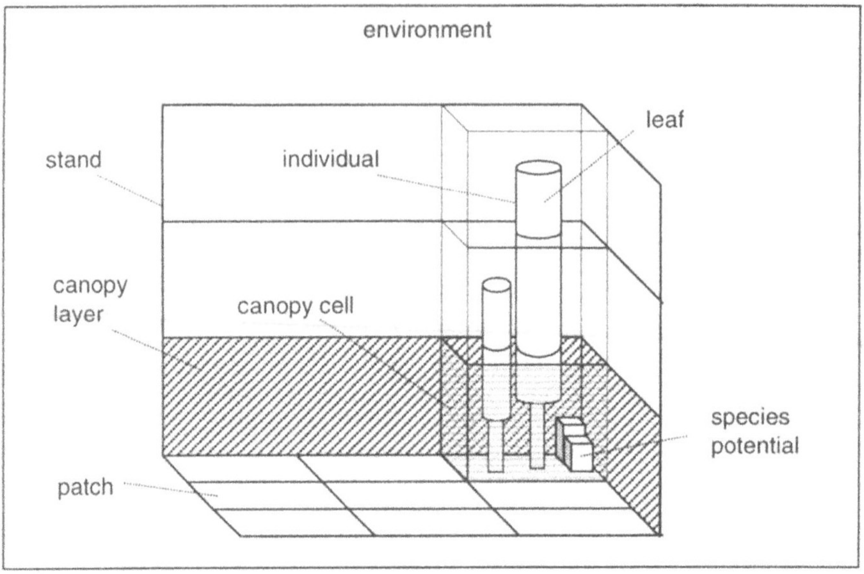

Fig 1. Schematic representation of the spatial structure of PICUS version 1.0.

The monthly potential evapotranspiration PET is calculated according to Thornthwaite and Mather (1957). When the soil moisture content decreases PET is modulated by a factor that depends upon the amount of water in the soil (Dunne and Leopold 1978). This relationship can be expressed as

$$AET = PET \cdot f\left(WC/_{FC}\right). \tag{2}$$

WC = water content of the soil
FC = field capacity of the soil
$f(\,)$ = some function of the term inside the parentheses

In this study f(WC/FC) represented a sandy loam. According to Zahner (1967), evapotranspiration for such soils is unlimited above a ratio of WC/FC = 0.5; below this threshold the ratio AET/PET equals WC/FC. As PICUS version 1.0 relies on monthly climate data, water input into the soil via snow melt or precipitation and water depletion from the soil are calculated iteratively within each month to characterize the site water balance more realistically. In preparatory tests we compared the water balance calculations of PICUS with the results of a detailed ecophysiologically-based water budget model which operates on a daily basis (Lexer 1995a) and found that the coincidence of the two models was closest with three iteration steps per month. The sequence of water input via precipitation or snowmelt and water output from the soil via evapotranspiration varies stochastically among the iterative steps. The radiation submodel of PICUS captures the essential effects of the radiation regime above and within the canopy. Diffuse and direct radiation above the canopy are modelled according to the methods given by Swift (1976), Garnier and Ohmura (1968), Frank and Lee (1966), Hungerford et al. (1989), Buffo et al. (1972), Brock (1981) and Liu and Jordan (1960). The effect of slope and orientation on incoming radiation as well as the shielding effect of the surrounding topography are explicitly included in the model.

To save CPU-time the complex calculations for the simulation of the radiation regime are run for 3 representative days per month in 10-minute time-steps each. For any crown cell within the canopy the proportion of impinging direct light is calculated for every time-step. Attenuation of direct light by cumulative leaf area in the sun´s direction is modelled according to Beer´s law. A fixed percentage of the attenuated light is "left" in every passed crown cell and scattered as diffuse light into all adjacent crown cells, assuming a spherical leaf area distribution. The diffuse light from above the canopy trickles vertically down the canopy and is attenuated according to a negative exponential function. The scattering of diffuse light into adjacent crown cells is computed as an iterative process. As a result each crown cell in the simulated forest contains information on the average amount of direct and diffuse available light as well as on the leaf biomass of individual trees on the corresponding parent patch. The amount of available light for an individual tree is calculated as the weighted average of the available light in each crown cell where an individual is present with leaf area. The weights are determined from the proportion of the particular tree´s leaf area in the crown cells. This preliminary, individual-specific, available light index is related to the corresponding available light index that an open-grown tree of the same species and size would have. Thus we preserve consistency with the modelling concept of a "constrained growth potential" which is derived from data on open-grown trees (compare Botkin 1993). The monocausal environmental response functions used in PICUS 1.0 are shown in Figure 2.

In the standard model version mortality is modelled as a stochastic process, where the probability of death for an individual tree depends on the assumption that 1 % of all individuals of a species can reach the species-specific maximum age under favourable environmental conditions. The annual probability of death for an individual increases to 0.368 if an individual fails to realize a specified minimum diameter increment for a number of successive years (Botkin 1993).

In PICUS the regeneration success of a species depends on the presence of mature seed-producing trees in the overstory and on the constraints of the abiotic environment. The seed shadow of each adult tree is modelled as a function of the size of the parent tree and the species´ seed characteristics. In addition seed propagation by animals is considered for zoochorous species.

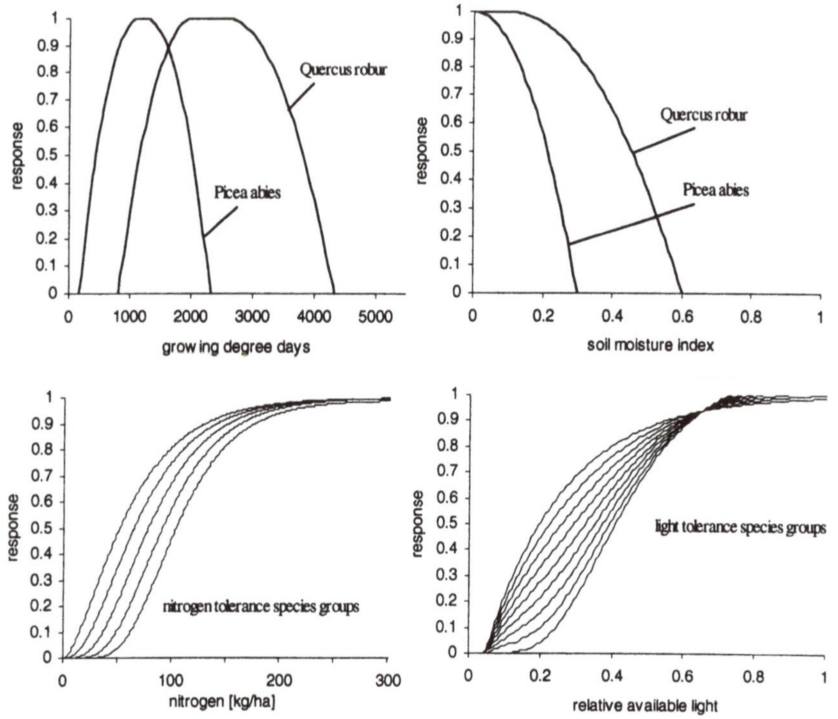

Fig. 2. Environmental response functions for selected species and species groups as used in PICUS 1.0.

2.2 The bark beetle risk-model

Originally the stand hazard model employed in this study was developed to estimate the risk of *Picea abies* stands with limited soil moisture supply suffering from bark beetle damage (Lexer 1995b, 1997). Stands predisposed to beetle damage by external influences such as windthrow and snow breakage were excluded from the model development. Risk models deal with highly stochastic host–insect systems. To reduce the implications of such stochastic effects in space and time the following recommendations have been given (Daniels et al. 1979; Stage and Hamilton 1981; Hedden 1981):

1. Separate the predictions of damage probability and damage intensity.
2. Use the periodic damage as the dependent variable.

Following these recommendations, in a first step the probability of a bark beetle infestation within a four-year-period for a given stand is estimated in a top-down approach. A uniform random deviate decides if a stand is subjected to bark beetle damage or not. Subsequently, for stands which are rated as "damaged", the damage intensity is estimated. This approach enables the use of different variables for the explanation of the damage probability and intensity. A logistic function (eq. (3)) was used to estimate the dichotomous outcome "damage" / "no damage" for a stand (Stage and Hamilton 1981).

$$P = \frac{1}{1+\exp\left(41.7253-0.0224\cdot SHI-1.6848\cdot\sqrt{AGE}-14.4867\cdot BAS-14.2009\cdot SVI\right)} \quad (3)$$

Predictor variables that were used in the risk model are the average age of *Picea abies* (AGE), the proportion of *P. abies* basal area (BAS), the stand vitality index SVI and the stand hazard index SHI1. SVI characterizes the proportion of *Picea* individuals with a decreasing diameter increment trend with respect to two successive 4-year periods (eq. (4)).

$$SVI = \frac{\sum_{i=1}^{n} N_{VIneg(i)}\cdot dbh_{(i)}}{\sum_{i=1}^{n} dbh_{(i)}} \quad (4)$$

$N_{VIneg(i)}$ = 1 for individuals with increment trend < 1 expressed as (increment period$_{(I-1)}$/(increment period$_{(I)}$), 0 for other individuals
dbh = diameter at breast height

SHI1 is the combined effect of the average number of dry days for the four-year risk-rating period (DD) and the stand edge effect (SEE), which characterizes the proportion of south- and west-oriented open stand edges of a stand (eq. (5)).

$$SH1 = DD\cdot SEE \quad (5)$$

DD = number of dry days (soil water potential below -2.0 bar) in the growing season April–October
SEE = stand edge effect, SEE = 1 for stands with no open south- and west-oriented edges

From eq. (3) it follows that stands on dry sites with a high proportion of *Picea abies* individuals with negative diameter increment trends are subjected to a higher risk of a bark beetle damage. Increasing stand age increases the risk of bark beetle damage. The intensity of bark beetle damage is subsequently predicted with eq. (6),

$$\%M = \frac{1}{1+\exp\left(4.1451-0.0251\cdot SHI2\right)} \quad (6)$$

where the stand hazard index SHI2 is the multiplicative combination of SHI1 and the proportion of *P. abies* basal area (BAS).

2.3 Coupling the models

Linking the risk model system to the forest succession model required some modifications to both models. The patch model had to provide the risk model with the stand specific variables SVI, SEE, BAS and DD. In order to extract the number of dry days from the water balance model included in PICUS, the state variable "soil water content" had to be linked to the corresponding soil water potential via a soil water characteristic curve. The calculation of the annual number of dry days (DD) involved the quasi daily values for the soil water potential interpolated from monthly values. The size of the simulated stands was fixed at 1 ha and no open south- or west-oriented edges were assumed to keep the stand edge effect SEE numerically at a value of 1. The variable SVI was calculated for each 10 x 10 m patch and aggregated to the stand level. The risk model variables BAS, DD and AGE had to be kept in arrays for four years, the stand vitality index SVI for eight successive years. In a moving time window covering four and eight years respectively the probability of a periodic bark beetle infestation was calculated using model (3). If a stand had been classified as damaged, bark beetle damage was possible four years after. It was assumed that only *Picea abies* individuals with dbh > 10 cm could be killed in the bark beetle risk-model. The reason behind this assumption was that in the course of our simulation experiments, any natural advance regeneration of *P. abies* had to be excluded from being rated as killed by bark beetles, as bark beetle damage does not generally occur in very young stands (Schmidt-Vogt 1977). In addition, the database for risk model development comprised stands with average diameters > 20 cm. Thus we avoided extrapolation beyond the range of stand characteristics represented in the calibration data set.

Each year the succession simulator PICUS simulates intrinsic and growth-related mortality. As the latter implicitly includes individuals which die due to density-dependent reasons and trees that die because of unfavourable environmental conditions, we had to assume that *Picea* individuals which had been rated as "dead" by the growth-related mortality algorithm essentially would have been infested and killed by bark beetles and thus would be included in the a posteriori bark beetle mortality estimate of eq. (6), if the following requirements were met:

1. a soil moisture response between 0.15 and 0.30 for *Picea abies*
2. the average air temperature had been > 7.0 °C and the average atmospheric water balance (eq. (7)) had been negative for 2 successive years.

$$AWB = \frac{P - PET}{PET} \tag{7}$$

AWB = atmospheric water balance
P = precipitation sum per year
PET = potential evapotranspiration per year

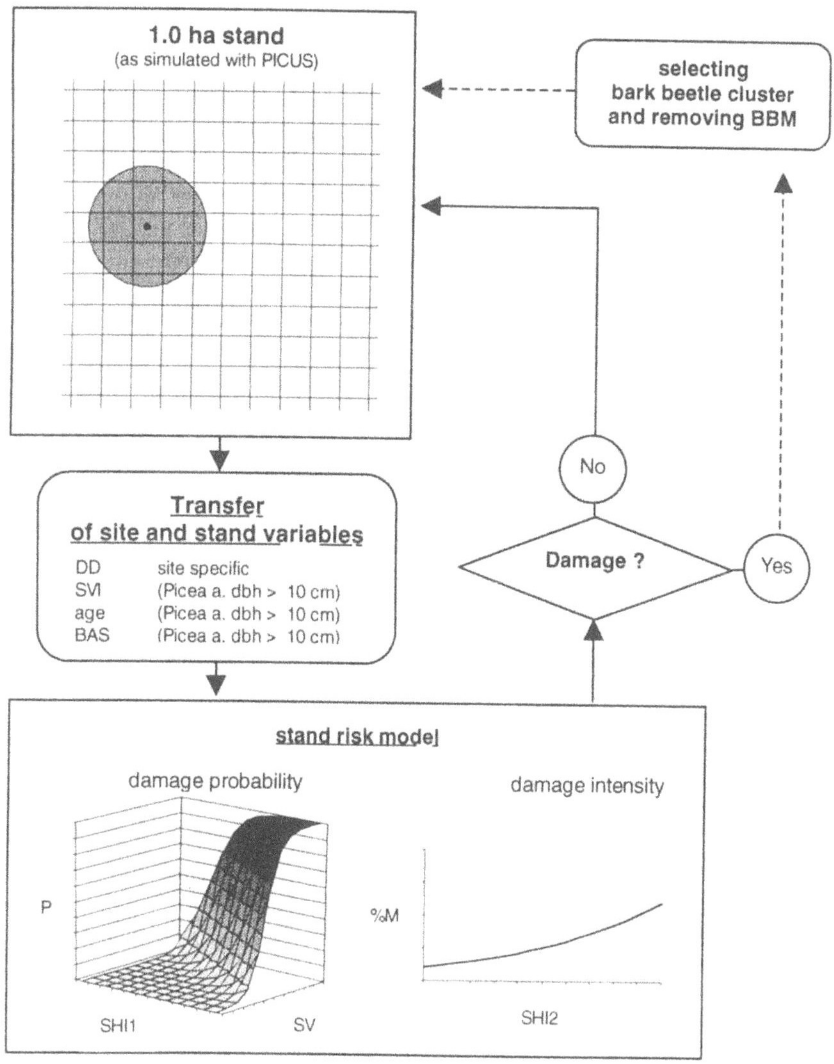

Fig. 3. Flow diagram of the modelling system.

All *Picea* individuals (dbh > 10 cm) which met these requirements were excluded from being killed in the stress mortality algorithm. The theoretical background for this constraint is that both suitable breeding material and an appropriate population of bark beetles have to coincide in space and time to cause damage (Waring and Pitman, 1980; Berryman, 1976). Water-stressed *Picea abies* trees provide the main source of breeding material for *Ips typographus* and *Pityogenes chalcographus* (Christiansen et al. 1987; Mullock and Christiansen 1986; Schmidt-Vogt 1977; Merker 1956). *P. abies* individuals with a soil moisture response below 0.1 were considered to die rapidly due to drought stress and thus would not provide suitable breeding habitat while trees with a soil moisture response of more than 0.3 were not considered to be seriously stressed. Warm and dry conditions, as indicated by air temperature and the atmospheric water balance, favour the physiological development of insects and subsequently the build-up of high population densities. Trees killed by bark beetles are neither evenly nor randomly distributed within an infested stand. The trees that are killed in a stand are usually aggregated in bark beetle spots. In our simulation experiment, the centre of a bark beetle spot was randomly drawn from the most vulnerable 20 % of all 100 m^2 – patches with respect to SVI. The rationale was that the probability of damage was considered to be highest on patches with the highest proportion of trees with low vigour. Once the centre of a bark beetle spot was identified, *Picea* individuals (dbh > 10 cm) were selected and killed within a maximum radius of 20 m. If the estimate of trees killed by bark beetles (BBM) was higher than the number of *P. abies* trees present within the selected spot, another centre was selected. Figure 3 presents a flow chart of the modelling system.

2.4 Simulation experiments

We had to design an array of simulation experiments to evaluate the simulated effects of bark beetle-induced mortality on total stand mortality rates and stand dynamics of low elevation *Picea abies* stands under different climate scenarios. We decided to choose ecoregion 6.2 (Kilian et al. 1994) as the "virtual" study area as the bark beetle risk model had been developed from data from that region. At elevations between 500 and 800 m a.s.l. *P. abies*, *Fagus sylvatica*, *Abies alba*, *Quercus petraea* and *Quercus robur* would be the most abundant naturally occurring tree species (Kilian et al. 1994). The present forests consist mainly of pure, even-aged *P. abies* and *Pinus sylvestris* stands. For our simulation experiments we initialized the forest model PICUS in accordance to a stand representative for the study area (Table 1).

Characteristic	value
stems/ha	470
mean dbh +/- std.dev.	33 +/- 5 cm
stand age	60
stand area	1.0 ha
water holding capacity	15 cm
available nitrogen	200 kg/ha

Table 1. Stand and site characteristics used for model initialization.

In a baseline experiment, we wanted to explore the effect of bark beetle mortality under current climatic conditions (baseline scenario). Climate data for the baseline scenario were generated once from 30-year averages and standard deviations for monthly temperature and precipitation data from the meteorological station St.Veit/Glan and were subsequently used for all model runs under the baseline scenario (Table 2). Scenario 1 was meant to characterise climatic conditions with an increased frequency of drought periods. To create such a climate record we embedded the instrumental data from the period 1986 – 1993 within the baseline climate data in 15-year intervals (Table 2). This approach was chosen because:

1. the period from 1986 to 1993 was characterized by several years with strongly reduced precipitation during the growing season
2. the bark beetle risk model (eqs. (3) and (6)) had been developed with data from that particular time period.

Scenario	Season	temperature [°C]	Precipitation [mm]
Baseline	Year	8.1 +/- 0.53	839 +/- 104
	Winter	1.2 +/- 0.6	294 +/- 76
	Summer	14.9 +/- 1.0	545 +/- 83
scenario 1	Year	8.1 +/- 0.48	803 +/- 98
	Winter	1.1 +/- 0.8	287 +/- 77
	Summer	15.1 +/- 0.6	516 +/- 84

Table 2. Characteristics (mean +/- standard deviation) of the climate scenarios used for the simulation experiments.

Both model versions were run ten times each under the two climate scenarios. For each scenario we compared by visual graphical comparison the averaged simulated species composition over 400 years, as predicted by the model versions with and without bark beetle mortality. The nonparametric Mann/Whitney U-test was used to test for differences in biomass/ha for the most abundant tree species between the model versions and climate scenarios in 100-year intervals. *Picea abies* mortality was summed over 100-year periods in three diameter classes to analyse the effect of bark beetle mortality on overall mortality rates of *P. abies*. To enable suitable comparisons, we related the growth-related and bark beetle mortality to the corresponding intrinsic mortality in order to account for changing stand structure over time and differences between model versions and climate scenarios, as intrinsic mortality is proportional to stem number/ha.

3. Results and discussion

3.1 Species composition

After 100 simulated years under the baseline climate scenario, the overall accumulated biomass and the species-specific biomass for *Picea abies*, *Abies alba*, *Fagus sylvatica*, *Quercus petraea* and *Quercus robur* were significantly different between the two model versions. In the bark beetle model version the accumulated biomass of *P. abies* never reached 400 tons/ha and was always lower than in the model version without bark beetle mortality, whereas the biomass of *A. alba* and *F. sylvatica* was increased (Figure 4). In the bark beetle version, *Quercus* spp. were present from year eight of the simulation onwards and steadily increased, whereas in the standard model version the *Quercus* spp. were not able to immigrate before year 20. From year 200 onwards, total accumulated biomass did not differ significantly between the model versions. In the model version with bark beetle mortality the proportion of *P. abies* was smaller whereas all the other main species showed increased biomass. After 400 simulated years no significant differences between the different model versions with respect to total accumulated biomass or biomass of *P. abies*, *A. alba*, *F. sylvatica* or *Quercus* spp. could be detected. The results of the Mann–Whitney U-test are shown in Table 3.

Species	year			
	100	200	300	400
Picea abies	**	**	**	
Abies alba	**	**		
Fagus sylvatica	**	**	*	
Quercus petraea	*			
Quercus robur	**	**	**	
Total biomass	**			

Table 3. Results of the Mann–Whitney U-test comparing the two model versions under the baseline climate with respect to biomass [t/ha]. - * = significant at α = 0.05, ** = significant at $\alpha = 0.01$.

It can be seen from Fig. 5 that under climate scenario 1 with an increased frequency of drought periods, the general pattern of simulated vegetation development is quite similar to those of the baseline scenario.

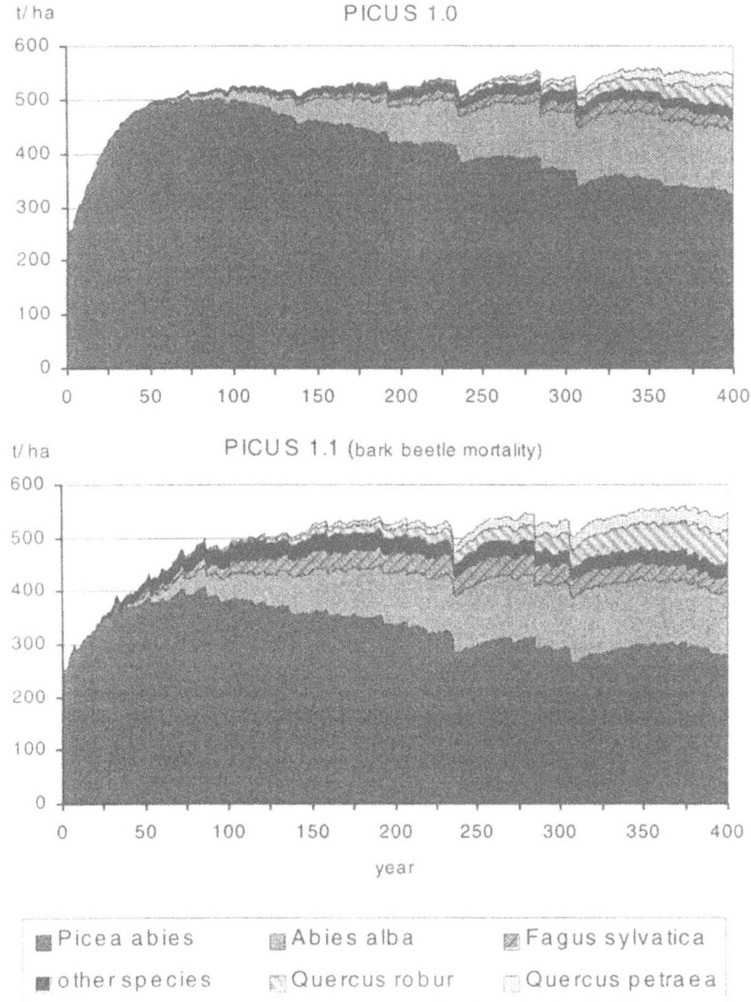

Fig. 4. Vegetation development under current climate (baseline scenario) as simulated by a model version without (left side) and with bark beetle induced mortality (right side), starting from a pure, even-aged 60-year-old *Picea abies* stand. (Average from 10 model runs for a 1.0 ha stand).

Again, *Quercus* spp. were able to immigrate much earlier if bark beetle mortality was explicitly considered. In the standard model version, *Quercus* spp. were present from year 20 onwards. The biomass of *Picea abies* declined steadily whereas the other most abundant species increased in biomass. Significant differences between the two model versions occurred mainly during the first 200 years after initialization (Table 4). After 400 simulated years, no significant difference in total accumulated biomass could be detected. Similar to the baseline scenario, the biomass of both *Quercus* spp. is significantly higher until year 300 for the bark beetle model version.

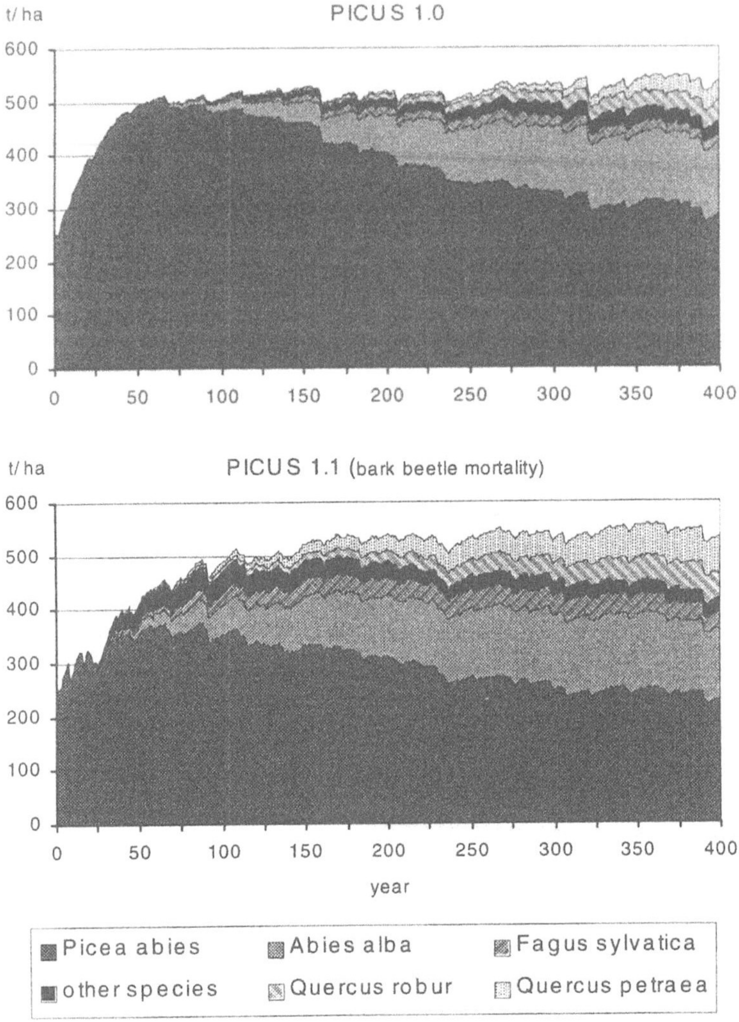

Fig. 5. Vegetation development under a climate scenario with increased frequency of drought periods (scenario 1) as simulated by a model version without (left side) and with bark beetle induced mortality (right side), starting from a pure even-aged 60-year-old *Picea abies* stand. (Average from 10 model runs for a 1.0 ha stand).

Species	year			
	100	200	300	400
Picea abies	**	**	**	*
Abies alba	**	**		
Fagus sylvatica	**	**	**	
Quercus petraea	*	**	**	
Quercus robur	*			
Total biomass				

Table 4. Results of the Mann–Whitney U-test comparing the two model versions under the climate scenario 1 (increased frequency of drought periods) with respect to biomass [t/ha]. - * = significant at $\alpha = 0.05$, ** = significant at $\alpha = 0.01$.

If we compare the results of each model version under different climate scenarios, no significant differences with respect to total and species-specific biomass could be found for the standard model version. In the time series for *Picea* biomass produced by the bark beetle model version, it is evident that mortality occurs more intensively than under current climatic conditions. Inspection of the graphs (Figure 4, Figure 5) reveals that the combined biomass of the *Quercus* species is higher for the bark beetle model version. However, significant differences were found for *Picea abies* at years 300 and 400 (lower biomass) as well as slightly increased biomass for *Quercus robur* at year 100 and reduced abundance for *Quercus petraea* at year 400 (Table 5).

Species	year			
	100	200	300	400
Picea abies			**	**
Abies alba				
Fagus sylvatica				
Quercus petraea	*			
Quercus robur				*
Total biomass				

Table 5. Results of the Mann/Whitney U-test comparing the results of the model version with bark beetle mortality under the baseline scenario and under climate scenario 1 (increased frequency of drought periods) with respect to biomass [t/ha]. - * = significant at $\alpha = 0.05$, ** = significant at $\alpha = 0.01$.

3.2 *Picea abies* mortality

Relative bark beetle mortality decreases over time in both climate scenarios. The reason for this consistent pattern is evident from Figure 6. The probability of bark beetle damage decreases substantially, mainly because of the decreasing relative *Picea* basal area.

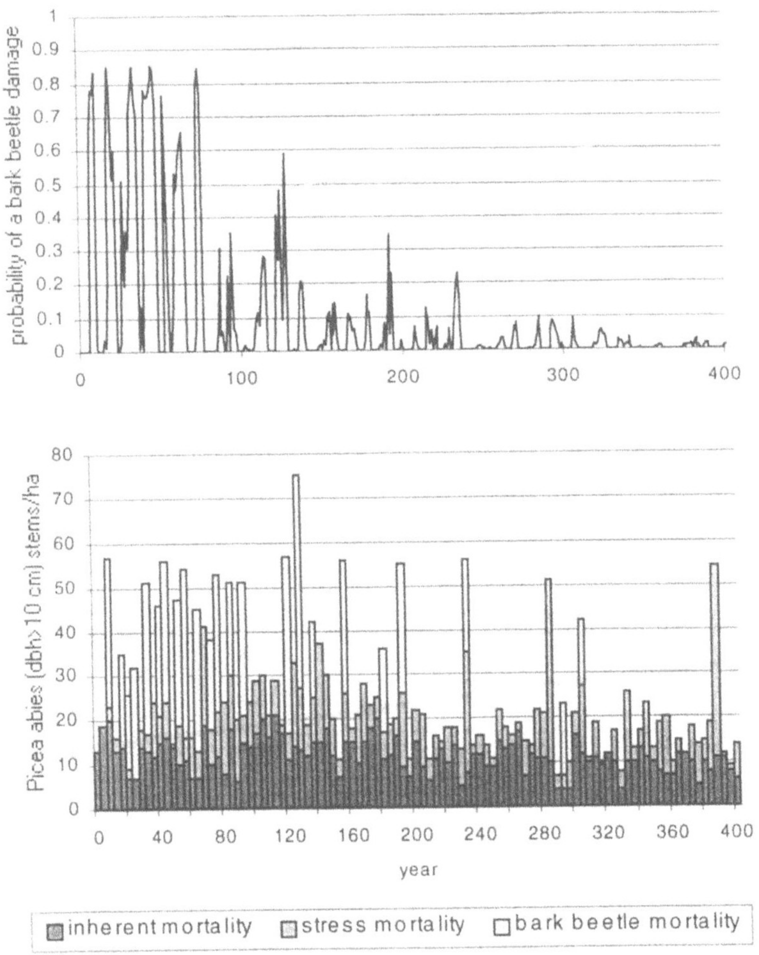

Fig. 6. Above: Development of bark beetle damage probability over time (baseline scenario). Below: Corresponding *Picea abies* mortality for four-year periods.

At first sight this might seem unusual, but the literature indicates that in mixed stands with a low proportion of *Picea abies* the risk of bark beetle infestation is lower due to improved habitat conditions for bark beetle antagonists (Nebeker et al. 1986). Under scenario 1 bark beetle mortality even approaches zero in the third period between years 300 to 400. However, the corresponding *P. abies* biomass proportion of 0.42 could not be validated by the literature and might not be an appropriate threshold for the non-occurrence of bark beetle infestations under the increased frequency of drought periods of scenario 1. Even low proportions of *P. abies* in mixed stands can be infested and killed if bark beetle population densities are sufficiently high (Thalenhorst 1958).

The bark beetle model version yielded significantly higher total *Picea abies* mortality rates in diameter classes (> 30 cm) in the first two 100-year simulation periods under current climate and in the first three periods under the climatic conditions of scenario 1. Growth-related mortality was generally not significantly different between model versions under current climate. Under the climatic conditions of scenario 1 with increased frequency of drought periods, growth-related mortality was significantly lower for the bark beetle model version in diameter classes (> 30 cm) from years 0–100 and in the diameter class (>10–30 cm) in the second period from years 100–200. This result partly corresponds with the expectation that growth-related mortality is negatively correlated with bark beetle mortality, as the latter reduces stand biomass and thus increases resource availability for the remaining individuals. Nevertheless, as bark beetle spots were concentrated in those parts of a simulated stand where tall trees with decreasing diameter increment trends were located and subsequently stem number/ha was low, the effect on growth-related mortality at a stand level is apparently small under current climate and only slightly increased under the conditions of scenario 1 (Table 6).

	base line				scenario 1			
diameter	Period				period			
[cm]	1	2	3	4	1	2	3	4
>10 – 30		*				*		
>30 – 50					**			
>50					*			

Table 6. Results of the Mann–Whitney U-test comparing the results of the two model versions under the baseline scenario and under climate scenario 1 (increased frequency of drought periods) with respect to relative stress-induced mortality. - * = significant at α = 0.05, ** = significant at α = 0.01.

It is interesting to note that neither bark beetle mortality nor stress-induced mortality were significantly higher under the increased frequency of drought periods of scenario 1.

4. Conclusions

Two points have essentially been addressed in this study:

1. The integration of a stand-level risk model with a forest succession model to account explicitly for bark beetle induced mortality,
2. An assessment of the effects of the spatial pattern of bark beetle mortality on simulated vegetation dynamics.

To the best of our knowledge, no similar study has been undertaken. One reason may be that the modelling of bark beetle mortality is difficult due to the underlying stochastic nature of the host – beetle system. The bark beetle risk model which was employed in this study integrates considerable knowledge on the interrelationship between bark beetles (Scolitidae and *Picea abies*. Moreover it links the probability of bark beetle damage not just to stand characteristics, but accounts explicitly for the effect of drought stress. Another important point is that variables at the stand level are considered in determining bark beetle mortality. Thus the risk of death for an individual tree is not exclusively modelled from individual-based variables. As with any empirical model, extrapolation beyond the range of conditions represented in the parameterization data set might yield erroneous results. In the case of the bark beetle risk model the underlying population density of bark beetles is not considered explicitly. A model of bark beetle population dynamics could be used to account for this potential source of error. Nevertheless, we confined the scenarios for our simulation experiments to climatic conditions similar to those represented in the parameterization data set.

Another problem in introducing explicitly formulated sources of tree mortality is the "black box approach" which is currently used to model mortality in forest gap models. As we did not totally alter the mortality logic of gap models, we had to segregate different reasons of mortality within this "black box", a task which turned out to be rather difficult. Instead of simply adding bark beetle induced mortality as another variable to the population dynamics system, we tried to separate bark beetle induced mortality which is included implicitly in the growth-related mortality estimates from other causes of stress-related mortality. The underlying assumptions to accomplish this task correspond well with qualitative knowledge about bark beetle - host tree interactions (cf. Berryman 1976). Our simulation experiment can thus be seen as an attempt to link mortality functions (eqs. (2) and (5)) developed from observational data with a rather qualitative and arbitrary model of tree mortality (Botkin 1993).

Conventional gap models cannot account for environmental patterns at scales larger than the modelled plot, which is approximately 800 m^2 in models of Central European forest ecosystems (Kienast 1987; Bugmann 1994). A spatially explicit model such as PICUS can generate a different species response to disturbances because it is not bound to the scale of the individually simulated plot. Urban et al. (1991) argued in a similar manner and demonstrated the importance of gap size on species composition. Despite the limited extent of the underlying simulation experiments, this study shows that if bark beetle mortality is considered explicitly in forest gap models, species composition and species abundance can be significantly different for periods of up to 300 years. Light intolerant species such as *Quercus petraea* and *Quercus robur* are able to immigrate earlier. In our case study, from simulation year 300 onwards no significant differences in *Quercus* spp. biomass were found between the two model versions. It has yet to be tested whether this is a consistent pattern for various combinations of site characteristics and climatic conditions. However, if the transient behaviour of forests in response to a climate change is to be predicted, then the results of conventional patch models may not be realistic.

During the last 20 years increasingly sophisticated gap models have been applied to the study of various aspects of population dynamics in forest ecosystems. Mortality has a considerable influence on regeneration patterns and stand structure. Our conclusion is that if forest patch models are to be used to project vegetation characteristics such as species composition or biodiversity indices for forest vegetation assemblages, then improvements are need in the way in which the spatial and temporal effects of tree mortality pattern are handled.

Acknowledgements. This study was conducted within a research project financed by the Austrian Federal Ministry for Forestry and Agriculture. We are grateful for their cooperation and support.

5. References

Berryman AA (1976) Theoretical explanation of mountain pine beetle dynamics in lodgepole pine forests. Environ. Entomol 5:1225 – 1233

Botkin DB (1993) Forest dynamics. An ecological model. Oxford University Press, Oxford

Botkin DB, Janak JF, Wallis JR (1972) Some ecological consequences of a computer model of forest growth. J Ecol 60:849–872

Brock TD (1981) Calculating solar radiation for ecological studies. Ecol Modelling 14:1–19

Buffo J, Fritschen L, Murphy J (1972) Direct solar radiation on various slopes from 0° to 60° north latitude. USDA, Forest Service, Res Pap PNW-142. Portland, OR

Bugmann H (1994) On the ecology of mountainous forests in a changing climate: A simulation study. Diss. ETH No. 10638

Christiansen E, Waring RH, Berryman AA (1987) Resistance of conifers to bark beetle attack: Searching for general relationships. For Ecol Manage 22:89–106

Daniels RF, Leuschner WA, Zarnoch SJ, Burkhart HE, Hicks RR (1979) A method for estimating the probability of southern pine beetle outbreaks. For Sci 2:265–269

Dunne T, Leopold LB (1978) Water in environmental planning. W.H. Freeman and Company, New York

Frank EC, Lee R (1966) Potential solar beam irradiation on slopes. USDA, Forest Service, Res Pap RM-18, Rocky Mtn. For. and Range Exp. Station, Fort Collins, Colorado

Garnier BJ, Ohmura A (1968) A method of calculating the direct shortwave radiation income of slopes. J Appl Meteorol 7:796–800

Hasenauer H, Stampfer E, Rohrmoser Ch, Sterba H (1994) Solitärdimensionen der wichtigsten Baumarten Österreichs. Österr Forstztg 3:28–29

Hedden RL (1981) Hazard-rating system development and validation: An overview. In: Hazard-Rating systems in forest insect pest management. Symposium Proceedings, General Technical Report WO-27, USDA.

Hönninger K, Lexer MJ (1997) Implementing an object-oriented and event-controlled forest succession model. Soc Am For 7th Symposium on Systems Analysis in Forest Resources (in press)

Houghton JT, Jenkins GJ, Ephraums JJ (eds) (1990) Climate change. – The IPCC climatic assessment. Report prepared for IPCC by Working Group 1. Cambridge University Press, Cambridge

Houghton JT, Meira Filho LG, Callander BA, Harris N, Kattenberg A, Maskell K (eds) (1996) Climate Change 1995. The Science of Climate Change. Cambridge University Press, Cambridge

Hungerford RD, Nemani RR, Running StW, Coughlan JC (1989) MTCLIM: A mountain microclimate simulation model. USDA, Forest Service, Intermountain Res Stat RP INT-414

Kienast F (1987) FORECE – A forest succession model for southern Central Europe. Oak Ridge National Laboratory, Oak Ridge, Tennessee, ORNL/TM-10575

Kilian W, Müller F, Starlinger F (1994) Die forstlichen Wuchsgebiete Österreichs. Eine Naturraumgliederung nach waldökologischen Gesichtspunkten. Berichte Forstliche Bundesversuchsanstalt Wien 82

Lee YJ (1971) Predicting mortality for even-aged stands of lodgepole pine. For Chron 47:29–32

Lexer MJ (1995a) Anwendung eines "big leaf"-Modelles zur Simulation des Bodenwasserhaushaltes in Fichtenbeständen (*Picea abies* (L.) Karst.). Centbl gesamte Forstwes 4:209–225

Lexer MJ (1995b) Beziehungen zwischen der Anfälligkeit von Fichtenbeständen (Picea abies (L.) Karst.) für Borkenkäferschäden und Standorts- und Bestandesmerkmalen unter besonderer Berücksichtigung der Wasserversorgung. Diss. Univ. f. Bodenkultur

Lexer MJ (1997). Risikoanalyse und Ableitung waldbaulicher Maßnahmen zur Beeinflussung des Borkenkäferrisikos in Fichtenbeständen. In: Müller (ed) Waldbau an der unteren Waldgrenze. Forstliche Bundesversuchsanstalt Wien, FBVA-Berichte Nr. 95, pp 79–89.

Liu BYH, Jordan RC (1960) The interrelationship and characteristic distribution of direct, diffuse and total solar radiation. Sol Energy 4:1–19

Merker E (1956) Der Widerstand von Fichten gegen Borkenkäferfraß. Allg Forst- Jagdztg 7:129–187

Monserud RA (1976) Simulation of forest tree mortality. For Sci 22:438–444

Moore AD (1989) On the maximum growth equation used in forest gap simulation models. Ecol Modelling 45:63–67

Mullock P, Christiansen E (1986) The treshold of successful attack by Ips typographus on Picea abies: A field experiment. For Ecol Manage 14:125–132

Nebeker TE, Houston DR, Hodges JD (1986) Forest pests: Influence of forest management practices on pest population dynamics and forest productivity. In: Hennessy TH, Dougherty PM, Kossuth SV, Johnson JD (eds) Stress physiology and forest productivity. Martinus Nijhoff Publishers

Paccala SW, Canham CD, Silander JA Jr. (1993) Forest models defined by field measurements: I. The design of a northeastern forest simulator. Can J For Res 23:1980–1988

Schmidt-Vogt H (1977) Die Fichte. Ein Handbuch in 2 Bänden. Band I. Verlag Paul Parey, Hamburg and Berlin

Shugart HH (1984) A theory of forest dynamics. The ecological implications of forest succession models. Springer, New York

Stage AR, Hamilton DA Jr. (1981) Sampling and analytical methods for developing risk-rating systems for forest pests. In: Hazard-Rating systems in forest insect pest management. Symposium Proceedings, General Technical Report WO-27, USDA.

Swift LW (1976) Algorithm for solar radiation on mountain slopes. Water Resour Res 12:108–112

Thalenhorst W (1958) Grundzüge der Populationsdynamik des großen Fichtenborkenkäfers Ips typographus L. Schriftenr. Forstl. Fak. Univ. Göttingen 21:1–126

Thornthwaite CW, Mather JR (1957) Instructions and tables for computing potential evapotranspiration and the water balance. Publ Climatol 10:183–311

Urban DL (1990) A versatile model to simulate forest pattern. A user's guide to ZELIG version 1.0. Environmental Sciences Department, The University of Virginia Charlottesville, Virginia

Urban DL, Bonan GB, Smith TM, Shugart HH (1991) Spatial applications of gap models. For Ecol Manage 42:95–110

Waring RH, Pitman GB (1980) A simple model of host resistance to bark beetles. Oregon State University Research Note 65

Zahner R (1967) Refinement in empirical functions for realistic soil-moisture regimes under forest cover. In: Sopper, W.E., Lull, W.H. (eds) Forest Hydrology. Pergamon Press, Oxford, pp 261–274

Impacts of climatic variability and extreme on forests: synthesis

Martin Beniston and John L. Innes

1. Introduction

The events associated with the El Niño episode during the winter of 1997–1998 have drawn the attention of the World to the importance of extreme climatic episodes. While model predictions tend to be viewed with extreme caution, a series of storms is immediately newsworthy because of its impact on human populations. Forests can also be severely affected by extreme events, as illustrated by the ice storm that affected the north-eastern USA and south-eastern Canada in January 1998. Clearly, such events need to be taken into account when assessing the future responses of forests to a change in climate.

2. Climatic variability

A much better understanding of climate variability in Europe is developing as a result of recent progress in the recognition of the North Atlantic Oscillation. Over the past several decades, countries surrounding the Atlantic Basin have experienced dramatic changes in climatic conditions. Since the mid-1960s there has been a steady increase in wintertime storminess in the northeastern Atlantic and the North Sea. During positive phases of this Oscillation, winds bring greater quantities of moist air to Europe, according to Hurrell (1997). Lack of snow in the Alps in the late 1980s and early 1990s, and its late arrival during the 1995/1996 season can to a large degree be attributed to the high positive values of the North Atlantic Oscillation index during these periods, as shown by Beniston (1997). At the same time northern European countries have experienced an upward trend in winter precipitation, to the extent that glaciers in many parts of Norway have exhibited significant increases in mass and length. On the contrary, over much of southern Europe and the Mediterranean, there has been a steady decline in precipitation, with a corresponding general retreat of alpine glaciers since the early 1970s. Surface temperatures in winter have increased significantly over the past 25 years from northern Europe across Eurasia, while temperatures over the Middle East, North Africa, Greenland and the Canadian Arctic have cooled. These changes in climate are strongly related to variations in the North Atlantic Oscillation (NAO), which has strengthened from its record low index state during the 1950s and 1960s to a historic maximum in the early 1990s. Superimposed on this trend have been large, quasi-decadal oscillations which have affected climatic conditions in many parts of Europe (see for example Weber et al., 1997).

In many instances, the temporal and temporal evolution of climatic variability and extremes can be ascertained from historical records; this provides some insight into whether today's generally warmer climate is associated with higher variability or more frequent extremes. Forest damage episodes during the 18th and the 19th centuries in Bohemia (Czech Republic), for example, have been analysed on this basis by Brazdil (this volume, Chapter 2). Problems of observation, data quality and homogeneity of climatic data need to be addressed, however, in order to make meaningful statements regarding the temporal and spatial trends in the frequency and intensity of extreme events. Despite these problems, Brazdil was able to argue that the effects of extreme climatic events have increased markedly during the 20th century.

With a view to looking at changes in variability in a changing climate over Europe, Déqué and Doblas-Reyes (this volume, Chapter 4) have applied the climate version of the French ARPGE-IFS General Circulation Model to conditions of doubling of atmospheric greenhouse gases. Significant changes in intra-seasonal and inter-annual variability in the future climate are observed in the model over most of Europe (for which the variable-resolution model has its highest spatial definition), compared to current climatic conditions.

Côté et al. (1998) have shown that by using a high-resolution Regional Climate Model (RCM), with initial and boundary conditions driven by either climatological anlayses or by a General Circulation Model, the spatial and temporal detail associated with variability is also enhanced. Simulations with the Canadian RCM over North America have enabled to assess the manner in which variability may change under enhanced greenhouse-gas conditions this century.

3. Long-term changes in average conditions and extremes

There a number of strategies which have been proposed for detecting changes in climate extremes and attributing these to anthropogenic forcing. There is increasing interest in analysing different regional ecotones for evidence of climatic change, as shown by Diaz and Graham (1997). For example, the distribution of potential vegetation types can be used to create regional climate indices, in order to evaluate changes in different climate parameters through time. Another approach consists in identifying regional climatic indices for areas which represent ecotonal boundaries or extreme conditions for current climate. It then becomes possible to determine whether these types of regions undergo changes in time, which may suggest expansion or contraction of such regions, and either amelioration or worsening of conditions in hitherto extreme environments. As a measure of shifts in extremes and variability in response to changes in mean climatic conditions, the climatic indicators suggested by the authors can be used to identify similar changes in General Circulation Models (GCM) for various scenario experiments.

In Switzerland, climate change has been observed in the instrumental record, in particular through temperature increases which are well above the global average (Beniston et al., 1994). Rebetez and Beniston (this volume, Chapter 3) show that minimum temperatures exhibit the most significant increases, particularly in winter, while maximum temperatures show less marked changes since the beginning of the century. Changes in mean temperature are accompanied by strong shifts in the extremes of the probability density functions of temperature distributions, particularly for the minima. The diurnal temperature range has decreased simultaneously with the observed warming this century. Additional analyses of the relationships between mean temperatures, variability and skewness indicate that warmer years are correlated with reduced variability as well as a reversal of the skewness of temperature distributions, compared to colder years; this is particularly the case in winter, since the number of extremely cold days has substantially diminished and been superseded by increases in warm extremes.

There has also been an increase in extreme precipitation events, particularly during autumn. Frei and Schär (1997) show that strong and relatively rare extreme events contribute a substantial fraction to the mean precipitation in the Alps; there is frequent occurrence of strong precipitation along the northern and southern Alpine rim. Spatial patterns of precipitation exhibit pronounced and non-synchronous seasonal and inter-annual variations. In many instances, vegetation may respond as much to climatic extremes as to changes in mean conditions; continued trends such as those found for the Swiss Alps may in time have profound impacts on forest ecosystems.

Seasonal snow-cover in mountainous regions is an important component of the regional hydrological cycle and constitutes a basic control on ecological systems. Long-term measurements of snow depth show that it is extremely variable, especially at middle elevations (1000-2000 m above sea level). Modelling of the snow-pack represents a valuable tool for the assessment of alpine snow-cover. A numerical model coupled with a meteorological analysis system has been used by Martin (this volume, Chapter 5) to generate a snow climatology of the French Alps and the Pyrenees. The simulations allow simple comparisons between regions (similar characteristics related to elevation, slope and aspect) which are not always possible with measured data. The model allows the sensitivity of the "snow climatology" to be analysed for changes in temperature and precipitation as projected by General Circulation Models for the next century; such information can then be used in vegetation models to assess the impacts on different species. Borel et al. (1997) also focused on snow–vegetation impacts modelling, by using a probabilistic vegetation model which is based on correlations between the spatial distribution of the principal plant communities and species, and the spatial distribution of climatic parameters for various time periods.

4. The influence of vegetation on climate

Impacts research commonly focuses on the effects of climatic change on a particular environmental or socio-economic system. However, the Earth system is more often characterized by complex feedbacks and interactions between its various elements than this simplified one-way forcing. Barron and Dutton (1997) note that while the geologic record exhibits marked changes in vegetation character and distribution associated with climatic changes, a number of numerical experiments today suggest that different vegetation distributions can significantly alter the global climate. Results from GCM experiments which incorporate vegetation feedback mechanisms highlight the fact that vegetation cannot be viewed simply as a response to climate change but often contribute to that change. The conclusions are based on a number of experiments using the GENESIS model developed at the National Center for Atmospheric Research in Boulder, Colorado, using a new soil–vegetation scheme devised by Cosgrove and Barron (1997). The experiments are designed to isolate the effect of vegetation changes on climate by specifying different vegetation types in the GENESIS simulations. The experiments include both idealised vegetation patterns and reconstructions from the geologic record. Vegetation changes in periods of the distant past are seen in the simulations to influence the globally averaged surface temperature by as much as 2°C.

Using a high-resolution RCM for the contemporary period, Walko et al. (this volume, Chapter 6) also show that at regional scales, vegetation properties can also directly influence climate on local to large regional scales. Using the RAMS RCM developed at Colorado State University (Ft. Collins, USA), and estimating what the natural landscape was prior to European settlement, it is shown what the climate would be for a given year using current observed weather at the lateral boundaries. By replacing this landscape with a USGS evaluation of current landscape, the authors have highlighted the importance of landscape in determining climate in the south-eastern United States.

5. Natural variability in forests

The workshop identified a number of outstanding problems. Many resource managers are being asked to ensure that changes to forests lie within the "range of natural variability". This is difficult, as there is often insufficient information to indicate what this range is. The difficulties were explained by Hughes (this volume, Chapter 7), who used 8000-year tree-ring series to show that the frequency of droughts in California has changed over the last 2000 years. The periods of drought identified in the tree-ring record for *Pinus longaeva* in the Great basin area of the western USA correlate with low stands in the water level of Mono Lake, California, and other water bodies. This work seems particularly significant as it demonstrates that the droughts that are often considered as extreme in the 20th century, such as the one in 1919–1923, are not the most extreme when a longer time-scale is considered. For example, of the 227 non-overlapping drier periods since 6000 BC, only six have occurred since AD 1700.

6. Extreme events

Drought plays an important role in forest dynamics. It has been responsible for pulses of tree mortality in the Argentinian Andes, according to Villalba and Veblen (this volume, Chapter 10). In the Canadian Rockies, Luckman and Kavanagh (this volume, Chapter 9) have identified times of extreme cold as a cause of tree death, with a sustained period of cold summers in 1696–1701 apparently being responsible for extensive tree mortality. Conversely, in the Czech Republic, wind, snow and frost are the main causes of forest damage, as indicated by Brazdil, (this volume, Chapter 2). These results reflect the differences in the climatic factors which limit trees in various parts of the world.

The impact of extreme events is determined to a greater or lesser extent by the antecedent conditions. For example, a given wind speed will have very different effects depending on whether the soil is water-logged or not (Combe, this volume, Chapter 13). This makes predictions based on linear extrapolations difficult, and is probably one reason why the modelling of extreme events has not been more successful. This conclusion also applies to the impacts of droughts; it is insufficient to look only at the intensity of a drought in relation to historical precedent. Instead, it is necessary to determine the incidence of droughts and other events prior to the drought in question, and then to determine the extent to which these predisposed the trees to drought damage.

7. Forest responses

The responses of forests to environmental changes are very uncertain. A variety of different climatically-induced stresses are important for forests, as emphasised by Innes (this volume, Chapter 1), and these are likely to have an impact on future changes in forest composition.. Whether or not this will be reflected in changes in the tree-line remains unclear as it is often difficult to determine the most important factors affecting the upper growth limit of trees (Körner; this volume, Chapter 14). Temperature may be important at some locations but at others, soil moisture may be critical, as indicated by Carrer et al. (this volume, Chapter 11). Alternatively, the length of snow-cover may be the limiting factor. Changes in the tree-line may be particularly important if these occur at the expense of alpine meadows, and there is much evidence to suggest that such changes are occurring (Peterson; this volume, Chapter 12). However, causal relationships of treelines in mountain regions are still not fully understood today. Körner (this volume, Chapter 14) notes that the often cited 10°C isotherm of the warmest month is an empirical relationship which holds for the Rocky Mountains and the Alps, where there is such a coincidence not generally found elsewhere. Global comparisons, if correlated with more adequate climatic terms, yield less regional deviation and hint at mechanisms that may have received insufficient attention. Körner has focused upon older suggestions that below-ground processes are crucial and that processes associated with photo-assimilation *per se* have less predictive value than previously believed.

The complex factors affecting some tree-lines are illustrated by Villalba and Veblen (this volume, Chapter 10). In Patagonia, the successful establishment of *Austrocedrus chilensis* is related to the presence of wet-cool summers persisting for a decade or more. Mortality, on the other hand, can be associated with the presence of a single dry-warm summer. Consequently, it appears to medium-term mean climatic conditions that control establishment and short-term extreme events that control mortality in this environment. Such a conclusion is not universally applicable. Peterson (this volume, Chapter 12) argues that although tree establishment occurs in discrete time periods, these periods are not synchronous among different regions of North America. In addition, tree establishment is spatially and temporally variable both between and within species at different locations.

Tree-ring width is a frequently-used method of determining the sensitivity of trees to climate, particularly in extreme situations. Several studies reported in this book indicate the importance of detailed analysis of climatic data, with a trend away from the use of mean monthly precipitation and temperature data in the calculation of the response functions. Other features of trees may also help to provide information on extreme events. For example, severe frosts are sometimes recorded as frost cracks in tree rings (Cherubini, 1997). However, when trees are lost to such extremes, the forest industry may be in a poor position to react, as demonstrated by events following the Vivian storm, which affected western and central Europe in February 1990 (Combe, this volume, Chapter 13).

While generalisations are often made about the response of forests to climate, there is increasing evidence that micro-site factors play a critical role in determining the responses of trees and other vegetation to climate. This was emphasised by Tessier et al. (this volume, Chapter 8) in their study of *Larix decidua* in the French Alps and by Saurer et al. (this volume, Chapter 16) working with *Fagus sylvatica* in Switzerland. Similar conclusions were reached by Schmelter (1997) working in northern Patagonia, by Luckman and Kavanagh (this volume, Chapter 9), working on the population dynamics of *Picea engelmannii* and *Abies lasiocarpa* in the Canadian Rockies, and by Peterson (this volume, Chapter 12) working in the mountains of the Pacific Northwest. In addition, Holten (this volume, Chapter 15) has stressed the importance of slope in determining the responses of alpine vegetation to climate change. The conclusion that can be drawn from these studies is that the same change in climate may have different effects in different stands and forests.

Although the climatic changes that have been forecast will be rapid, even greater changes have occurred in the past. Wick and Ammann (1997) have studied the effects of climatic changes at the end of the last glacial period. They used pollen analysis to show that the responses of animals and vegetation to such changes may be very rapid if no migration is involved. Conversely, when migration has to occur from refuge areas, substantial lag times may be apparent.

8. New methods for looking at climatic responses

Dendroclimatology has a long history in climatic research. Traditionally, the responses of tree growth to climate are determined by linear correlations between a number of mean monthly variables and tree-ring width, in a process known as response-function analysis. A major drawback of this method is that it does not allow for non-linear responses. These may be important: the sensitivity of a tree to climate can change through time, as a result of the normal ageing process and/or as a result of changes in environmental factors that influence the climatic response. Both Tessier et al. (this volume, Chapter 8) and Carrer et al. (this volume, Chapter 11) have used Neural Network techniques to overcome this problem. Carrer et al. use the method to show that for trees at the alpine tree-line there is not only a lower growth threshold to temperature, but also an upper one.

New methods for analysing the responses of forests to climate are being developed. Isotope analysis offers considerable potential, according to Saurer et al. (this volume, Chapter 16), as well as the monitoring of atmospheric pollutant species at high alpine sites. These are generally far removed from local effects (urban sources etc.) and therefore provide useful information on long-distance pollution which can affect tree growth and mortality (Schwikowski, 1997).

9. Modelling responses

Much of the information that we have today about future responses of forests to climate change has been derived by numerical modelling techniques. Recent developments in modelling have demonstrated the importance of extreme in events in determining forest development (Bugmann, 1997). Considerable progress has also been made in refining early models that dealt only with mean monthly temperatures. This was shown by Grote et al. (this volume, Chapter 17) who found that seasonal shifts in temperature conditions had a greater impact on modelled water and carbon balances than did changes in annual averages, and that precipitation changes were most important if they occurred in the dry season. Including soil moisture based on daily values, for example, results in radically different results from a model that used long-term mean monthly values (Lasch, this volume, Chapter 18; Grote, this volume, Chapter 17). Similarly models including moisture values based on precipitation and subsequent water storage also produce different results (Kräuchi and Baltensweiler, 1997). These results all confirm the empirical studies that site factors play a critical role in determining the responses of forests to climate change.

It is not only physical site factors that are important. Lexer and Hönninger (this volume, Chapter 19) have described the interactions between the results of a traditional climate gap model and the potential changes that occur when bark beetles are included. Including bark beetle effects significantly changes the forest species composition simulated by the models.

Although many changes in forests may occur, models suggest that these are most likely to occur near the upper tree-line. However, empirical evidence suggests that any changes in tree-lines may be very subtle (Peterson, this volume, Chapter 12). Other forests may be fairly resistant to change, such as the *Picea abies* plantations of the Swiss Plateau (Kienast, 1997). Such changes need to be looked at on a landscape scale, as the interactions between patches may be particularly important (Syke,s 1997).

10. Main conclusions

1. Tree-ring records reveal that mean temperatures are important for tree growth, but that tree rings contain much more information. This information reveals the importance of extreme events in determining the growth of individual trees.
2. Thc factors affecting tree growth vary between species and between sites. Consequently, a knowledge of the ecology of a particular species is needed before predictions are possible about how it will respond to climate change. Site factors need to be taken into account, as they will influence how a tree responds to a specific climatic change.
3. The main climatic events that are likely to impact on forests are wind storms, droughts, frosts, and the duration of snow-cover. These can have both direct (e.g. freezing injury) and indirect (e.g. by influencing bark beetle populations) effects.
4. The changes in climate which are projected by general circulation models on the basis of scenarios provided by the IPCC (Intergovernmental Panel on Climate Change) are not unique: similar (but not identical) very rapid changes have occurred in the past. Forests responded to these changes differently, making prediction about the impacts of the current changes difficult.
5. The causes of climatic variability are becoming increasingly clear. The Southern Oscillation in the Pacific Ocean has been acknowledged for some time, but the implications of the presence of a North Atlantic Oscillation are now being recognised.
6. Analysis of climatological records from Switzerland and elsewhere indicate changes in the patterns of extreme events. These include less extreme cold periods in winter and more intensive rainfall events in autumn.
7. The impacts of these changes on forests are currently being evaluated. Changes in temperature may affect the winter chilling requirements of some species. Changes in precipitation patterns will affect soil moisture conditions, with major implications for forest development. However, generalisation are not possible as the controlling climatic factors for trees vary between species and site and also through time.
8. Models of forest development are increasingly incorporating information on extreme conditions. A new generation of forest models is using information on

daily values for soil moisture availability, for example. The results generated by these models are very different from those derived from older, simpler models.

9. Simple models of changes in the distribution of trees based on mean temperatures are now seen as inadequate tools for the prediction of forest responses to climatic change. Changes in the distribution of forests will be affected by the availability of suitable sites for tree growth. For example, the upward movement of the treeline may be restricted by the absence of soils suitable for tree growth.

Forests are very sensitive to climate and respond to it in a complex fashion. At the same time, forests represent intricate ecosystems that are in a perpetual state of change. While past responses of forests to climate change provide an indication of what might happen if the climate changes in the way that has been suggested, future responses will be very difficult to predict. Scientists are making substantial progress in improving such predictions, particularly in relation to changes in climatic extremes and the impacts of these on forests, as was amply demonstrated by the Wengen-1997 workshop.

11. References

Barron, E. J., and Dutton, J. F., 1997: The role of vegetation-climate feedbacks in governing past climates. Wengen-97 Workshop on Past, Present and Future Climate Variability and Extremes: The Impacts of Forests. Wengen, Switzerland, September 1997. Abstracts Volume, p. 4

Beniston, M., 1997: Variations of snow depth and duration in the Swiss Alps over the last 50 years: Links to changes in large-scale climatic forcings. Climatic Change, 36, 281-300

Beniston, M., Rebetez, M., Giorgi, F. and Marinucci, R., 1994: An analysis of regional climate change in Switzerland. Theor. and Appl. Clim., 49, 135-159

Bugmann, H., 1997: Simulated impacts of interannual climate variability on past and future forest composition. Wengen-97 Workshop on Past, Present and Future Climate Variability and Extremes: The Impacts of Forests. Wengen, Switzerland, September 1997. Abstracts Volume, p. 29

Cherubini, P., 1997: Climatic influence on tree-ring patterns in a subalpine forest. Wengen-97 Workshop on Past, Present and Future Climate Variability and Extremes: The Impacts of Forests. Wengen, Switzerland, September 1997. Abstracts Volume, p. 27

Cosgrove, B. A., and Barron, E. J., 1997: A simplified dynamic vegetation model coupled to the GENESIS GCM. Wengen-97 Workshop on Past, Present and Future Climate Variability and Extremes: The Impacts of Forests. Wengen, Switzerland, September 1997. Abstracts Volume, p. 32

Côté, H, Laprise, R., Giguère,M., BergeronG., and Caya, D., 1997: Annual and seasonal variability of the climate of Western Canada as simulated by the Canadian Regional Climate Model under current and enhanced greenhouse gas scenarios. Wengen-97 Workshop on Past, Present and Future Climate Variability and Extremes: The Impacts of Forests. Wengen, Switzerland, September 1997. Abstracts Volume, p. 33

Diaz, H. F., and N. E. Graham, 1997: Monitoring climatic variability within ecoregions : a strategy for climate change detection. Wengen-97 Workshop on Past, Present and Future Climate Variability and Extremes: The Impacts of Forests. Wengen, Switzerland, September 1997. Abstracts Volume, p. 30

Frei, C., and Schaer, C., 1997: Spatial and temporal variations of strong precipitation in the Alpine region. Wengen-97 Workshop on Past, Present and Future Climate Variability and Extremes: The Impacts of Forests. Wengen, Switzerland, September 1997. Abstracts Volume, p. 21

Hurrell, J. W., 1997: Decadal variations in climate over the Northern Hemisphere: Impacts and possible causes. Wengen-97 Workshop on Past, Present and Future Climate Variability and Extremes: The Impacts of Forests. Wengen, Switzerland, September 1997. Abstracts Volume, p. 13

Kienast, F., 1997: Potential impacts of climate change on species diversity in forests - a modeling study. Wengen-97 Workshop on Past, Present and Future Climate Variability and Extremes: The Impacts of Forests. Wengen, Switzerland, September 1997. Abstracts Volume, p. 38

Kräuchi, N., and Baltensweiler, A. 1997: Relative abundance of tree species as affected by climatic variability. Wengen-97 Workshop on Past, Present and Future Climate Variability and Extremes: The Impacts of Forests. Wengen, Switzerland, September 1997. Abstracts Volume, p. 37

Schwikowski, M., 1997: Reconstruction of paleo atmospheric chemistry from alpine ice cores. Wengen-97 Workshop on Past, Present and Future Climate Variability and Extremes: The Impacts of Forests. Wengen, Switzerland, September 1997. Abstracts Volume, p. 6

Sykes, M., 1997: Forest dynamics and species range limits: modelling the response of Scandinavian forest landscapes to a changing climate. Wengen-97 Workshop on Past, Present and Future Climate Variability and Extremes: The Impacts of Forests. Wengen, Switzerland, September 1997. Abstracts Volume, p. 40

Weber, R.O., Talkner, P., Auer, I., Böhm, R., Gajic-Capka, M., Zaninovic, K., Brazdil, R., and Fasko, P., 1997: Twentieth-century changes of temperature in the mountai regions of Central Europe. Climatic Change, 36,327-344

Wick, L., and Ammann, B., 1997: Forest responses to rapid warming at the end of the Younger Dryas. Wengen-97 Workshop on Past, Present and Future Climate Variability and Extremes: The Impacts of Forests. Wengen, Switzerland, September 1997. Abstracts Volume, p. 2

Index

A

B

C

D

G

H

I

J

K

L

M

N

O

P

Q

R

S

T

U

V

W

X

Z

Lecture Notes in Earth Sciences

Vol. 1: Sedimentary and Evolutionary Cycles. Edited by U. Bayer and A. Seilacher. VI, 465 pages. 1985. (out of print).

Vol. 2: U. Bayer, Pattern Recognition Problems in Geology and Paleontology. VII, 229 pages. 1985. (out of print).

Vol. 3: Th. Aigner, Storm Depositional Systems. VIII, 174 pages. 1985.

Vol. 4: Aspects of Fluvial Sedimentation in the Lower Triassic Buntsandstein of Europe. Edited by D. Mader. VIII, 626 pages. 1985. (out of print).

Vol. 5: Paleogeothermics. Edited by G. Buntebarth and L. Stegena. II, 234 pages. 1986.

Vol. 6: W. Ricken, Diagenetic Bedding. X, 210 pages. 1986.

Vol. 7: Mathematical and Numerical Techniques in Physical Geodesy. Edited by H. Sünkel. IX, 548 pages. 1986.

Vol. 8: Global Bio-Events. Edited by O. H. Walliser. IX, 442 pages. 1986.

Vol. 9: G. Gerdes, W. E. Krumbein, Biolaminated Deposits. IX, 183 pages. 1987.

Vol. 10: T.M. Peryt (Ed.), The Zechstein Facies in Europe. V, 272 pages. 1987.

Vol. 11: L. Landner (Ed.), Contamination of the Environment. Proceedings, 1986. VII, 190 pages.1987.

Vol. 12: S. Turner (Ed.), Applied Geodesy. VIII, 393 pages. 1987.

Vol. 13: T. M. Peryt (Ed.), Evaporite Basins. V, 188 pages. 1987.

Vol. 14: N. Cristescu, H. I. Ene (Eds.), Rock and Soil Rheology. VIII, 289 pages. 1988.

Vol. 15: V. H. Jacobshagen (Ed.), The Atlas System of Morocco. VI, 499 pages. 1988.

Vol. 16: H. Wanner, U. Siegenthaler (Eds.), Long and Short Term Variability of Climate. VII, 175 pages. 1988.

Vol. 17: H. Bahlburg, Ch. Breitkreuz, P. Giese (Eds.), The Southern Central Andes. VIII, 261 pages. 1988.

Vol. 18: N.M.S. Rock, Numerical Geology. XI, 427 pages. 1988.

Vol. 19: E. Groten, R. Strauß (Eds.), GPS-Techniques Applied to Geodesy and Surveying. XVII, 532 pages. 1988.

Vol. 20: P. Baccini (Ed.), The Landfill. IX, 439 pages. 1989.

Vol. 21: U. Förstner, Contaminated Sediments. V, 157 pages. 1989.

Vol. 22: I. I. Mueller, S. Zerbini (Eds.), The Interdisciplinary Role of Space Geodesy. XV, 300 pages. 1989.

Vol. 23: K. B. Föllmi, Evolution of the Mid-Cretaceous Triad. VII, 153 pages. 1989.

Vol. 24: B. Knipping, Basalt Intrusions in Evaporites. VI, 132 pages. 1989.

Vol. 25: F. Sansò, R. Rummel (Eds.), Theory of Satellite Geodesy and Gravity Field Theory. XII, 491 pages. 1989.

Vol. 26: R. D. Stoll, Sediment Acoustics. V, 155 pages. 1989.

Vol. 27: G.-P. Merkler, H. Militzer, H. Hötzl, H. Armbruster, J. Brauns (Eds.), Detection of Subsurface Flow Phenomena. IX, 514 pages. 1989.

Vol. 28: V. Mosbrugger, The Tree Habit in Land Plants. V, 161 pages. 1990.

Vol. 29: F. K. Brunner, C. Rizos (Eds.), Developments in Four-Dimensional Geodesy. X, 264 pages. 1990.

Vol. 30: E. G. Kauffman, O.H. Walliser (Eds.), Extinction Events in Earth History. VI, 432 pages. 1990.

Vol. 31: K.-R. Koch, Bayesian Inference with Geodetic Applications. IX, 198 pages. 1990.

Vol. 32: B. Lehmann, Metallogeny of Tin. VIII, 211 pages. 1990.

Vol. 33: B. Allard, H. Borén, A. Grimvall (Eds.), Humic Substances in the Aquatic and Terrestrial Environment. VIII, 514 pages. 1991.

Vol. 34: R. Stein, Accumulation of Organic Carbon in Marine Sediments. XIII, 217 pages. 1991.

Vol. 35: L. Håkanson, Ecometric and Dynamic Modelling. VI, 158 pages. 1991.

Vol. 36: D. Shangguan, Cellular Growth of Crystals. XV, 209 pages. 1991.

Vol. 37: A. Armanini, G. Di Silvio (Eds.), Fluvial Hydraulics of Mountain Regions. X, 468 pages. 1991.

Vol. 38: W. Smykatz-Kloss, S. St. J. Warne, Thermal Analysis in the Geosciences. XII, 379 pages. 1991.

Vol. 39: S.-E. Hjelt, Pragmatic Inversion of Geophysical Data. IX, 262 pages. 1992.

Vol. 40: S. W. Petters, Regional Geology of Africa. XXIII, 722 pages. 1991.

Vol. 41: R. Pflug, J. W. Harbaugh (Eds.), Computer Graphics in Geology. XVII, 298 pages. 1992.

Vol. 42: A. Cendrero, G. Lüttig, F. Chr. Wolff (Eds.), Planning the Use of the Earth's Surface. IX, 556 pages. 1992.

Vol. 43: N. Clauer, S. Chaudhuri (Eds.), Isotopic Signatures and Sedimentary Records. VIII, 529 pages. 1992.

Vol. 44: D. A. Edwards, Turbidity Currents: Dynamics, Deposits and Reversals. XIII, 175 pages. 1993.

Vol. 45: A. G. Herrmann, B. Knipping, Waste Disposal and Evaporites. XII, 193 pages. 1993.

Vol. 46: G. Galli, Temporal and Spatial Patterns in Carbonate Platforms. IX, 325 pages. 1993.

Vol. 47: R. L. Littke, Deposition, Diagenesis and Weathering of Organic Matter-Rich Sediments. IX, 216 pages. 1993.

Vol. 48: B. R. Roberts, Water Management in Desert Environments. XVII, 337 pages. 1993.

Vol. 49: J. F. W. Negendank, B. Zolitschka (Eds.), Paleolimnology of European Maar Lakes. IX, 513 pages. 1993.

Vol. 50: R. Rummel, F. Sansò (Eds.), Satellite Altimetry in Geodesy and Oceanography. XII, 479 pages. 1993.

Vol. 51: W. Ricken, Sedimentation as a Three-Component System. XII, 211 pages. 1993.

Vol. 52: P. Ergenzinger, K.-H. Schmidt (Eds.), Dynamics and Geomorphology of Mountain Rivers. VIII, 326 pages. 1994.

Vol. 53: F. Scherbaum, Basic Concepts in Digital Signal Processing for Seismologists. X, 158 pages. 1994.

Vol. 54: J. J. P. Zijlstra, The Sedimentology of Chalk. IX, 194 pages. 1995.

Vol. 55: J. A. Scales, Theory of Seismic Imaging. XV, 291 pages. 1995.

Vol. 56: D. Müller, D. I. Groves, Potassic Igneous Rocks and Associated Gold-Copper Mineralization. 2nd updated and enlarged Edition. XIII, 238 pages. 1997.

Vol. 57: E. Lallier-Vergès, N.-P. Tribovillard, P. Bertrand (Eds.), Organic Matter Accumulation. VIII, 187 pages. 1995.

Vol. 58: G. Sarwar, G. M. Friedman, Post-Devonian Sediment Cover over New York State. VIII, 113 pages. 1995.

Vol. 59: A. C. Kibblewhite, C. Y. Wu,Wave Interactions As a Seismo-acoustic Source. XIX, 313 pages. 1996.

Vol. 60: A. Kleusberg, P. J. G. Teunissen (Eds.), GPS for Geodesy. VII, 407 pages. 1996.

Vol. 61: M. Breunig, Integration of Spatial Information for Geo-Information Systems. XI, 171 pages. 1996.

Vol. 62: H. V. Lyatsky, Continental-Crust Structures on the Continental Margin of Western North America. XIX, 352 pages. 1996.

Vol. 63: B. H. Jacobsen, K. Mosegaard, P. Sibani (Eds.), Inverse Methods. XVI, 341 pages, 1996.

Vol. 64: A. Armanini, M. Michiue (Eds.), Recent Developments on Debris Flows. X, 226 pages. 1997.

Vol. 65: F. Sansò, R. Rummel (Eds.), Geodetic Boundary Value Problems in View of the One Centimeter Geoid. XIX, 592 pages. 1997.

Vol. 66: H. Wilhelm, W. Zürn, H.-G. Wenzel (Eds.), Tidal Phenomena. VII, 398 pages. 1997.

Vol. 67: S. L. Webb, Silicate Melts. VIII. 74 pages. 1997.

Vol. 68: P. Stille, G. Shields, Radiogenetic Isotope Geochemistry of Sedimentary and Aquatic Systems. XI, 217 pages. 1997.

Vol. 69: S. P. Singal (Ed.), Acoustic Remote Sensing Applications. XIII, 585 pages. 1997.

Vol. 70: R. H. Charlier, C. P. De Meyer, Coastal Erosion – Response and Management. XVI, 343 pages. 1998.

Vol. 71: T. M. Will, Phase Equilibria in Metamorphic Rocks. XIV, 315 pages. 1998.

Vol. 72: J. C. Wasserman, E. V. Silva-Filho, R. Villas-Boas (Eds.), Environmental Geochemistry in the Tropics. XIV, 305 pages. 1998.

Vol. 73: Z. Martinec, Boundary-Value Problems for Gravimetric Determination of a Precise Geoid. XII, 223 pages. 1998.

Vol. 74: M. Beniston, J. L. Innes (Eds.), The Impacts of Climate Variability on Forests. XIV, 329 pages. 1998.

Vol. 75: H. Westphal, Carbonate Platform Slopes – A Record of Changing Conditions. XI, 197 pages. 1998.

Vol. 76: J. Trappe, Phanerozoic Phosphorite Depositional Systems. XII, 316 pages. 1998.

GPSR Compliance
The European Union's (EU) General Product Safety Regulation (GPSR) is a set of rules that requires consumer products to be safe and our obligations to ensure this.

If you have any concerns about our products, you can contact us on

ProductSafety@springernature.com

In case Publisher is established outside the EU, the EU authorized representative is:

Springer Nature Customer Service Center GmbH
Europaplatz 3
69115 Heidelberg, Germany

www.ingramcontent.com/pod-product-compliance
Ingram Content Group UK Ltd.
Pitfield, Milton Keynes, MK11 3LW, UK
UKHW020133290926
15471UKWH00003B/463

9783662172742